W9-BEJ-894

Lifelong Motor Development

Lifelong Motor Development

Second Edition

Carl P. Gabbard
Texas A & M University

Boston, Massachusetts   Burr Ridge, Illinois   Dubuque, Iowa
Madison, Wisconsin   New York, New York   San Francisco, California   St. Louis, Missouri

WCB/McGraw-Hill

*A Division of The **McGraw·Hill** Companies*

Book Team

Editor   *Scott Spoolman*
Publishing Services Coordinator   *Peggy Selle*
Permissions Coordinator   *LouAnn Wilson*
Visuals/Design Production Manager   *Janice Roerig-Blong*
Production Manager   *Beth Kundert*
Publishing Services Manager   *Sherry Padden*
Visuals/Design Freelance Specialist   *Mary L. Christianson*
Marketing Manager   *Pam Cooper*
Promotions Manager   *Mike Matera*

Basal text: Times Roman
Display type: Universe, Times Roman
Typesetting system: Quark XPress for Macintosh
Production Services: Edwards Brothers
Printing and Binding: Edwards Brothers

President and Chief Executive Officer   *Thomas E. Doran*
Vice President and Executive Publisher   *Edgar J. Laube*
Vice President/Production and Business Development   *Vickie Putman*
Vice President/Sales and Marketing   *Bob McLaughlin*

President and Chief Executive Officer   *G. Franklin Lewis*
Senior Vice President, Operations   *James H. Higby*
Corporate Senior Vice President and President of Manufacturing   *Roger Meyer*
Corporate Senior Vice President and Chief Financial Officer   *Robert Chesterman*

Consulting Editor   *Aileene Lockhart*

The credits section for this book begins on page 443 and is considered an extension of the copyright page.

Cover design and partial interior design by Kay Fulton design

Cover image © Claudia Kunin/Tony Stone Images

Copyedited by Laurie McGee

Library of Congress Catalog Card Number: 94–79727

ISBN 0–697–23333–2

Printed in the United States of America

10   9   8   7   6   5   4   3   2

Contents

--->

PART T W O   Biological Growth and Development

Chapter 2
Heredity and Neurological Changes

Chapter 3
Physical Growth Changes

-->

PART *THREE* Perception and Information Processing

- ->

PART *F O U R* Motor Behavior across the Life Span

Chapter 8
Early Movement Behavior

Chapter 9
Motor Behavior during Early Childhood

Chapter 10
Motor Behavior during Later Childhood and Adolescence

Chapter 11
Motor Behavior in the Adult Years

--➤

PART *F I V E* Assessment

--➤

PART *S I X* Sociocultural Influences on Motor Development

Preface

J ust as motor development is a continuous process of change, so must the textbooks that contain its body of knowledge be updated with some regularity. In essence, the timely publication of the second edition of *Lifelong Motor Development* is a manifestation of the need to stay as current as possible in a field that is expanding at a phenomenal rate. Although the second edition reflects several changes from the initial version, it may still be described as a *comprehensive* and *up-to-date, science-based* discussion of physical growth, development, and motor behavior from conception through the adult years.

SCIENCE

Above all else, it was important that the second edition of *Lifelong Motor Development* be an extremely up-to-date presentation of multidisciplinary research in life-span motor development. To achieve this goal, hundreds of references from the various scientific fields that contribute to our understanding of motor development were reviewed. The result for the student of motor development is a blend of classic and cutting-edge research information.

READABILITY

The second major intent was to improve the readability of the text. This was accomplished, with the help of dedicated reviewers, by thoroughly reviewing each section for organization, transitions, and general readability. One significant improvement is the elimination of excess referencing while still providing key credits. In addition, more tables and figures are provided to clarify textual material.

SELECTED HIGHLIGHTS OF CHANGE

- In general, each section was revised to provide current thought in the science of motor development across the life span.

- A major change is introduced in chapter 1 with the inclusion of the latest theoretical perspectives on human development: *contextualist* and *developmental biodynamics* approaches. These themes are referred to (with examples of research) in several chapters. In addition, the discussion of theoretical views now includes *biological theories of aging*.

- To stimulate critical thinking while reviewing the various theoretical perspectives and related material throughout the text, the timely issue of *nature-nurture* is given special emphasis in regard to the influence of heredity and the environment on development.

- Expanded coverage of information processing (chapter 7) with a new section under Programming, *Developmental Theories of Motor Control*. This section complements the latest theoretical views with discussion of organizing *perception into action* (the brain and body behavioral connection).

- New material on theory and development of *motor (functional) asymmetries* complements the chapter 8 discussion of early movement behaviors.

- The chapter on motor assessment (12) has been updated to include the latest revisions of recommended tests (e.g., Bayley II, Denver II, and Peabody Developmental Scales).

- Chapter 13 now includes information on *self-esteem* and expanded coverage of *sociocultural theories of aging*.

- Several new and revised tables and figures complement the material; this is especially evident in chapter 9 (Motor Behavior during Early Childhood).

- Appendix A (literature retrieval/relevant journals) has been updated with the latest in computer-assisted resources to aid the student in motor literature searches.

- The supplemental learning activities section (Appendix B) has been expanded to stimulate the application of theory and inquiry.

BASIC ORGANIZATION

Lifelong Motor Development presents a topical and chronological approach to the study of human development. While there are good arguments for using either format entirely, the intent is to provide the student with a multidimensional perspective. The topical sections offer the student a rich feel for the interrelatedness of the lifelong developmental process. On the other hand, conceptualizing life-span development within a framework for study and placement of significant events necessitates a time-related continuum. This text provides such a developmental framework; however, it is based on the premise that while change may appear phasic (stagelike), its underlying processes are continuous. It is the continuous nature of development that requires topical discussions.

The basic organization of the text is not different from the first edition. For first-time users (in brief), *Lifelong Motor Development* is divided into six parts and two appendixes. Part One is a single chapter presenting a multidisciplinary overview of lifelong human development with emphasis on basic developmental principles, terms, issues, and theoretical approaches. Also provided is a conceptual model of the phases of motor behavior used as the framework for the material presented in Part Four, "Motor Behavior across the Life Span." Part Two is devoted to topical discussions on the body of information related to lifelong biological growth and development. Chapters 2, 3, and 4 provide information on the various hereditary, neurological, and physical characteristics that, together with experience, form the bases for motor behavior across the life span. Chapter 5 deals with factors and conditions that may affect the course of biological growth and development.

Chapters 6 and 7 (Part Three) present comprehensive topical discussions of lifelong perceptual development and information-processing characteristics (including the latest theoretical thoughts). The chapters in Part Four (8 through 11) provide a chronological (phase model) description of motor behavior (performance) characteristics across the life span. Coverage begins with early movement behavior (chapter 8), then extends through early childhood (chapter 9), later childhood and adolescence (chapter 10), and finally to motor behavior during the adult years that is characterized by peak performance and with older age, regression (chapter 11).

Part Five (chapter 12) offers a broad perspective of the diversity of motor assessment techniques and discusses the considerations for selecting and implementing a wide variety of assessment instruments. Part Six (chapter 13) presents a discussion of the influence and importance of sociocultural factors on motor development from a lifelong perspective.

Finally, two appendixes are included. Appendix A tells how to conduct both computer-assisted and manual literature searches on topics relevant to this text. Appendix B provides a variety of supplemental learning projects designed to augment the chapter objectives with practical "hands-on" experience. Chapter references are provided to allow for review of related text information.

PEDAGOGICAL FEATURES

An important goal in creating this text was to make it a *teachable* resource with an *effective learning system*. As with the first edition, each of the six parts of the text begins with an overview of its content. Included in the introduction of each chapter are *chapter objectives* and *key terms*. The objectives reflect the conceptual framework used by identifying the important facts, topics, and concepts to be covered. Marginal notations and page numbers are provided to aid the reader in identifying text material related to chapter objectives. Feedback from first-edition users and reviewers suggest that these characteristics are quite useful in developing course materials. For example, in some instances instructors may wish to designate selected (rather than all) chapter objectives for course coverage.

Key terms appear in boldfaced type in the text and pinpoint the words of greatest importance to understanding the broader concepts of each chapter; these and other significant terms are highlighted throughout. A chapter *summary* and *suggested readings* are presented at the end of each chapter.

And finally, a much desired outcome of the second edition of *Lifelong Motor Development* is that the reader will feel the excitement associated with studying and understanding human motor development from a life-span perspective.

ACKNOWLEDGMENTS

Lifelong Motor Development (2nd ed.) is the outcome of much more than my efforts. Foremost, I am grateful for the continuous and diligent efforts of the scientific community in providing the knowledge base. Much of the inspiration for creating the second edition came from comments from current users, colleagues, and students. I also owe a special debt of gratitude to Brown & Benchmark, more specifically, Scott Spoolman and his editorial staff for their dedication to making this text the very best available in lifelong motor development. I also wish to express considerable gratitude to the reviewers whose suggestions play a vital role in any success that the text encounters. The reviewers were

Sherry L. Folsom-Meek
Mankato State University

Yvonne M. Becker
Augustana University College

Stevie Chepko
Springfield College

Kathleen Williams
University of North Carolina at Greensboro

Robert Kraft
University of Delaware

Once again, a final note of loving thanks for the support, love, and above all, patience of my wonderful daughters and wife. And, to all things, give thanks to the Lord.

Lifelong Motor Development

PART ° one --

----------► *chapter*

1   Introduction to the Developmental Perspective

P art One presents an overview of lifelong human development. Because developmental psychology is the basic body of knowledge from which an understanding of growth and motor behavior is derived and appreciated, much of this section will focus on its fundamentals. Basic developmental principles, terms, and theoretical approaches and issues are provided to aid the reader in forming an eclectic perspective of human development and motor

An Overview of Lifelong Human Development

behavior across the life span. Along with the basic elements of developmental psychology, chapter 1 provides the initial introduction to the ways in which motor development parallels other human experience across the life span. Highlighting this section is a conceptual model of the phases of motor behavior illustrating how the chronological stages of life-span development integrate with the evolution and regression of motor performance. This model provides the foundation (conceptual framework) for much of the material presented in subsequent chapters.

Introduction to the Developmental Perspective

OBJECTIVES

Upon completion of this chapter, you should be able to

1. Define the term *motor development* and describe its association with the life-span perspective. **5**
2. Discuss the multidisciplinary approach to studying motor development. **5**
3. Briefly describe the five major objectives of the developmental specialist. **5**
4. Define the general terms that are unique to the fields of human and motor development. **6**
5. Discuss and support eight major observations associated with human development. **9**
6. Outline the periods of life-span development. **10**
7. List the phases of motor behavior and briefly describe their characteristics. **11**
8. Discuss the purpose and identify primary strategies used in conducting research in the scientific study of life-span development. **13**
9. Identify and briefly discuss the main controversial issues in human development. **16**
10. Name and describe the major theoretical views on human development. **19**
11. Describe the various biological theories of advanced aging. **28**

KEY TERMS

motor development	phases of motor behavior
life-span perspective	phase
multidisciplinary approach	nature-nurture
heredity	continuity-discontinuity
maturation	stages
growth	critical period
development	modifiability-reversibility
motor behavior	biological-maturation approach
cephalocaudal development	psychosocial approach
proximodistal development	learning-behavioral approach
phylogenetic behaviors	cognitive-developmental approach
ontogenetic behaviors	information-processing approach
differentiation	contextualist (ecological) approach
integration	developmental biodynamics approach
aging	dynamical systems
periods of life-span development	biological theories of advanced aging

In general terms, **motor development** is the study of movement behavior and the associated biological change in human movement across the life span. From another perspective it may be viewed as the process of change in motor behavior resulting from the interaction of heredity and the environment. Basic to a comprehensive understanding of this field of specialization is knowledge related to the characteristics and principles of *growth* (change in size), *development* (change in level of functioning), and *motor behavior* (performance). Traditionally this area of inquiry was studied by developmental psychologists primarily interested in the childhood stages of development. In more recent years, however, the scope of motor development has been extended in recognition that the developmental process is continuous and observable from conception to the final stage in human life. The basis for accepting this **life-span perspective** is formed on the theoretical notion that the developmental process extends beyond puberty and young adulthood. Significant physiological and motor behavior changes occur during older adulthood and are important to our understanding of the full scope of human development. Hughes and Noppe (1991) state, "To examine only isolated segments of the life span is equivalent to studying isolated scenes from a film or play" (p. 6).

■ *Objective 1.1*

Along with the promising trend to view motor development from a *life-span perspective* has emerged the practice of studying behavioral change using an integrated **multidisciplinary approach.** It is generally accepted that behavior in any domain (i.e., cognitive, affective, psychomotor) is the product of many influences. To have a fuller understanding of human development, one should consider the full range of possible influences. Perhaps the strongest support for this point of view has been seen among those individuals interested in child development from a total-development perspective. While working both independently and cooperatively, professionals from such fields as developmental psychology, exercise physiology, medicine, biomechanics, physical education, and sociology have provided data adding to our understanding of total human development and behavior.

■ *Objective 1.2*

Developmental psychologists and motor development specialists seek to accomplish five major objectives: (a) to determine the common and characteristic changes in behavior, function, and appearance across the life span; (b) to establish when it is that these changes occur; (c) to describe what causes these changes; (d) to determine whether change can be predicted; and (e) to find out whether these changes are individual or universal.

■ *Objective 1.3*

   Excellent commentaries by Clark and Whitall (1989) and Thelen (1989) offer contemporary discussions on the description and history of motor development (see Suggested Readings).

As with any specialized field of science or education, motor development has established its own terminology and adopted words from related disciplines. The

GENERAL TERMINOLOGY

■ *Objective 1.4*   study of motor development cuts across several disciplines and subareas within the study of movement behavior and therefore uses a considerable amount of the general terminology of these related fields. Familiarity with this terminology will be important in developing a clear understanding of the developmental perspective.

Prerequisite Terms

Heredity refers to a set of qualities that are fixed at birth and account for many individual traits and characteristics. The cells of normal humans possess 46 chromosomes arranged in 23 pairs. These chromosomes are made up of thousands of genes that influence such traits as eye and hair color, personality, intelligence, height, muscle fiber type, and general body build. Although these traits are strongly influenced by genetic structure, they may be modified by environmental factors.

The term **maturation** is often used interchangeably with the words *growth* and *development*. Maturation, however, is a more distinctive process, actually referring to the qualitative functional changes that occur with age. Maturation is also used to describe the successive tissue changes that take place until a final form is achieved. These changes are associated with the progression of an individual from one level of functioning to a higher level. Primarily innate (i.e., genetically determined) and resistant to external influences, maturation is a fixed order of progression; the time intervals may vary but the sequence of appearance of characteristics generally does not. The development of locomotion, for example, follows a consistent order (sit, walk, run) and an approximate age of appearance. The rate at which these motor capabilities are attained may differ among individuals, but the sequence generally remains fixed.

Environment is the circumstances, objects, or conditions by which one is surrounded.

Experience refers to conditions within the environment that may alter or modify various developmental characteristics through the learning process.

Learning is defined as the relatively permanent change in performance that results from practice or past experience.

Readiness is the combination of maturation and experience that prepares an individual to acquire a skill or understanding.

Adaptation is the process of altering one's behaviors to interact effectively with the environment. The term is often used to describe the complex interplay between the individual and the environment. The developmental aspects of maturation and the individual's experience interweave to create behavior.

Motor Development Terms

As previously mentioned, *motor development* is both a process and scientific field of study. As a process, it may be viewed as the lifelong change in motor behavior resulting from the ongoing interaction between heredity and the environment. The study of motor development focuses on the biological changes associated with movement behavior across the human life span. More specifically it is the study of lifelong growth, development, and motor behavior (performance).

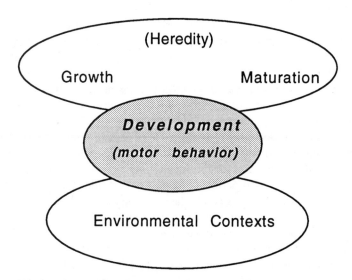

Figure 1.1
The interrelated nature of human development.

Growth, often used interchangeably with *development* and *maturation,* refers to observable changes in quantity; in this context, growth represents an increase in body size. Although maturation may be a factor in the growth process, environmental factors may also contribute.

Development is a term that can be applied to several human behaviors. Basically, development refers to the process of change in the individual's level of functioning. Those changes may be either quantitative or qualitative in nature. Development is a product of growth as well as of heredity, maturation, and experience.

Motor behavior is the product of growth and development characteristics and refers to observable changes in movement or the learning and performance of a particular movement or motor skill. In the study of the motor behavior aspect of motor development, the developmental descriptions of fundamental and sport skill performance abilities are documented. Figure 1.1 presents a graphic illustration of the interrelated nature of human development and the influences on motor performance.

In its purest sense, the term *motor* refers to the underlying biological and mechanical factors that influence movement (or observable action) even though the terms movement and motor are frequently used interchangeably. As noted earlier, motor (or movement) performance is influenced by several integrated factors. Learning, integrated with growth and basic developmental factors, is a primary influence.

Learning was earlier defined as a relatively permanent change in performance resulting from practice or past experience. Although learning cannot be observed directly, it can be inferred from a person's motor performance. The term *motor learning* refers to a relatively permanent change in the performance of a motor skill, also resulting from practice or experience. Another term used frequently in the movement literature is *motor control,* referring to the area of study concerned primarily with the underlying processes involved in movement and how various movements are controlled. The focus of motor control study is on understanding specific neural, physical, and behavioral aspects of human movement.

Cephalocaudal and *proximodistal* are terms used to describe the orderly and predictable sequences of growth and motor control. **Cephalocaudal development** is growth that proceeds longitudinally from the head to the feet. Complementing this physical growth is a gradual progression of increased muscular control moving first from the muscles of the head and neck to the trunk and then to the legs and feet. **Proximodistal development** is growth that proceeds from the center of the body to its periphery; that is, growth and motor control occur in the trunk region and shoulders before the wrists, hands, and fingers. These developmental trends are clearly observed in young children who exhibit greater coordination in the upper torso than in their legs and feet (cephalocaudal). Similarly, children will exhibit gross shoulder movements in their first attempts to draw, long before their ability to make fine motor cursive forms is developed.

Phylogenetic and *ontogenetic* behaviors are movements that are acquired naturally or must be learned. **Phylogenetic behaviors** tend to appear somewhat automatically and in a predictable sequence. Such behaviors as reaching, sitting, and walking are presumably resistant to environmental influences and are common to the human species. **Ontogenetic behaviors** such as writing, swimming, and cycling are specific to the individual and therefore are influenced by learning and the environment.

Differentiation and *integration* are processes associated with neurological development and increased functional complexity. **Differentiation** is the process by which structure, function, or forms of behavior become more specialized. In general, it is the progression of motor control from gross, poorly controlled movements (like those displayed by young infants) to more precise, complex forms of motor behavior commonly exhibited by older individuals. For example, the way newborns reach for and grasp (a small object) is described as *corralling* because they primarily use the shoulders and arms. By the end of the first year more controlled manual movements, primarily involving the thumb and forefinger, take the place of the corralling movement.

Integration, on the other hand, refers to the intricate interweaving of neural mechanisms of various opposing muscle groups into a coordinated interaction. Using the example of reaching and grasping again, integration is observable as young children mature and demonstrate ever-improving eye-hand coordination in their movements. The differentiation and integration processes occur concurrently in the course of development. These terms will be discussed in greater detail in chapter 2.

The scope of motor development includes the study of biological change over the life span; therefore the term *aging* has relevance to our understanding of the developmental process. **Aging** has been characterized as the diminished capacity to regulate the internal environment (e.g., cellular, molecular, organismic structures), resulting in a reduced probability of survival. It may also be thought of as the process of growing old regardless of chronological age.

___

To grasp a comprehensive understanding of human development, certain fundamental and predictable observations must be taken into consideration. The following general characteristics of human growth and development provide specific relevance to the study of motor development. Selected characteristics will be discussed in more detail in subsequent sections of this chapter.

1. *Heredity and environmental context play vital interactive roles in development.* Human development stems from biological (e.g., maturation) and environmental (e.g., culture, learning, practice) influences. This "nature-nurture" relationship provides the resources for the individual to achieve his or her potential, the interplay of these factors being unique to the individual.
2. *Although development appears stagelike, its bases are underlined by processes that are likely continual and cumulative.* Some theories describe development as progressing through a series of stages (usually related to age), suggesting qualitative differences in how individuals move, think, feel, and so on. Others describe development as change arising from quantitative increments that are gradually transforming. It is most likely, however, that development involves a complex weaving of gradual and more abrupt observable changes.
3. *Development follows a definite and predictable pattern.* This fact is exhibited somewhat clearly in regard to the orderly growth and development patterns of such behaviors as intelligence, speech, and motor control (i.e., proximodistal and cephalocaudal development).
4. *Development is aided by stimulation.* Although most development occurs as a result of maturation and environmental influence, much can be done to stimulate the process to reach its full potential. Stimulation, especially during initial learning and development periods when the individual is at a point of readiness, can be very effective.
5. *Although there may be certain critical periods during early development when the individual is particularly responsive to specific influences, the human body has a unique resiliency to overcome many adverse experiences or lack thereof (plasticity).* Critical periods have been described in relation to, for example, emotional bonding with a parent and nutrition needs at specific times during early development. Unless the particular case is extreme, the human body can, in many situations, overcome (or "catch up" on) such conditions to function normally.

SIGNIFICANT OBSERVATIONS ABOUT DEVELOPMENT

■ *Objective 1.5*

6. *Early foundations are critical.* Patterns of behavior that are established during the early periods of life determine to a large extent the quality of existence and performance in later years. This assumption has continued to gain support in light of recent evidence and predictions related to health intervention techniques and practices during the early years.

7. *There are social expectations for every stage of development.* Every cultural group expects its members to master certain skills at various points during the life span. Such milestones are described as "developmental tasks" (Havighurst, 1972), specific learning problems that arise at particular stages of life that individuals must accomplish to meet the demands of their culture. Some developmental tasks arise as a result of maturation, such as learning to walk before a specific age, whereas others, such as learning to read and throw a ball, develop from cultural pressures.

8. *All individuals are different.* Therefore, no two people can be expected to react in the same manner to similar stimuli. Science provides evidence that every person is biologically and genetically different. This is true even in the case of identical twins; in fact, individuals are not only different from one another, but as they move from youth to old age, the differences increase.

PERIODS OF LIFE-SPAN DEVELOPMENT

■ *Objective 1.6*

Closely associated with the life-span approach to studying the various aspects of motor development are designated periods (stages) representing approximate chronological age behavior. The following stages are used as the framework from within which to address each aspect of motor development.

Prenatal (***conception to birth***). This is where the story of human development begins. The *prenatal period* is one of immense physical change that begins with genetic transmission and continues through a number of cellular and structural variations. Development of the embryo and fetus are genetically predetermined, but there are several environmental influences on prenatal development. Significant stages of growth and development within this period are the *embryonic* period (up to 8 weeks) and the *fetal* period (8 weeks to birth).

Infancy (***birth to 2 years***). The first month of *infancy* is known as the *neonate* period. This period of development has grown popular as an area for specialized study because of the possibilities for observing initial motor responses (outside the womb) and infant survival characteristics. Infancy is a time of extensive dependency on adults. It also marks the beginning of many motor and psychological activities such as language, symbolic thought, and sensorimotor coordination.

Childhood (***2 to 12 years***). Due to the large number of developmental milestones that are reached during this period, *childhood* is divided into two stages:

1. *Early childhood (2 to 6 years).* This period corresponds roughly to the time in which the child prepares for and enters school. This period represents a

significant stage in the development of fundamental motor skills, perceptual-movement awareness, and the ability to care for oneself.

2. *Later childhood (6 to 12 years).* Sometimes called the elementary school years, this is a period of fundamental motor skill refinement and the mastery of certain academic skills. Physical growth slows substantially and thought processes are usually more concrete than in the adolescent period.

Adolescence (*12 to 18 years*). The term *adolescence* comes from the Latin verb *adolescere,* meaning "to grow into maturity." With this period begin the physical changes relating to one of the major landmarks in human development: puberty. Some of the dramatic changes associated with adolescence are accelerated growth in height and weight, the appearance of secondary sex characteristics, the ability to reproduce, and deepening of the voice. This is also a period in which the degree of logical and abstract thought increases, as well as a concern about identity and independence.

Adulthood (*18 years and older*). The period of *adulthood* has traditionally been subdivided into three stages:

1. *Young adulthood (18 to 40 years)*
2. *Middle-age (40 to 60 years)*
3. *Older adulthood (60 years and older)*

With the increased interest in the physiological and motor behavior change that takes place with aging has emerged a greater focus on the characteristics of adults who are 30 to 70 years old and older.

Figure 1.2 describes the phases of life-span motor behavior and the relationship of that behavior to specific age-related stages (or periods). Inherent in the notion of a **phase** in this context is the relatively continual/overlapping characteristics of human development. Developmental stages are included to complement the overlapping relationship as related to motor behavior characteristics normally associated with chronological age categories.

More-detailed discussions related to characteristic behaviors associated with each of the phases will follow in later sections of this text. The following is a brief overview of the categories of motor behavior across the life span.

Reflexive/spontaneous movement phase. The *reflexive/spontaneous movement phase* is that span of motor behavior which begins at about the third fetal month and continues on after birth into the first year of life. Movements at this phase mirror the relative immaturity of the nervous system. As the system matures, reflexes (involuntary motor responses) and spontaneous movements (stereotypic rhythmic patterns of motion that appear in the absence of any known stimuli) are gradually phased out as voluntary control increases.

PHASES OF MOTOR BEHAVIOR

■ *Objective 1.7*

Figure 1.2

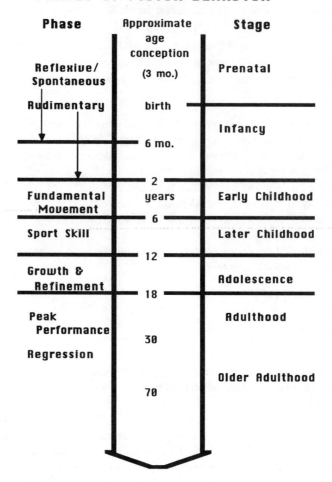

PHASES OF MOTOR BEHAVIOR

Rudimentary phase. The *rudimentary phase* corresponds with the stage of infancy (from birth to 2 years). Rudimentary behavior is voluntary movement in its first form. Motor movements are determined by maturation and appear in a somewhat predictable sequence. Motor control generally develops in a cephalo-caudal/proximodistal order and is characterized by such behaviors as crawling, creeping, walking, and voluntary grasping.

Fundamental movement phase. The *fundamental movement phase* is acknowledged as a major milestone of early childhood (2 to 6 years) and of life-span motor development. Considered an outgrowth of rudimentary behavior, this phase witnesses the appearance of a number of fundamental movement abilities: perceptual-motor awareness (e.g., body awareness, balance); locomotor skills (e.g., running, jumping); nonlocomotor skills (e.g., twisting/turning, stretching/bending); and manipulative skills (e.g., throwing, kicking). These abilities establish the foundation for efficient and more complex human movement in later phases of development. More than 30 characteristic movement skill abilities that emerge during the early childhood period have been identified.

Sport skill phase. The motor skills and movement awareness abilities that the child acquires during the fundamental movement phase gradually become more refined and, in many instances, adapted to sport and recreational activities in the *sport skill phase.* The primary stimulus during this phase is the individual's increased interest in sport skill events and the ability to learn and practice these movements.

Growth and refinement phase. Growth occurs during all periods of development, but perhaps the most significant motor behavior change is seen at the time of puberty and the accompanying growth spurt that generally marks the first stage of adolescence. This is the *growth and refinement phase.* As the levels of hormones rise in the body, changes in muscle and skeletal growth provide a new dimension within which acquired motor skills can be asserted. During the later stages of adolescence, sex differences (mainly favoring males in regard to physical size) become more apparent due primarily to the increased amount of androgen hormones.

Peak performance. Most sources have identified the time of *peak performance* (peak physiological function and maximal motor performance) to be between 25 to 30 years of age. As a general rule, females tend to mature at the lower end of the range (22 to 25) and males at the upper end of the range (28 to 30 years). This is especially evident in three of the most influential factors in motor performance: strength, cardiorespiratory function, and processing speed.

Regression. Although considerable variation among individuals is apparent after 30 years of age, most physiological and neurological factors decline at a rate of about 0.75% to 1% a year. During this phase the phenomenon known as "psychomotor slowing" (primarily in speeded tasks) appears, and the earlier developmental process of differentiation begins to reverse, resulting in similar performance characteristics among the very young and the older adult. The *regression* of motor behavior is generally characterized by decreases in cardiovascular capacity, muscle strength and endurance, neural function, flexibility, and increases in body fat. Though exercise training has not been shown to retard the aging process, it does allow the individual to perform at a higher level.

I question this.

The study of motor development has emerged as a full-fledged scientific endeavor only in recent years. Although researchers and the medical profession have been concerned with motor development as long ago as the days of Hippocrates, little scientifically designed inquiry was conducted until the 1930s. During that period, psychologists (e.g., Bayley, Shirley, Gesell, Ames, McGraw) interested primarily in child development began to establish the data base upon which the life-span perspective was later formed. Motor development specialists then began to make contributions to the literature led by the pioneering efforts of such individuals as Wild, Espenschade, Eckert, Rarick, Glassow, and Halverson.

RESEARCH IN MOTOR DEVELOPMENT

■ *Objective 1.8*

Today researchers across several disciplines are actively involved in the scientific study of growth and motor behavior across the developmental continuum. The science of motor development is supported by millions of dollars in research funds annually, and the results of that research are disseminated through numerous highly respected technical journals and numerous instructional periodicals. Through the efforts of researchers in basic inquiry, educators and parents gain information with which to implement effective educational programs and understand the characteristics of human behavior. Excellent articles by Thelen (1989) and Lockman and Thelen (1993) discuss in some detail the "(re)discovery" of motor development as a field of scientific inquiry (see Suggested Readings).

Why Study Motor Development?

The importance of studying motor development is tied closely to our need to understand the diverse nature of life and self. From conception until death, physical growth and motor behavior are an integral part of human development. To fully understand human behavior, we need to gather knowledge of all the determinants of existence. Kaluger and Kaluger (1984) have identified four types of factors as primary determinants of human growth and development: physiological, psychological, environmental, and metaphysical. Some important reasons for studying motor development are summarized in the following statements:

1. To interact effectively with other individuals we need to be aware of what they can and cannot do—motorically, cognitively, and socially.
2. If we understand what constitutes the normal range of motor development, we can better understand and guide individuals who may be developing abnormally.
3. We can use our understanding of motor development to improve health and motor performance.
4. Our understanding of motor development makes a contribution to a more comprehensive body of knowledge that, in turn, enables us to better understand ourselves.

Methods of Study

As mentioned, the study of motor development has evolved into a legitimate and contributing field of research. It is important to have some knowledge of the way researchers gather their data for drawing conclusions about the way motor development takes place. Although a discussion of the numerous experimental and statistical techniques is beyond the scope of this text, it will be helpful to have a general understanding of those research designs and data collection methods that are used most frequently.

Investigators use the scientific method to answer questions about human development. The investigator first formulates a hypothesis, which is a hunch, guess, or prediction about some aspect of motor development. The hypothesis (or research question) is then tested by collecting evidence in the form of data and analyzing the information to determine whether it supports or rejects the original

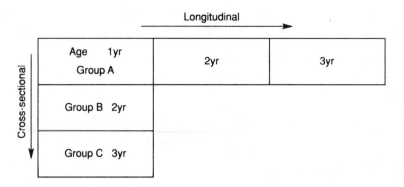

Figure 1.3
Cross-sectional and
longitudinal designs.

assumption. In most cases, the study is repeated with different subjects to further confirm the original findings.

The primary types of research designs used in motor development investigations are longitudinal, cross-sectional, sequential, and cross-cultural.

Longitudinal Design　One of the most popular and perhaps the most reliable type of design is the *longitudinal design.* In this method, data are collected on the same individual over an extended period of time, usually across several years. One strength of this design is that development is observed directly and not implied, as with the cross-sectional and cross-cultural methods. It may be one of the best designs, but the method is not without weaknesses. Over a long period of time, subjects may move away, testing personnel may change, and the cost of the study in time and money may be a problem. Repeated testing of the same subjects may also be a problem. The individual's performance may improve with repeated test sessions because of increasing familiarity with the task (the "practice effect") rather than because of increased maturity. The shorter the span of time and the fewer the number of measurement sessions, the easier it is to conduct a longitudinal study.

Cross-Sectional Design　The vast majority of research studies in motor development have a *cross-sectional design.* Investigators observe age differences by selecting individuals representing different age groups and measuring their behaviors. Figure 1.3 compares cross-sectional and longitudinal designs. In the cross-sectional study, subjects are chosen carefully so that the individuals in the different age samples are as nearly alike as possible in all ways other than age. The primary value of this design is that it permits a developmental change to be detected in a fairly short period of time. However, it has its limitations too. Researchers cannot be certain that the behavioral differences are due to age because all individuals grow up under different circumstances (e.g., education, nutrition, habits of physical activity, etc.).

Sequential Design　The *sequential design* offers some improvement over the longitudinal and cross-sectional strategies. This method includes elements from each of the other two designs in that it allows the researcher to study several different-aged samples over a period of years. An advantage of the design is that subjects with different characteristics (e.g., age, background) can be compared at the same chronological age to identify any existing behavioral differences. Another is that a study of this type can be conducted in a relatively short span of time.

Cross-Cultural Design   A study with a *cross-cultural design* compares individuals among different cultures. A comparison of fitness and throwing ability among samples of children in the United States and the Soviet Union would be one example of this method.

Regardless of the design used, data must be collected. Systematic observation is perhaps the most popular method for collecting data and is the basic tool of any science. If researchers wish to focus on naturally occurring spontaneous behavior in a "natural" setting (e.g., at home, at play, at school) the method is known as *naturalistic observation. Structured observation* requires that the researcher manipulate the environment in some way, as is the case when the study is conducted in a laboratory setting or in a familiar environment in which new elements are presented.

In other techniques the individual is questioned directly. The *case history* provides an in-depth analysis of background information on the individual. The *interview* generally does not provide as much background information but is especially valuable if the goal is to obtain a lot of information in a short period of time. In the interview, researchers actively probe the individual's ideas, feelings, and motives; the data come in the form of words rather than observed actions, as is also true of the case history. The *survey* (questionnaire) is similar to the interview except the researcher seeks written responses to a series of questions. One of the most commonly used methods of collecting information about motor development is through the use of *standardized tests.* These tests are widely available and offer researchers a source of objective data. The chapter on assessment includes a detailed discussion of motor development tests and makes recommendations for their use.

To design an investigation that conforms to the scientific method, the researcher must make a series of choices. Experts agree that there is no single "right" investigative method. The choice should depend on the research question under investigation, the available resources, and the investigator's judgment.

CONTROVERSIAL ISSUES IN DEVELOPMENTAL PSYCHOLOGY

A number of controversial issues have generated spirited debate in the field of developmental psychology over the last 70-year period. In fact, most theories of development were formulated to address and have been evaluated based on their accommodation to the critical questions: What roles do nature (heredity) and nurture (environment) play in development? Is development a continuous process or stagelike? Are there critical (sensitive) periods in development? Can progress of undesirable characteristics related to advanced aging be slowed down or reversed? Although these issues have been generalized to human development as a whole, each question has direct relevance to our understanding of motor development.

■ *Objective 1.9*

Perhaps the most enduring issue of child development has been the **nature-nurture** question. This topic has fueled controversy among theorists since the early days of psychology and continues to stimulate debate to date. In essence, the issue is a debate over whether development is the result of individuals' genetic endowment (nature) or the kinds of experiences they have had (nurture). In other words, it is the issue of "heredity versus the environment." For example, do children naturally crawl at 6 months and walk at 12 months because of what they have learned (from the environment) or because of some biological unfolding process? These are the types of questions that theories are based on.

Nature-Nurture

From a motor development perspective, if environmental factors are the chief determinants of motor behavior, a structured educational setting during infancy may stimulate a successful athletic career. To the other extreme, if genetic elements predominate and athletes are "born" and not "made," early educational experiences would do little to influence motor behavior. A more contemporary view recognizes that heredity and the environment are intertwined and that human behavior is influenced by an interaction of the two factors. In essence, genes and environment are necessary for an organism to exist. The emphasis of more and more research efforts has been on seeking to understand the interactive mechanisms that mediate development, rather than on the proof or disproof of extreme positions. As you read the upcoming section, Theoretical Views, you are encouraged to note how each approach addresses this universal question of nature versus nurture.

Everyone generally agrees that with development, an individual's behavior and abilities change. However, a controversy exists on how best to explain these changes. The notion of **continuity** reflects the view that development involves gradual, cumulative change from conception to death. Developmentalists who emphasize this view suggest that puberty, for example, though it appears to be an abrupt, discontinuous event, is actually a gradual process occurring over several years. The changes that occur are believed to be continuous and based upon "quantitative" improvements. For example, as neural and muscle components gradually increase, various functional abilities may then appear (e.g., walking, exhibiting strength, coordination).

Continuity-Discontinuity (Stages)

From another perspective, **discontinuity** is the view that development is much more stagelike. That is, it undergoes a series of stagelike transformations during which underlying processes and structures exhibit rapid reorganization followed by a period of relative stability (Bukatko & Daehler, 1995). For example, theorists such as Piaget and Erikson describe development in terms of the individual's progress through a series of distinct **stages,** or periods, when common developmental milestones occur. The concept of developmental stages suggests "qualitative" differences in how individuals behave (e.g., think, move, feel) at certain periods in their lives. Complementing this, development appears to undergo rapid changes as one stage ends and a new one begins, which is then followed by a relatively stable period with minimal transition.

Readers of developmental literature will come across other terms that have similar meanings as the concept stage; such as *sequence* and *phase*. These terms may also describe qualitative transitions over time but generally are not strictly affixed to specific age levels. Although a sequence or phase may be associated with behaviors usually exhibited during a specific age period, the basic premise of a phase (or sequence) is that a specific behavior may be accomplished at varying age levels but generally in the developmental order identified.

Most, if not all, theorists agree that there are continuities and discontinuities in the course of human development. The debatable question is which processes occur in stages or by plateaus (phases). The issue appears to be one of identifying which developmental processes are controlled by continuous underlying constructs or mechanisms (continuity) and which processes are influenced only when experiential circumstances, events, or environmental forces interact with the individual (discontinuity). From a contemporary motor behavior perspective, Thelen and Ulrich (1991), in reference to dynamical systems theory, emphasize that one of the primary challenges of this line of research is to "account for discontinuities in performance arising from processes that are themselves continuous" (p. 2).

Critical Periods

A **critical period** refers to an optimal time for the emergence of certain behaviors and is a point in development when an individual is unusually receptive to influences by environmental or other mitigating factors. Some supporters of this concept also believe that specific forms of stimulation must be present during the critical period for normal development to proceed. For example, suppose that there is an optimal period between 4 and 8 months of age for the infant to acquire basic reach-grasp skills. If this assumption is correct, infants who do not have an opportunity to learn the skill during that critical period may never be as proficient as if they had acquired it during the optimal time period. Some experts also believe that the failure to form a strong emotional bond with a significant adult during infancy could affect emotional stability later in childhood.

Not everyone agrees that critical periods are useful for all domains of development. Although it seems reasonable that children may be influenced either positively or negatively from lack of certain experiences, there is difficulty in determining the extent of the problem. Some children have the astonishing ability to display considerable resilience to adverse conditions. Consider the child born into a low socioeconomic condition with lack of proper nutrition, parental guidance, education, and enrichment experiences. Such children may still grow up to be professionals, athletes, and productive citizens (Werner & Smith, 1982). Perhaps the crucial challenge is to identify the most vital experiences or deprivations and the circumstances that allow the individual to recover from any adverse impacts of these factors.

From another perspective, a critical (sensitive) period refers to a span of time when the individual may be especially "sensitive" to specific influences, such as during possible exposure to teratogens, which are agents (e.g., drugs, stress, alcohol) that can cause fetal malformation. The embryonic and fetal periods have been reported as especially sensitive to teratogens (Little, 1992).

Certain aspects of advanced aging, such as the decline in vision and hearing, loss of muscular strength, and general psychomotor slowing are undesirable. Declines in memory function, though somewhat controversial, are another one of the often-mentioned areas of regression with advancing age. One of the really exciting possibilities of development, however, is that certain undesirable aspects may be reversed under some circumstances or, if not reversed, modified to be less detrimental in their effects. Can certain characteristics of regression be altered or removed through appropriate intervention, such as training and educational programs that modify diet or health-related lifestyle practices? Scientific speculation suggests that even purely biological deterioration may someday be reversible by means of techniques developed through research and a better understanding of the physiological mechanisms underlying the aging process. Although this **modifiability-reversibility** of biological decline is an exciting prospect, little scientific evidence exists as yet that would suggest how much change is possible. Aging takes its toll on almost every facet of physiological function, but the quality of life among individuals can be extremely variable. Current research findings do not indicate that exercise training can actually "retard" the aging process, but evidence does suggest that regular physical activity and proper diet can improve the quality of life. More discussion on this concept will be given in chapter 11.

Modifiability-Reversibility

In the field of developmental psychology, theories are used to describe various patterns of behavior and to explain why those behaviors occur. From this perspective, a *theory* may be defined as a set of concepts and propositions that allow the theorist to organize, describe, and explain some aspect of behavior. To gather a comprehensive and meaningful understanding of motor development, it is essential to know something about the way individuals learn and react to their environment. Theories that have earned respect for describing some aspect of behavior serve to enrich our knowledge of the complexities of human development.

In the early years of developmental psychology much of the information reported, especially on motor development, was of a "descriptive" nature. That is, researchers viewed and explained development through a rather systematic cataloging of growth norms and age-related behaviors. Some researchers for example, would ask, What is the age when infants sit, stand, and walk, or express a two-word vocabulary? because one of the primary views was that maturation (i.e., maturational theory) rather than learning (or the environment) was the main impetus for developmental change. On the other hand, traditional learning/behavioral psychologists focused less on normative characteristics (explained as

THEORETICAL VIEWS

■ *Objective 1.10*

a product of maturation) and more on learning principles. In more recent years, contemporary views of development have taken a much more comprehensive and explanatory perspective of developmental change. These theories consider the environmental *context* in which development occurs and try to explain the *mechanisms* and *processes* accounting for change (growth, development, and behavior), thus attempting to provide a more comprehensive explanation of the dynamic interaction between the organism and its environment.

What follows is a brief description of seven traditional and contemporary theoretical approaches to the study of human development. Although there is wide diversity among the views and some were not formulated around motor behavior, a common thread running through all approaches is the issue of "nature-nuture." While reviewing each approach the reader is encouraged to reflect on this issue by determining its emphasis in development; that is, does the approach place more importance on the role of biological factors? environmental influences? or is it more of a balance of nature-nuture?

In addition, no dialogue of life-span development would be complete without comment on theories of advanced aging. Although most treatises on this subject, and the discussion that follows, focus on the biological aspects of advanced aging, evidence also suggests that an individual's longevity and quality of life are influenced by social factors. Chapter 13 will discuss some views on how an individual's perspective on aging and social experiences may influence the aging process. It is hoped that with this information, the reader will develop a better understanding of the aging process as influenced by nature and nurture.

Biological-Maturation Approach

Gesell's *maturation theory* played a significant role in the evolution of the study of child development during the 1930s and 1940s. Gesell was one of the pioneers of systematic observation; his timetables detailing when most children achieve certain developmental milestones are still widely used today. Basically Gesell's theory contends that development is the result of inherited factors and requires no stimulation from the external environment. The maturational view of development emphasizes the emergence of patterns of development of organic systems, physical structures, and motor capabilities under the influence of genetic forces. Gesell believed that growth of the intellect and of the motor functions was tied closely to growth of the nervous system, a perspective that underlies our fundamental understanding of phylogenetic and ontogenetic motor behavior. Unless certain neurological and biological characteristics are sufficiently developed (matured), it may be futile to practice certain phylogenetic skills. On the other hand, learning and experience (practice) are necessary for the existence of ontogenetic behavior. For example, maturational theory is illustrated in those developmental charts that show how human locomotion develops in the infant from rolling to walking. Most specialists today would agree that this aspect of development can not be accelerated by special training. Unfortunately, perhaps, Gesell extended his maturational view of development to form a more general perspective of human development. Using an analogy of parents tending plants, parents may "plant" and tend their children, but whether they turn out to be hardy is determined primarily by inheritance, not by care (Clarke-Stewart, Perlmutter, & Freedman, 1988).

Erikson (1963) expanded Freud's theory of psychosexual development and focused on the social and cultural aspects of human behavior. Through his *psychosocial theory,* Erikson proposed that individuals' lives are shaped by the way in which they cope with their social experiences. Like Piaget, Erikson viewed development as occurring in distinct stages throughout a person's life span. He identified eight stages (from birth to death) in the development of personality, stressing that each stage is characterized by some emotional or interpersonal problem that must be resolved by the individual before the stage can be completed. Erikson also stressed that personality development is highly dependent upon the characteristics of the relationship between the child and parent (or other adult). Psychosocial theory supports the notion that what happens early in life is of utmost importance to the development of a healthy personality. Parents and significant others (such as teachers) play a major role in the development of an individual's self-concept and conscience. Erikson's opinion on the importance of physical activity in development is rather implicit; however, he does point out the importance of success-oriented movement experiences as a way of reconciling the developmental crises through which each person passes (Gallahue, 1989).

Psychosocial Approach

John Watson (1928) gave birth to behaviorism, from which several behavioral-learning theories have been proposed. Watson believed that children were the products of their parents' training. According to *behavioral learning theory,* the primary motivator of development and behavior is the environment, which manipulates basic drives and needs by offering incentives for "appropriate" behavior and punishment for "undesirable" behavior. Another aspect of this view is that behavior can be influenced in this way only after a process called *conditioning* has taken place. Whereas Watson adhered to a model of *classical conditioning* (i.e., a conditioned stimulus paired with an unconditioned stimulus will, over time, produce a particular response), Skinner (1974) developed what is known as the operant conditioning model.

Learning-Behavioral Approach

Operant conditioning contends that human behavior is controlled by environmental reinforcers and that behavior can therefore be altered by controlling the reinforcement. *Reinforcement* in this context refers to the favorable consequence that follows a response, thus increasing the chances that the response will occur again. *Operant learning* is a form of training in which reward and punishment are used to shape behavior. This type of learning has been demonstrated in many behavior modification experiments. Several psychologists, following the lead of Skinner, have applied the principles of operant learning to the study of child behavior.

Bandura (1977, 1986, 1989) and Mischel (1984) are the main architects of contemporary arguments associated with observational learning and social learning theory. *Observational learning* contends that virtually all of what individuals (children) learn comes from observing others. Bandura suggests that through observational learning (also called *modeling* or imitation), we cognitively represent the behavior of others and then are likely to adopt this behavior ourselves. For example, a boy may adopt the competitive manner of his father (as seen while golfing) and display such behavior when observed with the boy's peers. Positive behaviors may also be adopted, for example, after observing peers or family members participating in healthy physical activity.

The application of behavioral learning theory plus cognitive processes to the development of learning is known as *social learning theory.* This view emphasizes behavior, the environment, and cognition as key factors in development, with each factor influencing the other factors. According to this theory, because we think, we are able to foresee probable consequences of our actions (and alter our behavior). Therefore the individual has the ability to control behavior and, if desired, resist environmental influence.

Both the behavioralist and social learning views may be criticized for lack of elaboration on the biological determinants of children's development.

Cognitive-Developmental Approach

Of the theories of cognitive development, Jean Piaget's efforts have perhaps had the single greatest effect on modern developmental psychology. Piaget (1963) believed that the individual discovers solutions to problems primarily through interaction with the environment. According to *Piaget's theory of cognitive development,* the child is not a passive recipient of events (as the behaviorists propose), but, rather, the child seeks out experiences. Piaget studies focused on the structure of cognitive thought and its orderly sequence of development. Though Piaget did not deny the significance of maturation in development of the intellect, he did not view all of development as the unfolding of biological processes as maturational theory contends. Piaget instead viewed the developmental process as the interaction of biological maturation and environmental (the more contemporary term being *contextual*) experience. The researcher believed that an innate force drives children to actively pursue cognitive *equilibrium* (balance). For example, when a child feels comfortable with what she thinks, that is, when she is able to assimilate information in the environment into her existing cognitive structures, she achieves a level of equilibrium. The general notion of equilibrium as noted in the later works of Piaget (Piaget, 1985) has been mentioned by contemporary motor developmentalists in reference to dynamical systems theory (Thelen & Ulrich, 1991).

According to Piaget, cognitive structures are developed as the result of organization and adaptation. *Organization* is the building of simple perceptual and cognitive processes into higher-order mental structures. *Adaptation* is the process of continuing change that takes place as a result of the individual's interaction with the environment. To explain the process of adaptation, Piaget uses the terms schema, assimilation, and accommodation. A *schema* is the basic unit of an organized pattern of behavior (cognitive structure), or of representations of actions or concepts. A schema may be used just as it was originally programmed into memory or it may be altered in some way to meet the demands of a changing environment. The process of *assimilation* is at work when individuals deal with the environment through their current schema. *Accommodation* refers to the "modifying" process that takes place when the structure of basic schema is changed to meet the demands of the environment, such as in the performance of a novel task. Thus, adaptation is the balance between assimilation and accommodation, which provides a temporary state of equilibrium. The notion of the organization of similar cognitive functions into schema was formalized into schema theory (Schmidt, 1975, 1977) to become one of the foremost explanations of the way motor skills are acquired.

One of Piaget's lasting contributions to developmental psychology was the notion of stages. Even though researchers today do not always agree on the meaning of the term "stage," Piaget's broad distinctions among basic periods of development from infancy through adolescence continue to be used. Piaget proposed that these particular periods of cognitive development are based on qualitative, structural changes that have a fixed order and cannot be skipped. Piaget identified four stages of intellectual development:

1. *Sensorimotor period (0 to 2 years).* Infants learn through their sensory impressions, motor activities, and the interactions of the two.
2. *Preoperational period (2 to 7 years).* The individual continues to discover both the environment and self through movement activities. Children are not yet able to think using logical operations to transform information, but they have developed the ability to think in images and symbols and can form mental representations of objects and events.
3. *Concrete operations (7 to 11 years).* Children begin to think in a logical manner, and although they cannot yet abstract, they are capable of thinking in terms of the concrete or actual experience.
4. *Formal operations (over 11 years).* Adolescents and adults can think abstractly and use logical and systematic problem solving in cognitive processing. The individual is able to create a hypothesis to account for some event and then test it in a deductive fashion.

Many of Piaget's writings make references to movement activities and the importance of play. Piaget believed that through play (both structured and free), the child has the opportunity to test novel physical, cognitive, emotional, and social behaviors that cannot otherwise be accommodated in the real world. Once the behaviors are tested through play, they become part of memory; hence, from a cognitive perspective, play is a medium for intellectual development.

Although many contemporary researchers agree with Piaget's basic suggestions, in recent years several revisions, extensions, and new interpretations of the theorist's works have appeared. Two areas of controversy center on the assumed qualitative differences in thinking between children and adults and the stagelike character of the transformations. For example, many researchers suggest that Piaget underestimated how quickly children can learn certain cognitive strategies (e.g., T. Flavell, 1985; Meltzoff, 1988). Other neo-Piagetian theorists have proposed extended models that offer a picture of development that is more continuous and gradual, rather than stagelike (Case, 1985; Fischer & Pipp, 1984). Arlin (1975, 1977) extended Piagetian theory into adulthood (another frequently noted limitation) by proposing a fifth stage of cognitive development termed "problem finding." The theorist suggests that rather than just possessing the ability to answer complex questions, adults seek out challenges and operate at a higher level of critical thinking by creating problems and discovering questions to ask.

Information-Processing Approach

Since its introduction in the 1970s *information processing* has been one of the dominant approaches in the field of cognitive psychology and motor behavior (Shea, Shebilske, & Worchel, 1993; Schmidt, 1988). The information-processing approach analyzes cognitive activities in terms of successive stages of information processing such as attention, perception, memory, thinking, and problem solving. Scientists who support this approach have drawn close analogies between the human mind and a computer. Like computers, human minds have a limited capacity for processing information. With development, changes in the capacities of cognitive structures, including critical strategies, occur. The primary challenge for researchers is to provide an understanding of how individuals' thinking changes as they develop (across the life span). Richard Schmidt, one of the leading scientists in the field of motor behavior, notes that the information-processing perspective studies movement from the point of view of the human as a processor of information; it focuses on storage (memory), coding, retrieval, and programming, the result being production of an outcome (movement). Details of the information-processing perspective (a simplified model) and its relevance to lifelong development will be discussed in chapter 7.

Contextualist (Ecological) Approach

Since about the mid-1980s the *contextualist approach* has had a profound influence on the study of human development. This view emphasizes the role of the environmental *context* and recognizes the importance of the historical period in which the individual develops (Kleinginna & Kleinginna, 1988; Morris, 1988). Closely associated with this approach is ecological theory, which gets its label from the branch of biology dealing with the relation of living things to their environment and to one another. The basic premise is that human development is inseparable from the environmental contexts in which the individual develops.

One of the most extensive models of the contextual approach is the *ecological systems theory* proposed by Bronfenbrenner (1986, 1989, 1993). This perspective emphasizes the broad range of situations and contexts individuals may encounter. These characteristics continue to be affected and modified by the individual's contextual surroundings. Bronfenbrenner describes the settings or environments (contexts) in which individuals develop as five distinct systems: the microsystem, mesosystem, exosystem, macrosystem, and chronosystem (see fig. 1.4).

In brief, the *microsystem* is the setting in which the individual lives, the contexts being the person's family, peers, school, and neighborhood. The

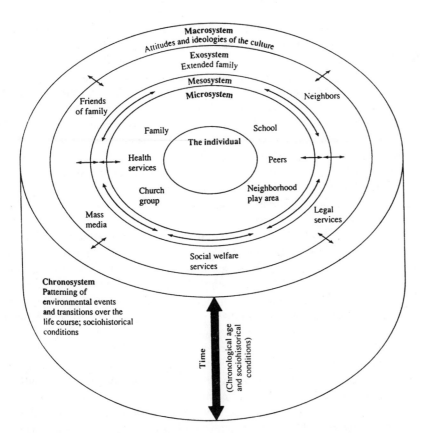

Figure 1.4
Bronfenbrenner's ecological theory of development.

mesosystem is concerned with the interrelations among the various settings within the microsystem; for example, the relation of family to school experiences, or family to the community. When experiences in another social setting in which the individual does not have an active role influence what the individual experiences in an immediate context, the setting is the *exosystem*. An example of this is city government (parks), which is responsible for the quality of play and recreational opportunities. The *macrosystem* involves the culture in which the individual exists. In many instances, behavior patterns, beliefs, and other products of a particular group are passed from generation to generation. In many cultures, sport and dance are an integral part of the setting, thus providing the individual with the refinement of such behaviors. The final contextual influence is the *chronosystem,* which involves the patterning of environmental events and transitions over the life span; that is, the sociohistorical contexts. For example, females of today are much more likely to participate in athletic endeavors than they were 40 years ago. Although the individual with his or her biological and psychological makeup is placed at the core of this model (Bukatko & Daehler, 1995), its main criticism has been the failure to adequately account for the interaction between genetic and contextual history (Santrock, 1995).

Needless to say, from a motor development perspective, the contextualist approach has great promise and direct implications for providing a more comprehensive explanation of motor behavior. As in the general field of human development, several motor developmentalists have come to realize that a comprehensive study of behavior should include an understanding of both the biological and environmental contexts in which motor behaviors emerge or are performed.

Pioneering efforts to include the ecological perspective in the study of motor development were introduced through E. J. Gibson's (1982, 1988) and J. J. Gibson's (1979) ecological theory of perception. In brief (further elaboration is given in chapter 6), to illustrate the active and meaningful nature of perception (the monitoring and interpretation of sensory information), the researchers described the concept of *affordances,* that is, things that the "environment" offers or provides to the individual (perceiver). For example, a soccer ball affords the child the opportunity to develop the skill of kicking. In essence, affordances are the stimulus for action. They are used to describe the reciprocal relation between the organism and the environment that is required to perform a given action. Parts of this theory provide the foundation for one of the most dynamic theoretical approaches to the contemporary study of motor development—developmental biodynamics (the behavioral study of perception and action), to be described in the latter part of this section.

Other emergent lines of motor development research have also adopted the contextualist approach. For example, Burton & Davis (1992), in reference to their ecological model of motor behavior referred to as ecological task analysis (ETA), suggest that qualitative and quantitative aspects of movement emerge from three sets of constraints: characteristics of the *performer* (biological factors), *environment,* and *task.* These same constraints have also been suggested as a model for classifying factors that may influence developmental change, that is, they interact to constrain the control of motor tasks (Newell, 1986).

The contextual approach has also been adopted in aging research. To illustrate, in her work with age-related changes in posture and movement (e.g., walking, reaching, lifting objects), Woollacott (1993) states that "an inevitable accompaniment of the aging process for many older adults is a restriction in their ability to move independently within the *context* of constantly changing task demands and environmental *contexts*" (p. 56). In essence, though these individuals can manage (walk) independently within their home environment, when presented with a different set of contextual factors such as standing on a moving bus, walking up and down stairs, or moving across unfamiliar surfaces, their level of mobility may be significantly hindered. This same general scenario is applicable to the infant learning to walk.

Inquiries into other contextual factors such as classroom culture (e.g., teacher's behaviors) and contexts of other motor-skill learning environments (e.g., home, community) (Branta, 1993) also show promise in providing a more comprehensive understanding of motor development and behavior. Figure 1.5 presents a graphic illustration of the contextual model for studying human motor development.

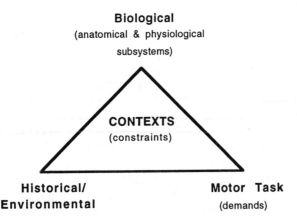

Biological

(anatomical & physiological subsystems)

CONTEXTS

(constraints)

**Historical/
Environmental**

Motor Task

(demands)

Figure 1.5
Contextual model for studying motor development.
Source: Data from Newel, 1986.

**Developmental
Biodynamics
Approach**

One of the most exciting and promising approaches to the comprehensive study of motor development has emerged along with the arrival of the 1990s. Sparked by advances in the neurosciences, biomechanics, and behavioral sciences, *developmental biodynamics* represents an interdisciplinary attempt to integrate promising theories and findings related to the behavioral study of *perception* and *action* (Lockman & Thelen, 1993). In essence, the primary attempt is to describe and "explain" the intimate connection between the brain and body. This approach represents a vast improvement from the more traditional maturational approach, which described change in a descriptive series of developmental milestones. Although this approach is just beginning to address the challenges associated with a developmental synthesis of the brain-body phenomenon, a few researchers have already made significant contributions toward a better understanding of the problem. A thorough scientific examination of the topic is beyond the scope of this section and text (further discussion is provided in chapter 7), but a brief description of the basic tenets of the approach should be useful in understanding the merits of an interdisciplinary view of the research problem.

Lockman and Thelen (1993) note that Bernstein (1967) introduced the term "biodynamics" in his writing on the genesis of motor coordination. The researcher's basic premise was that coordination and control emerged from continual and intimate interactions between the nervous system (brain) and the periphery (body). The central question Bernstein asked was, How can the brain control "individually" all the diverse elements (e.g., neurons, muscles, joints) and multiple linkages associated with the production of movement? To address the problem, Bernstein introduced the notion of *synergies*—classes of movement patterns involving collections (groups) of muscle or joint variables that act as basic units in the regulation and control of movement. Bernstein also proposed that there must be an inseparable relation between the brain and associated movement components (e.g., hands, arms, legs), and they must act and develop collectively. This basic premise set the agenda for the efforts of different disciplines to address the interface between perception and action.

This basic theme—interaction between perception and action—and variations have been addressed using, for example, mathematical techniques of nonlinear dynamics, principles of biomechanics, neural modeling, and, as previously mentioned, considerations of the context in which the development and behavior occurs. In addition to computer modeling and the use of robotics and animals, experimental observations have included humans, usually infants, performing a variety of upper limb (reaching, grasping), locomotor (walking), and reflexive/spontaneous movements. For a detailed description of contemporary studies associated with developmental biodynamics, refer to the special section of *Child Development* introduced by Lockman and Thelen (1993).

An applied branch of developmental biodynamics that has stimulated considerable inquiry in recent years is **dynamical systems** research (Thelen & Ulrich, 1991). Originating from Bernstein's problem with brain and body control, Kugler, Kelso, and Turvey (1980, 1982) introduced the dynamical systems approach, which focuses on the developmental perspective, that is, the "emergence," or unfolding, of motor behavior. Most traditional views on how human movement evolves have been based on the descriptive-maturational perspective, that is, the orderly and sequential emergence of behaviors linked to development of the nervous system (e.g., spontaneous/reflexive behavior, crawling, walking). Although this type of research has contributed to the understanding of when motor skills emerge and what they look like, one criticism has been that it falls short in explaining *how* (i.e., the processes of change) motor behavior emerges. The dynamical systems perspective suggests that coordination and control emerge as a result of the dynamical properties of the muscle collectives (i.e., *coordinative structures*). This view of development emphasizes the importance of the *dynamic* and *self-organizing properties* of the motor system to the individual's developing motor competencies. Along with biological considerations, this perspective (as a feature of biodynamics) suggests that change is stimulated by characteristics of the environment and demands of the specific task (contexts). Since its inception dynamical systems research has done much to provide a better understanding of how motor behavior emerges. Further elaboration on this important perspective will be given in chapters 7 and 8. Table 1.1 gives a summary of the theoretical approaches and their positions on nature-nurture discussed in the previous sections.

Biological Theories of Advanced Aging

■ *Objective 1.11*

As noted earlier, (advanced) aging may be defined as the diminished capacity to regulate the internal environment (e.g., cellular, organismic structures), resulting in a reduced probability of survival. Biological aging is commonly equated with biological decline (regression)—such as losses in eyesight, muscle and bone mass, strength, and flexibility—and a general psychomotor slowing of the body. Although the body is continuously aging from conception to death, after about age 30, regression in some form typically develops. There are numerous theories, subtheories, and lines of research on aging, with no single hypothesis answering all of the questions proposed. The following discussion briefly describes some of the more general explanations of biological aging. For a more detailed

Table 1.1 Theoretical Positions on Nature-Nurture

Approach	Description
Biological-Maturation	Heredity is more important
Psychosocial	Balanced biological-environmental perspective
Learning-Behavioral	Environment is more important
Cognitive-Developmental	Interaction between nature-nurture. Maturation sets limits on developmental pace, but experience is vital
Information-Processing	Not addressed, but hardware-software metaphor suggests interaction
Contextualist (Ecological)	Strong environmental view
Developmental Biodynamics	Emphasizes biological dynamics with some consideration of environmental context

discussion, refer to Birren and Schaie (1990) and Christiansen and Grzybowski (1993) in the Suggested Readings.

✓ **Genetic theories.** Also referred to as *cellular clock* and *programmed* theories, these views suggest that aging is under the direction of the DNA in human genes. Evidently there are proliferative and antiproliferative genes that, in theory, provide checks and balances of biological systems. Antiproliferative genes result in decreased cell division, representing a weakening agent to the body. Once these genes become more active, regression develops. It seems clear, based on several lines of supportive research, that some sort of genetic program determines the aging process, but our understanding of the specifics remains unclear.

✓ **Wear-and-tear theories.** The relatively simple premise of this theory is that aging is a result of long-term, accumulated damage to various vital bodily systems. In the course of daily living, the body suffers some damage but repairs itself under relatively normal conditions. With advancing aging, this ability gradually becomes less effective. Although some aspects of this notion appear reasonable, such as wear due to prolonged exposure to sunlight and other damaging elements, there is no evidence that hard work causes early death. To the contrary, vigorous exercise is a factor in predicting longer, not shorter life.

✓ **Cellular garbage/mutation theories.** These theories suggest that with advancing age, the body accumulates cellular garbage (waste products) due to destructive mutation, resulting in a decline in functioning. Two of the most noted views on this approach are the *free-radical* and *cross-linkage* theories. Free radicals are highly unstable molecules that react with other molecules in a way that may damage cells and diminish bodily functions. The normal process of cellular metabolism produces some free radicals, others are derived from the environment. Although the body can protect itself against some free radicals, with time the number builds up and results in signs of aging. Related to this view is the cross-linking theory, which focuses on the processes by which different types of molecules are joined together permanently (cross-linked), resulting in deterioration such as wrinkling and atherosclerosis. Another substance connected with the wear-and-tear of the body over time is *lipofuscin*. As the body ages, this inert granulated pigment accumulates in nerve, cardiac, and skeletal cells, filling spaces and interfering with function. It has been suggested that lipofuscin may be created during metabolism, and it is probably an effect rather than a cause of aging.

Immune system theories. With aging, the immune system's response declines, thus the body's ability to fight many infections diminishes. It is suggested that many symptoms of aging are a result of the accumulated effect of past and present disease that the system was or is unable to control. From another perspective, some researchers suggest that with aging, the body loses its ability to distinguish between its own specific materials and threatening foreign organisms (e.g., bacteria, viruses, fungi). Thus it attacks and annihilates its own tissues at a gradually progressing rate. Although these theories do not address aging of the immune system itself, it stands to reason that life expectancy would increase if the power of the system were sustained.

Hormonal theories. One possible cause for decline in the immune system's effectiveness and the subsequent aging effects has been atrophy of the thymus gland, which influences immune functioning via thymic hormone. Its progressive, age-related loss has been linked with declines in functioning of certain immune cells. Other examples of hormonal influence on characteristics associated with advanced aging are estrogen decline and bone loss in women and loss of growth hormone, which is related to a decline in muscle mass, bone density, and body fat and the occurrence of thin, wrinkled skin. Individuals who have been treated with growth hormone (discussed in chapter 5) have shown improvements in these areas.

SUMMARY

Motor development is a field of specialized study that examines patterns of growth, development, and motor performance (behavior) across the life span. With the development of a lifelong perspective on motor development has emerged the multidisciplinary approach to studying associated behavioral change. This approach has added significantly to our understanding of total human development and behavior. The major objective of the developmentalist is to identify and describe the characteristics associated with change across the life span.

Several general and specific terms are commonly associated with the study of growth and motor development. Growth refers to an observable change in quantity (size); development represents change in the individual's level of functioning. Motor behavior is the product of growth and development and describes observable changes in movement or the learning and performance of a movement or motor skill. All characteristics and behaviors are linked to other factors, including heredity, maturation, and experience. Important terms associated with developmental patterns include: cephalocaudal/proximodistal (the orderly and predictable sequences of growth and motor control); phylogenetic/ontogenetic (behaviors acquired naturally or learned); and differentiation/integration (specialization and coordinated interaction). Aging is another significant aspect of the lifelong motor development process. Although aging is continuous from conception, it is most often characterized by some type of diminished capacity.

In the study of human development, several fundamental and predictable observations have relevance to motor development. In brief, heredity and environmental context (with social expectations) are important; development appears stagelike, but its bases are underlined by processes that are likely continual and cumulative; development follows a definite and predictable pattern; development is aided by stimulation, with early foundations being critical; though critical periods are apparent, the human body is quite resilient; and, each individual is unique.

Life-span development covers five primary periods: prenatal (conception to birth), infancy (birth to 2 years), childhood (2 to 12 years), adolescence (12 to 18 years), and adulthood (18 years and older). Interrelated with these periods are the seven phases of motor behavior: reflexive, rudimentary, fundamental movement, sport skill, growth and refinement, peak performance, and regression. These phases provide an outline for the study of lifelong growth, development, and motor performance.

Motor development is a scientific and applied discipline. Through motor development research we learn to better understand the capabilities of individuals in order to guide their development and improve their health and performance. The scientific method of studying human development and performance may employ longitudinal, cross-sectional, sequential, or cross-cultural research designs. Of these designs, the longitudinal method (tracking the same group across several years) is preferable in many ways; its limitation is the extended period of time it may take to complete a study. The most popular method for collecting information and the basic tool of any science is systematic observation. This can be conducted in either a natural or structured setting. Case studies, interviews, surveys, and standardized tests are other widely used methods of collecting information on human development.

Several controversial issues are associated with the study of human development. Among the most debated issues have been: nature-nurture (the relative importance of heredity/maturation or the environment on development), continuity-discontinuity (whether changes are gradual and continuous or sudden and stagelike); critical periods (whether there are optimal times for the emergence of certain behaviors and specific periods when the body is most susceptible to teratogens); and modifiability-reversibility (whether certain undesirable behaviors and physical characteristics can be reversed or minimized).

Much of our present understanding of human development is derived from various traditional and contemporary theoretical models. Whereas earlier approaches to explaining development were more descriptive, using a systematic cataloging of norms and age-related behaviors, more-contemporary approaches attempt to explain the mechanisms and processes accounting for change. Included in the comprehensive study of lifelong development are considerations of the environmental context in which development occurs and explanations of advanced aging.

SUGGESTED READINGS

Birren, J., & Schaie, K. (Eds.). (1990). *Handbook of the psychology of aging* (3rd ed.). New York: Academic Press.

Bornstein, M. H., & Lamb, M. E. (Eds.). (1992). *Developmental psychology: An advanced textbook* (3rd ed.). Hillsdale, NJ: Lawrence Erlbaum.

Bronfenbrenner, U. (1993). Ecological systems theory. In R. H. Wozniak (Ed.), *Development in context.* Hillsdale, NJ: Erlbaum.

Bukatko, D., & Daehler, M. W. (1995). *Child development* (2nd ed.). Boston: Houghton Mifflin.

Christiansen, J. L., & Grzybowski, J. M. (1993). *Biology of aging.* St. Louis: Mosby.

Clark, J. E., & Whitall, J. (1989). What is motor development? The lessons of history. *Quest, 41,* 183–202.

Crain, W. (1992). *Theories of development* (3rd ed.). Englewood Cliffs, NJ: Prentice-Hall.

Lockman, J. J., & Thelen, E. (1993). Developmental biodynamics: Brain, body, behavioral connections (Special section introduction). *Child Development, 64,* 953–959.

Plomin, R., & McClearn, G. E. (Eds.). (1993). *Nature-nurture and psychology.* Hyattsville, MD: American Psychological Association.

Santrock, J. W. (1995). *Life-span development* (5th ed.). Dubuque, IA: Brown & Benchmark.

Shaffer, D. R. (1993). *Developmental psychology* (3rd ed.). Pacific Grove, CA: Brooks/Cole.

Thelen, E. (1989). The (re)discovery of motor development: Learning new things from an old field. *Developmental Psychology, 25*(6), 946–949.

PART ○ *two* -

- - - - - - - - - → *chapter*

Part Two provides the reader with information related to the structural and functional elements of human development and motor behavior. Although each of the component chapters in this section addresses the process of physical growth, the specific

Biological Growth and Development

topics have been placed in a broader context to provide a more explicit interpretation of developmental activity. Keep in mind that these topics are strongly interrelated: Chapters 2 and 3 focus on the hereditary, neurological, and physical aspects of human development that are typically manifested as structural growth; chapter 4 builds upon that information in discussing change related to functional processes; finally, chapter 5 describes the variety of other factors that may influence growth and development, such as prenatal care, nutrition, environmental agents, hormones, and physical activity.

chapter

2

Heredity and Neurological Changes

OBJECTIVES

Upon completion of this chapter, you should be able to

1. Define the term *heredity* and describe its primary characteristics. **37**
2. Identify the three primary functions of the nervous system. **39**
3. Outline the basic structure of the central and peripheral nervous systems and briefly describe their functions. **39**
4. Explain the basic structure and function of a neuron and identify the various types of neurons. **41**
5. Identify the terms *motor unit* and *motor pathways.* **42**
6. Describe how nerve impulses are conducted. **45**
7. Outline and describe the sequence of early developmental changes in the central nervous system. **46**
8. Identify the primary developmental changes that occur in the brain. **53**
9. Explain brain lateralization in terms of both brain structure and its effect on motor behavior. **55**
10. Outline and briefly describe the primary neurological changes that occur with aging. **55**

KEY TERMS

heredity
genotype
phenotype
central nervous system (CNS)
peripheral nervous system (PNS)
motor unit
nerve conduction velocity
cell proliferation

integration
differentiation
myelination
myogenic behavior
neurogenic behavior
cell death
brain lateralization

Heredity is the total set of characteristics biologically transmitted from parent to offspring. From the moment of conception, two primary factors interact to determine the developmental patterns of the individual: genetic makeup and the environment. The environment, even during prenatal development, modifies and interacts with heredity to shape the individual and to control the extent to which maximal potential will be realized. The *nature-nurture* debate suggests the significant influence of these two factors. Today behavioral scientists know that nature (heredity) and nurture (the environment) both contribute continually and inseparably to an individual's development. There is no organism in which one part is influenced *only* by genes and another segment *only* by the environment. The capacity of a living thing to respond to experiences in the environment is as genetically determined as the maturation of its neurological system.

A *gene* is the basic unit of heredity found within a chromosome. Each human cell contains 46 *chromosomes* (in 23 pairs) in its nucleus. Each chromosome consists of more than 1,000 genes, strung together in a chainlike formation. Genes are made up of molecules of *DNA* (deoxyribonucleic acid) packed tightly into the cell nucleus. During normal cell division, or *mitosis,* the DNA "unzips," breaking its rungs and forming two separate strands (fig. 2.1). In this fashion, all 46 chromosomes of the single cell reproduce themselves.

The DNA molecule contains the genetic code that determines what hereditary information is to be transmitted from one generation to the next. The code contains "instructions" for creating an individual with a sex type and basic human characteristics like head, limbs, and hair (rather than fins or feathers). Only one of the 46 pairs of chromosomes determines whether the new human being is male or female. The female has two X chromosomes; the male has one X and one Y chromosome. Females always contribute an X chromosome to the offspring, so it is the male who determines the sex of the next generation. The code also provides the directions for creating a new individual with his or her own unique combination of characteristics such as intelligence, height, blood type, and eye color. To get from blueprint to creation, information in the DNA molecules is transcribed into molecules of *RNA* (ribonucleic acid). These molecules take the code from the DNA into the cytoplasm of the cell, which contains the raw material for making protein. It is out of the proteins, made according to instructions from the RNA, that the functioning parts and capabilities of each human being are developed.

The **genotype** is all of an individual's genetic inheritance. The actual expression of the genotype as the person's visible characteristics and behavior is referred to as the **phenotype.** The phenotype depends not only on the individual's genetic makeup but also on all the environmental elements that affect the person from the moment of conception. The observable characteristics of a phenotype include physical traits such as height, weight, eye color, skin pigmentation, and psychological characteristics such as intelligence, creativity, and personality. Heredity also influences aging, puberty, time of tooth eruption, muscle fiber type ratio, and the development of skeletal age. Although body weight and aging are

Figure 2.1
Model of the DNA
molecule.

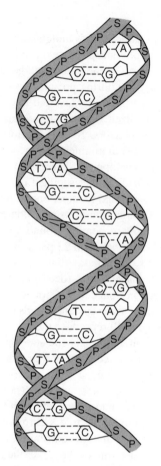

primarily the product of heredity, both can be significantly affected by environmental factors such as diet and exercise.

Much of our information about the genetic component of behavior has come about through the comparison of identical and nonidentical twins. Because identical twins have the same genetic material, variability in their behavior is due, in theory, primarily to environmental factors, such as learning. Only a few studies of twins have been conducted in the fields of motor development and exercise physiology. For example, Scheinfield (1939) found that sensory discrimination, reaction time, and general motor coordination were among those facets of human behavior determined primarily by heredity. More recent work (Brooks & Fahey, 1985; McArdle, Katch, & Katch, 1991) also suggests that the genetic component is the most important factor determining individual differences in aerobic and endurance performance. Although the answers are far from complete, data such as these do suggest that selected motor performance activities are significantly influenced by genetic factors.

Of all of the systems of the human body the nervous system is one of the most important. Everything that takes place consciously or unconsciously, voluntarily or as a reflex, has its primary initiation within the nervous system. Growth, development, and motor behavior all depend on the efficient functioning of this system. The effectiveness of a motor response is significantly influenced by the quality and capability of the nervous system and the brain. According to Guyton (1991), the nervous system has three primary functions: (a) a sensory function, (b) an integrative function, which includes the memory and thought processes, and (c) a motor function.

A basic review of the anatomy of the nervous system is important to the further discussion of the neurological changes that occur across the life span. The nervous system has two major parts: the **central nervous system (CNS),** which consists of the spinal cord and brain, and the **peripheral nervous system (PNS),** which is made up of all the nerve fibers that enter or leave the brain stem and spinal cord to supply the sensory receptors, muscles, and glands. Essentially, the PNS represents the lines of communication, whereas the CNS is the center of coordination and the mechanism that determines the most appropriate response to incoming impulses.

Central Nervous System   The structures of the central nervous system (CNS) basically function to transmit information about the environment and the body to the brain where it is recorded, stored in memory, and compared with other information (fig. 2.2). The CNS also carries information from the brain to muscles and glands, thus producing motor responses and the body's adaptations to environmental demands. The following discussion provides a brief description of selected parts of the CNS deemed most relevant to the study of motor development.

Spinal Cord   The spinal cord has an essential role in the input and response phases of information processing and motor behavior. Its primary function is to act as a transmission pathway; it carries to the brain all sensory information from the body and all motor commands sent down from the brain to muscles (and glands). The spinal cord also has an important function in reflex behavior.

The Brain   The brain is the principal integrative area of the nervous system. It is the location where memories are stored, thoughts are developed, emotions are generated, and complex control of motor behaviors is performed.

The *brain stem* is that part of the brain primarily responsible for several involuntary (reflexes) and metabolic functions. The brain stem also sets the rhythm of breathing and controls the rate and force of breathing movements and heartbeat.

Several important fiber tracts pass both upward and downward through the brain stem, transmitting sensory signals from the spinal cord mainly to the thalamus and motor signals from the cerebral cortex back to the spinal cord. The

NEURO-LOGICAL CHANGES

■ *Objective 2.2*

Basic Structure and Function

■ *Objective 2.3*

Figure 2.2
Anatomy of the central
nervous system.

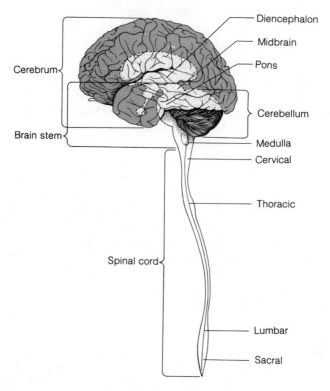

major structures of the brain stem are the pons, midbrain, medulla, diencephalon, and reticular formation. The *medulla* contains a number of sensory tracts for carrying information to the brain and motor tracts for carrying information to the muscles and glands. The medulla serves primarily to regulate vital internal processes such as respiration, blood pressure, and heart rate. The ventral and dorsal parts of the *pons* contain several nerve tracts that allow for coordination and involuntary influences on automatic movement and posture. The *midbrain* is involved with reflex movements caused by visual and auditory stimulation. The *diencephalon* consists of two areas: the thalamus and hypothalamus. The *thalamus* is an important integration center through which most sensory information passes. The *hypothalamus* is the structure where neural and hormonal functions work to create a constant internal environment (body temperature). The *reticular formation* plays an important role in attention and activation of the individual for cognitive and motor activity.

The *cerebral cortex* is the outermost layer of the cerebrum and is composed of an estimated 75% of the total neurons in the CNS. It is the functional head of the nervous system in its responsibility for higher-order critical thinking and information processing. Basically the cortex mediates: (a) the reception and interpretation of sensory information, (b) the organization of complex motor behaviors, and (c) the storage and use of learned experiences. The principal functional area of the cortex controlling motor activities is called the *motor area*. The motor area occupies the posterior half of the frontal lobe and consists of the motor cortex, premotor cortex, and Broca's area. The *motor cortex* controls the specific fine motor muscles throughout the body (e.g., of the fingers, feet, toes, lips). The *premotor cortex* elicits the coordinated movements that involve either sequences of individual muscle movements or the combined movements of a number of different muscles simultaneously. It is in this area that much of the individual's knowledge of learned skilled movements is stored. *Broca's area* has the primary function of controlling the coordinated movements of the larynx and mouth to produce speech.

The *basal ganglia* area of the brain is made up of a group of nuclei located in the inner layers of the cerebrum. The basal ganglia integrates the sensory motor centers and is involved with unconscious behavior, such as the maintenance of muscle tone required in upright posture. A major function of the basal ganglia is to control very fundamental gross body movements, whereas the cerebral cortex plays prominently in the performance of more precise movements of the arms, hands, fingers, and feet.

The *cerebellum* is an important part of the motor control system. Even though it is located far away from both the motor cortex and the basal ganglia, it interconnects with both of these through special nerve pathways. It also interconnects with motor areas in both the reticular formation and spinal cord. Its primary function is to determine the coordinated sequence of muscle contraction during complex movements. Its functions are also associated with vestibular awareness (balance), postural adjustments, and reflex activity.

Peripheral Nervous System   The peripheral nervous system (PNS) is a branching network of nerves, extensive to the degree that hardly a single cubic millimeter of tissue anywhere in the body is without nerve fibers. The PNS is divided into two systems: the somatic and the autonomic. The *somatic system* controls all the skeletal muscles (contracted through voluntary initiation); the *autonomic system* is primarily responsible for regulating the automatic functioning of the smooth muscles of the internal organs such as the heart, liver, lungs, and endocrine glands. The activities of the autonomic system are seldom subject to voluntary control.

PNS nerve fibers are of two functional types: *afferent fibers* for transmitting sensory information "into" the spinal cord and brain, and *efferent fibers* for transmitting motor impulses back from the CNS to the peripheral areas, especially to the skeletal muscles. The peripheral nerves that arise directly from the brain itself and supply mainly the head region are called *cranial nerves*. The remainder of the peripheral nerves are *spinal nerves*.

Neuron   A *neuron* (nerve cell) is the basic structural unit of the nervous system. There are billions of neurons in the nervous system; the average neuron is a complex structure with thousands of physical connections with other cells. Nerve impulses travel along neurons to relay information from one cell to another (and through the nervous system). The junction between two cells across which the

■ *Objective 2.4*

information must pass is called a *synapse*. The three basic parts of the neuron are the cell body (soma), dendrites, and the axon (fig. 2.3).

The *cell body* (*soma*) is the metabolic center of the cell. It contains the nucleus, which is responsible for regulating the various processes of the cell. Neuron cell bodies are located mostly within the central nervous system. Clusters of cell bodies are called nuclei. A *dendrite* is a nerve fiber that extends from the cell body. Each neuron can have anywhere from one to thousands of dendrites. Dendrites are the "receiving" part of the neuron, serving the important function of collecting information and orienting it toward the cell body. Although many dendrites branch from the cell body of a neuron, there is only one axon. The *axon* is the nerve cell structure that carries information away from the cell body to other cells. As previously noted, nerve impulses travel along neurons to transmit information. An impulse is an electrochemical process that travels in a chainlike sequence from the dendrite or cell body of one neuron to its axon to the dendrite or cell body of another neuron, and so on, through the length of the nerve tract. The functional connection between the axon and another neuron is the *synapse*.

Neurons are generally classified according to their function. *Afferent neurons,* also referred to as sensory neurons, carry nerve impulses from the sensory receptors into the spinal cord or brain of the central nervous system. *Efferent neurons* transmit impulses from the central nervous system to the muscles and glands. Efferent neurons passing impulses to muscles are commonly called *motoneurons*. Over 95% of all neurons are classified as *interneurons*. This type of neuron originates and terminates solely within the CNS.

Along with the neuron, the other basic types of nervous system tissues are supporting and insulating cells. These cells have the important function of holding neurons in place and preventing signals from spreading between the neurons. In the CNS these cells are collectively called *neuroglia* and referred to specifically as *glial cells*. In the PNS they are referred to as Schwann cells. *Schwann cells* wrap myelin sheaths around the large nerve fibers, thus insulating the pathway for an electrochemical nerve impulse.

■ *Objective 2.5*   **The Neuromuscular (Motor) Unit**   Each motoneuron axon branches into several synaptic terminals, and each of these terminals provides the nerve supply to a muscle fiber. A neuron and all the muscle fibers innervated by it are referred to as a **motor unit** because all the muscle fibers contract as a unit when stimulated by the motoneuron (fig. 2.4). Each of the muscle fibers making up the small delicate muscles of the eye may be supplied by a motoneuron, but larger postural muscles may only have one motoneuron to supply as many as 150 muscle fibers.

Motor Pathways   Nerve impulses travel through two major motor pathways on their way to producing muscular contraction or relaxation. The *pyramidal system* is made up of those neurons that originate in the cerebral cortex (premotor area), the axons of which descend through a direct channel to the spinal cord (fig. 2.5). The neurons of the cortex that send their axons down the pyramidal tract provide a direct channel from the cortex to the spinal neurons, which in turn

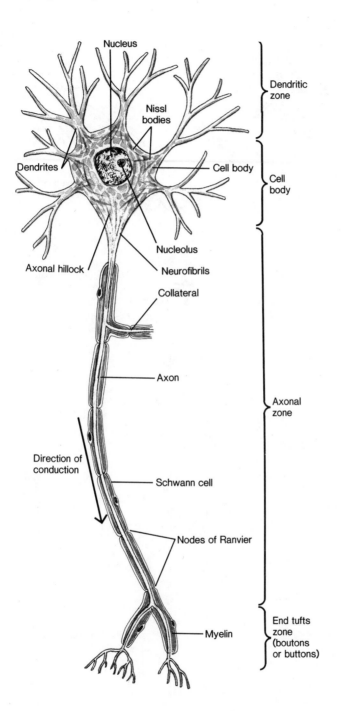

Nucleus

Nissl bodies

Dendrites

Cell body

Nucleolus

Axonal hillock

Neurofibrils

Collateral

Axon

Direction of conduction

Schwann cell

Nodes of Ranvier

Myelin

Dendritic zone

Cell body

Axonal zone

End tufts zone (boutons or buttons)

Figure 2.3
The basic parts of a neuron.

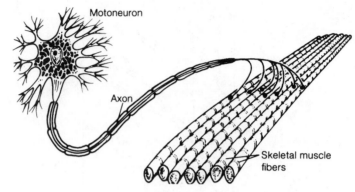

Figure 2.4
A motor unit.

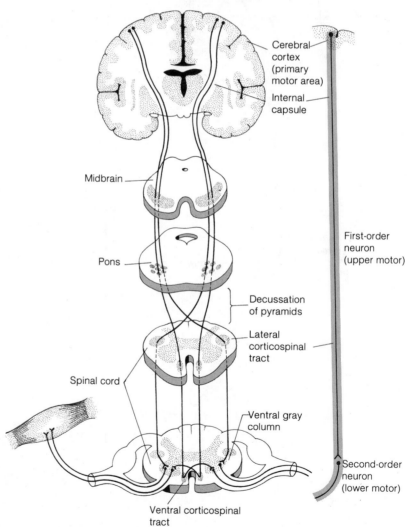

Figure 2.5
The pyramidal motor pathways.

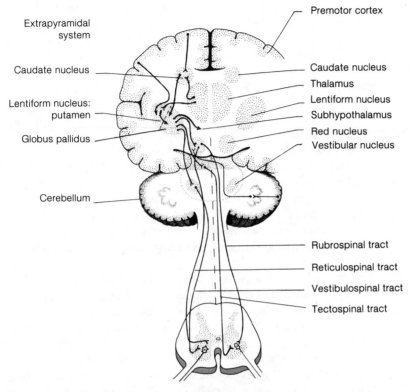

Extrapyramidal system

Caudate nucleus

Lentiform nucleus: putamen

Globus pallidus

Cerebellum

Premotor cortex

Caudate nucleus
Thalamus
Lentiform nucleus
Subhypothalamus
Red nucleus
Vestibular nucleus

Rubrospinal tract

Reticulospinal tract

Vestibulospinal tract

Tectospinal tract

Figure 2.6
The extrapyramidal motor pathways.
From A. B. McNaught and R. Callander, *Illustrated Physiology,* 4th edition. Copyright © 1983 Churchill Livingstone, Edinborough, Scotland. Reprinted by permission.

cause either muscle contraction or relaxation. Functions of the pyramidal system are primarily associated with specific movement patterns.

The *extrapyramidal system* includes all of the motoneuron axons not included in the pyramidal system. These neurons originate in the motor area of the cortex as well as in other cortical areas. Instead of traveling a direct pathway to the spinal neurons as happens in the pyramidal tract, several axons on the extrapyramidal tract are interrupted synaptically to terminate in various subcortical structures, especially the basal ganglia, cerebellum, and thalamus (fig. 2.6). The extrapyramidal system is primarily associated with general, rather than specific, movement patterns.

Impulse Conduction Neurons send information through their axons in the form of brief impulses, or "waves," of electricity in the form of single electrical clicks called *action potentials.* **Nerve conduction velocity** is the speed at which information travels; this rate is greatly affected by the presence or absence of a material called *myelin* around the axon. *Myelin* is a fatty material that forms a sheath around many axons both within and outside the CNS. The myelin sheath is formed from a type of insulating cell called a *Schwann cell.* These cells wrap around the axon and form a jelly-roll-type structure that serves as an effective insulator of electrical currents. The myelin sheath is interrupted every millimeter or

■ *Objective 2.6*

so by gaps called *nodes of Ranvier* (fig. 2.3). The neuron membrane is active only at the nodes, so the impulse is conducted when the action potential "jumps" from one of these nodes to another. This type of impulse conduction is known as *saltatory conduction* and is significantly faster than conduction in nonmyelinated axons.

Saltatory conduction also requires less metabolic energy, enabling myelinated axons to fire nerve impulses at higher frequencies for longer periods of time. Research has also indicated that there is higher conduction velocity, lower threshold, shorter latency, and higher amplitude coincident with the time of appearance and the degree of myelination (Timiras, Vernadakis, & Sherwood, 1968). Another factor in conduction velocity is the size of the axon. Basically, the larger the cross-sectional diameter, the faster the speed of conduction. Apparently axon conduction velocity is related to the urgency of the information that it is called upon to transmit. Axons with greater speed potential are concerned with the control of movement, especially in the mediation of rapid reflexes; axons that transmit visceral information are small, generally unmyelinated, and slow.

Early Developmental Changes

■ *Objective 2.7*

Normal CNS development appears to follow a rather exact sequence of integrated biological events; these include the processes of cell proliferation, migration, integration, differentiation, myelination, and cell death (modified from Williams, 1983; Cowan, 1978). Basically the developmental process begins with immature neurons (cell proliferation). These immature cells become specified as to their function and location within the system. When their location has been determined, the different cell types migrate to various sites and integrate with other cells. At the site of integration, neurons begin to elaborate axons and dendrites in preparation for establishing the functional connection (synapse) between cells (differentiation). In their final stage of morphological development, most nervous system pathways become coated with myelin to allow them to transmit impulses more effectively. Many neurons die during early development of the nervous system (cell death), which is believed to be a normal part of the development process.

It should be reemphasized that while the developmental sequence is rather exact, it is also a very complex and intricate process. The sequence described only represents highlights in the process. The following discussion elaborates in greater detail the events that occur within the sequence of early neurological development.

Cell Proliferation and Changes   Neurons first appear in the brain during the second prenatal month and virtually all of the **cell proliferation** process (growth in number) is completed by birth. Not only do neurons multiply very rapidly during early development, but they also grow in size as well. The period of dramatic increase in neuron size seems to occur from the sixth prenatal month through the first year of life. Goldman-Rakic et al. (1983) note that the peak period of growth is about the time of birth.

Beginning around the third trimester of gestation and continuing into at least the fourth year is a period of rapid brain growth and development called the *brain growth spurt*. The general period also includes development in terms of cell proliferation, myelination, dendritic and synaptic growth, and refinement of certain enzyme systems. Dobbing (1976) suggests that the brain has a "once only" opportunity for laying down its foundation for optimal development. It is also important to note that the timing of the growth spurt cannot be altered; the time line for development is predetermined genetically. The growth spurt represents a period of vulnerability for elaboration of neural process and potential brain function. This fact points to the importance of prenatal care and an optimal postnatal environment, especially during the first two years. Further discussion on factors that may influence growth and motor development is presented in chapter 5.

The genetic structure of the cell nucleus plays a primary role in controlling cell proliferation and differentiation, so its development is important. Mandel et al. (1974) note that the amount of DNA in the CNS increases 15 to 20 times during the period from four weeks after conception to the time of birth. The rate of increase after birth is much smaller (2 to 3 times) and the amount of DNA reaches adult levels by the end of the first year. Prenatal increases in RNA are similar to those of DNA but the increases after birth are more rapid, increasing by about 10 times before maturity.

Along with the process of cell proliferation several developmental changes occur in the axon and dendrite structures. Once the neurons have migrated to their final location in the CNS, they begin to elaborate their axon and dendritic structures in readiness to accept impulses through synaptic interconnections. Again, neurons may be categorized as sensory (afferent sending signals to the CNS); or motor (efferent sending signals from the CNS to the muscles); and central or interneurons (originating and terminating in the brain or spinal cord). It is most likely that this specification is biochemically set. The axons of sensory neurons must often travel relatively long distances to reach their synaptic conjunction. Thus these axons generally must also possess a high degree of biochemical specificity. According to Cowan (1978), this specificity seems to be present early in the developmental process and cannot be altered.

Terminal targets of motor neurons do not appear to be as specific as those for sensory neurons, but complex specificity is still evident in the growth process. The growth of the axon of a motor neuron is guided somewhat by the chronological order in which it matures and differentiates as well as by its biochemical properties. When a motor neuron innervates a muscle, the axon makes contact and induces biochemical specificity into the muscle, thus allowing the two structures to match up biochemically. Once specificity of the neuromuscular junction is established, sensory and central neurons form their synaptic connections.

One of the major events in cortical neuronal differentiation is the elaboration of dendritic structures. Dendrites are important because they are the main receptors for the neuron. At the synaptic junctions the dendrites of each neuron can

----

Figure 2.7
Representation of
dendritic development.

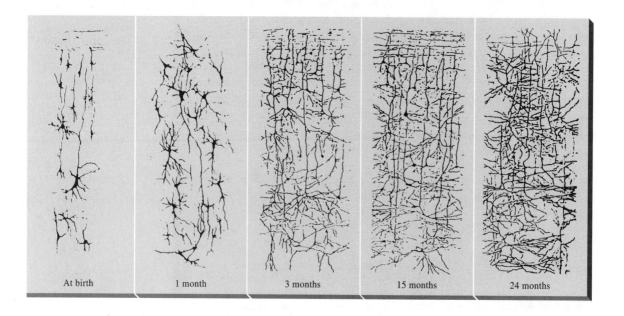

| At birth | 1 month | 3 months | 15 months | 24 months |

receive signals from literally thousands of other neurons. It has been estimated that dendrites of cortical neurons provide more than 95% of the targets for transmitting information through the system. Not until approximately the eighth fetal month do the first signs of thick dendrites with conspicuous spines start to appear on cells in the visual cortex. Even though the motor cortex is noted as being more advanced in dendritic development than the visual cortex at that time, the appearance of dendritic spine development is not as evident. Williams (1983) describes the period from halfway through the seventh fetal month to six months after birth as a span of critical importance in dendritic development. At around 30 weeks of gestation, the dendritic spines show obvious immaturity in being few in number and irregular in shape, but by 8 months (postnatal) the number of well-formed, lollipop-shaped, stubby spines has multiplied. Figure 2.7 represents the various stages of dendritic development in large pyramidal neurons (Lund, 1978).

Integration and Differentiation   As cells begin to multiply, migrate to their final location, and elaborate, they also undergo the concurrent processes of integration and differentiation. **Integration** refers to the intricate interweaving of neural mechanisms of various opposing muscle groups into a coordinated inter-action. **Differentiation,** on the other hand, is the process by means of which structure, function, or forms of behavior become more specialized. These terms also refer to the progression of motor control from gross, poorly controlled movements to precise, complex motor behavior. Differentiation cannot occur until *synaptogenesis* has taken place (i.e., the synapses between neurons have formed). The effectiveness of information delivery between neurons and thus among widespread parts of the nervous system depends on the number and indi-vidual strength of the synapses formed during development.

An average of about 6,000 small knobs called *presynaptic terminals* lie on the surfaces of the dendrites and soma of a motor neuron in adults; approxi-mately 80% to 90% are on the dendrites (Guyton, 1991). Estimates for the aver-age number of neuronal connections range from 5,000 to 100,000, depending on the type and function of the neuron. Figure 2.8 illustrates the basic parts of a motor neuron with presynaptic terminals.

Initially synapses develop on dendritic shafts and later on the dendritic spines and neuron somas. By approximately 8½ weeks the fetus has an estimated five synapses per square millimeter of nerve tissue. Quality synapses are usually not seen in the cortex until approximately 23 weeks following conception. Ap-parently, proliferation of synapses occurs so rapidly during early development that the number of neural connections peaks at about age 2 (Goldman-Rakic et al., 1983). Based on the presence of synapses in the fetus and their relationship to functional maturity of the brain, it appears that cortical activity is just begin-ning to develop toward the end of gestation. This type of data has led researchers to believe that some aspects of neuronal circuitry are established quite early in the life span (Cragg, 1974; Molliver, Kostovic, & Van Der Loos, 1973).

Animal studies have also provided some insight into the influence of sen-sory stimulation on synaptogenesis (J. Lund & R. Lund, 1972; R. Lund & J. Lund, 1972). Apparently the development of some synapses is independent of sensory stimulation, whereas others do not form without appropriate sensory input. Study of selected parts of the rat's visual system has identified three dis-tinct stages of synaptic development. The first (prefunctional) stage consists of initial synaptic contact and requires no visual stimulation. Stage 2 is the first functional stage, which begins immediately after the eyes open and during which myelination of the optic nerve and transmission of visual information take place. Synapses appear to proliferate and elaborate through the visual system at the same time. Stage 3 is the second functional stage and is identified by the addi-tional formation of intrinsic synapses and the refinement of response characteris-tics of neurons. This stage depends on patterned vision and is the period during which the development of synaptic connections is completed. Researchers have observed the characteristics of the first two stages in visually deprived animals but have not identified stage 3 traits in such animals. Cragg (1974) found that when afferent input (sensory stimulation) to the brain was cut off, the number of synapses normally developed was reduced by 50% in just six weeks. This type of data has led researchers to conclude that a lack of patterned stimulation will not

Figure 2.8
A typical motor neuron
with presynaptic
terminals.

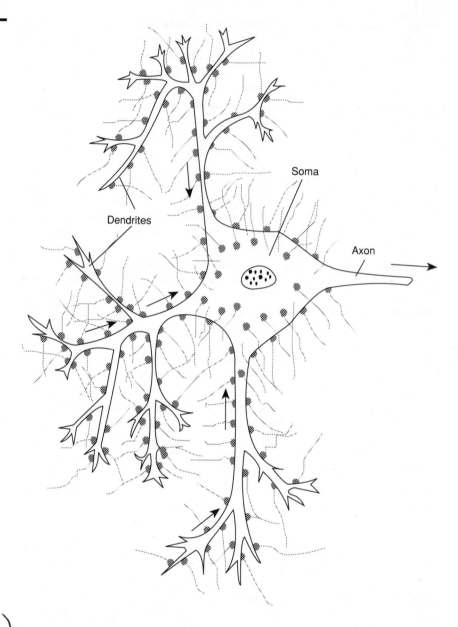

only reduce the number of total synapses formed but may decrease the level of
specificity or refinement of responses in the neuron itself.

Myelination   The developmental process of **myelination** has been one of the
most extensively examined indicators of neurological growth. Much of the early
growth in brain size and weight can be attributed to myelination. Evidence has

already been presented to show that neurons have slower transmission rates before myelination, are more prone to fatigue, and are more limited in their rate of repetitive firing. The relative rates of myelination in different areas of the brain give a rough estimate of when these areas reach adult levels of functioning (e.g., Bronson, 1982). The degree of myelination is also closely related to maturation (and readiness) in acquisition of motor skills during early childhood. The ages at which myelination begins and ends seem to vary from one brain structure to another, as does the time required for the myelination process itself.

The formation of myelin begins in the spinal cord about halfway through fetal development, continues through adolescence and adulthood, and in some areas of the brain, perhaps into old age. Figure 2.9 provides an approximate time line of the myelination process in selected structures and pathways within the CNS. Both sensory and motor roots begin myelination four to five months before birth. The motor roots are the first parts to develop myelin; the sensory structures exhibit a rapid increase about one month later. Motor mechanisms of the spinal cord appear to be fully myelinated and functional by the end of the first month after birth. Once again, the sensory mechanisms lag behind somewhat and do not show myelin growth in the cord until approximately six months after birth.

The primary sensory tracts to the cortex appear to mature at slightly different times during the life span. The myelination of auditory pathways begins as early as the fifth prenatal month, and the visual pathways begin around the time of birth. However, once the developmental process of the visual pathways begins, it proceeds very rapidly and is completed sometime during the first five months of postnatal life. Auditory myelination, on the other hand, is not completed until approximately 4 years of age. The higher somesthetic pathways related to touch show myelin growth about the eighth prenatal month, and by birth these pathways are myelinated through to the cortex. Evidence of this level of development is exhibited by newborns who are normally quite sensitive to touch stimuli. Significant growth continues in the somesthetic pathways until the child is between 1 and 2 years of age.

The pyramidal tract, the major efferent pathway from the motor cortex, begins myelination a month before birth. Although the pathway does not achieve full maturity for about two years, it is probably functional by 4 to 5 months, when intentional (voluntary) motor behavior can be observed in the infant (Bushnell, 1982; McDonnell, 1979). The part of the brain that integrates information (cerebral commissures) exhibits a rapid increase in myelin about the third month after birth and this process continues until approximately age 10. The reticular formation, associated with the attention processes, begins a period of rapid myelination at birth and continues to mature in an individual until sometime after the second decade. This observation has led to the belief that the capability of a person to selectively attend to a task is still being modified until early adulthood. Last to undergo myelination are the areas of the brain associated with memory. These structures, called association areas, begin to show significant myelin growth that

Figure 2.9
An estimated time line of
the myelination process.

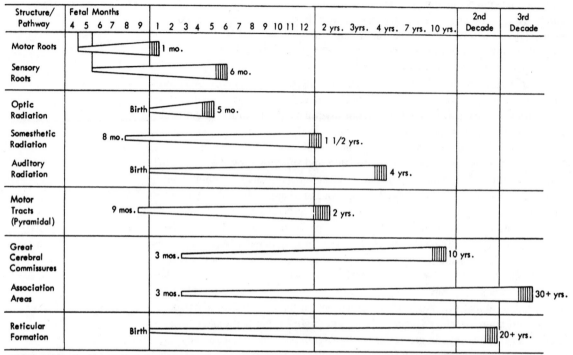

*Width and length of bars indicate increasing density of myelination; blacked-in areas at end of bars indicate approximate age range of completion of myelination process.

starts around the third month after birth and continues into and beyond the third decade of life.

The development of myelin and the characteristics of early motor behavior present an interesting developmental parallel. Prior to the initiation of voluntary motor control, movement is exhibited in the form of *myogenic, neurogenic,* and *reflex* behaviors. **Myogenic behavior** is movement that is the result of direct muscle stimulation rather than stimulation that arrives at the muscle through some intervening neural structure. Myogenic movement can sometimes be elicited in the fetus prior to the appearance of myelin.

After the appearance of myelin in the spinal cord at about the fourth fetal month, movement can be elicited through the motor neuron's connection with the muscle. This type of behavior is affected by neural structures and is called **neurogenic behavior.** At approximately 6 months (prenatal), myelination is more complete in the spinal cord and *reflex arcs* appear, meaning that the fetus can receive sensory input and reflexively translate the information into a behavior or motor response. Touch stimulation can elicit such movement behaviors as the primitive grasp reflex of the hands and the Babinski reflex of the feet. By the eighth fetal month, the cord is close to fully myelinated and the direction of growth proceeds upward toward the higher brain regions such as the medulla, midbrain, and thalamus. At this point several important developmental motor responses can be observed: respiratory movements, primitive sucking, the Moro reflex, tonic neck reflexes, and other righting reflexes.

At birth and during the neonate period, and as the motor pathways from the cortex are myelinated, voluntary motor behavior becomes possible. Since the somatosensory pathways are the most advanced of the sensory tracts, the infant responds readily to tactile stimulation on almost any part of its body. Although myelination of the optic tracts trails somewhat behind at this point, the neonate is capable of visual fixation and simple tracking movements as well as some form and depth perception awareness. As previously noted, myelination of the auditory pathways develops at a slower rate than the other sensory tracts; therefore, in comparison, the infant's responses are less mature (e.g., exhibiting primarily gross reactions to a sharp, loud sound) during the first month of life.

Cell Death   Although we may never be able to get an honest count of neuronal loss, researchers do believe that there is considerable **cell death,** or natural elimination of neurons, during early development and as we age. During early development some structures may lose 40% to 75% of neurons initially generated. Cell death of neurons in this instance is believed to be a normal part of the developmental process of establishing synaptic connections (i.e., differentiation). It has been hypothesized that proliferating cells compete with each other for a limited number of synaptic sites or for some function that is vital to their existence (Berry, 1982; Cowan, 1978).

Changes in Brain Size   The adult brain weighs about 3½ pounds, and although it makes up only about 2.5% of the total body weight, it requires 15% of the body's blood supply and about 25% of all the oxygen consumed. From two to eight weeks following conception the nervous system begins to develop as a long, hollow tube on the back of the embryo. As the system develops, brain size increases into a mass of neurons, losing its primitive tubular appearance. At birth, the infant's brain weighs about one-fourth of its adult weight. After birth, nerve cell size increases, other supporting cells called neuroglia are formed, and the myelin develops, causing the brain to double in volume. By age 3 the brain has reached nearly 90% of its adult size (Trevarthen, 1983) and by age 6 has basically achieved its full size. In contrast, total body weight at birth is just 5% of young adult size and by age 10 is only 50% of the weight that will eventually be attained. The part of the brain most fully developed at birth is the midbrain. As previously mentioned, the midbrain is the part of the CNS that controls much of the early reflex behavior. The midbrain, pons, and medulla occupy approximately

■ *Objective 2.8*

8% of the total brain volume at 3 fetal months of age, but by birth this proportion has fallen to around 1.5%. During the first decade of life, these percentages increase slightly due to fiber tract growth.

Growth and differentiation of the cortical regions of the brain are landmark features in the functional maturity of the CNS. As previously noted, the cortex is composed of an estimated 75% of the total neurons found in the CNS; this is an estimated 10 to 20 billion (Nowakowski, 1987). Yet in actual size, it accounts for only a very small portion in forming the outermost layer of the cerebrum—about one-fourth inch in thickness. While thickness of the cortex normally reaches adult levels by approximately $2\frac{1}{2}$ years of age, functional maturity of the cortical areas in general is usually not achieved until around age 6 (Cratty, 1986). There are indications that growth and differentiation of specific areas continue into at least early adulthood (Tanner, 1978; Williams, 1983).

Williams (1983) describes three major phases in the early development of the cerebral cortex: (a) general cortical differentiation (week 6 to week 16 of fetal development) in which the cortex becomes distinguishable from other subcortical structures and the six-layered appearance of the adult cortex appears; (b) divisional cortical differentiation (week 8 to week 32 of fetal development) during which different parts of the cortex become structurally defined. Although the cellular integrity of the cortex is still fairly immature, the major lobes (frontal, parietal, occipital, and temporal) and prominent fissures (Sylvian and central) are fairly well defined; and (c) local cortical differentiation (week 24 of fetal development on into the postnatal years) during which the cortex begins to present an orderly arrangement and organization. At approximately week 28 of gestation the external configuration and patterns of the fissures approximate those of the adult brain. By $2\frac{1}{2}$ months after birth some parts of the cortex (e.g., the motor cortex) have reached adult levels of thickness.

The rate of development among the cerebral lobes is quite varied. Each lobe has its own rate of development, each area in each lobe has its own developmental rate, and each layer of those areas have different rates of development. In order of increasing maturity, the occipital lobe (visual functions) matures first, the parietal lobe (somatosensory functions) next, and the temporal (auditory and memory functions) and frontal (memory and motor functions) lobes are the slowest to reach full maturity. In general, by 6 months of age the parts of the cortex that control hearing and visual processes are already developed, although the child probably lacks interpretive capabilities due to inadequate development of association areas. The association areas are responsible for integrating sensory information with memory data and organizing deliberate motor responses. The area of the frontal lobe that controls fine movements of the hands may undergo periods of aggressive development and refinement during the first 15 months following birth and again after 6 years of age. The developmental "lag" (characteristic of development in other abilities as well) between 15 months and 6 years probably represents a time when the system is laying down the neurological framework that will ultimately suppress weaker characteristics (Rabinowiz, 1974).

Following the basic principle of cephalocaudal development, motor control of the legs generally evolves later than the capabilities for upper body movements. It should be noted, however, that while this is the more typical pattern of development, other recent studies have reported well-coordinated leg movements in young infants about the same time as upper body control (Thelen, 1979, 1985; Thelen & Bradley, 1988). The cerebellum, the area of the brain primarily responsible for controlling coordinated motor responses, lags behind the midbrain, areas of the spinal cord, and cerebral cortex in development.

Brain Lateralization and the Corpus Callosum   Although the two hemispheres of the brain appear symmetrical, in functional terms they are quite different. Each hemisphere has its own specialized functions (i.e., functional asymmetries), a characteristic known as **brain** (hemispheric) **lateralization**. In most humans, the left hemisphere is associated with the governance of language, logic, and sequential processing, whereas the right hemisphere is specialized for nonverbal, visuospatial functions (e.g., music awareness, map reading, figure drawing). Of more direct relevance to motor behavior is that the cortex of the right hemisphere controls muscular activity in, and receives sensory input from, the left half of the body (Figure 2.10); whereas the left hemisphere has a complementary role in conscious movement on the right side of the body (Seeley, Stephens, & Tate, 1992).

Dividing and connecting the two hemispheres of the brain is a tough band of myelinated tissue called the *corpus callosum*. One of its primary functions is to provide the link for shared information between the two hemispheres, in essence allowing the right hand to know what the left hand is doing. Thus, the corpus callosum is important for the functional integration of the two cerebral hemispheres and possibly for the manifestation of functional asymmetries (Leporé, Ptito, & Jasper, 1986). The callosum undergoes marked growth in overall size and myelination during postnatal development and reaches adult proportions by approximately age 10 (Larroche & Amakawa, 1973). With regard to sex differences in cerebral lateralization of function, some research has indicated that neither the male brain nor the female brain is more asymmetrically organized (Hahn, 1987). This includes actual callosum area and its growth curve during development (Bell & Variend, 1985; de Lacoste, Halloway, & Woodward, 1986; Peters, 1988). Further discussion of this topic will be provided in chapters 8 and 9 in relation to theoretical and applied considerations of functional (motor) asymmetries (i.e., the development of handedness, footedness, and eye preference).

The loss of neurons with aging, is accepted as a significant part of the aging process (Moushegian, personal communication, 1985). It is estimated that this loss is about 5% to 10% until we reach our seventies and that neural loss may accelerate after that time. The brain loses about 7% of its weight from the time when we are younger adults. This amount could be substantially more if some form of organic brain disease or other secondary aging element (e.g., drugs, diet,

■ *Objective 2.9*

Neurological Changes with Aging

■ *Objective 2.10*

Figure 2.10
The concept of lateral
dominance of limb
control.

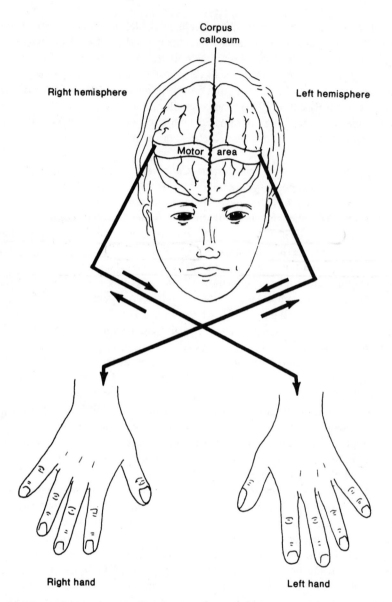

alcohol, health of the cardiovascular system) is present. It is generally believed that the brain has such remarkable recovery capabilities that, even though significant neuron loss may occur, the brain may only lose a small portion of its ability to function (Labouvie-Vief, 1985; Morgan, 1992).

In addition to a significant loss of neurons, other important changes appear to be occurring within the neurons themselves. One of the most critical of these changes involves the gradual shrinking and withering away of the dendrites and axons (i.e., interconnectivity among neurons) while other mutations may include the cell body. Process deficits normally associated with advanced aging include a decrease in neuron connectivity, delay in new axon growth and myelination, decrease in neuron excitability, and an estimated twofold increase in synaptic delay (Scheibel, 1992; Williams, 1990). Several of the characteristics listed have been associated with the older adult phenomenon known as *psychomotor slowing* (to be discussed in greater detail in chapters 7 and 11).

In general, aging may be constituted by a proliferation of abnormal material and functions referred to as vacuoles, plaques, and tangles. *Vacuoles* are thick granules surrounded by fluid, and although they are most common in diseased brains, they showed up in three-fourths of the brains checked from people over 80 years of age (Adams, 1980). *Senile plaques,* collections of debris consisting of cells, silicone, and macrophagic elements, are frequently found in the frontal and occipital lobes of the brain. *Neurofibrillary tangles,* tangled clumps of double-helical strands of protein, appear in older brains, but no one clearly understands their effect. Tangles tend to be particularly prominent in the cerebral cortex as a whole and in the thalamus, basal ganglia, and spinal cord. Eckert (1987) notes that plaques and tangles appear to some degree in the brains of 99% of all individuals older than 80 years of age. Senile plaques and neurofibrillary tangles have been found in brains of people who had primary degenerative dementia (i.e., Alzheimer's disease). Studies have also noted that as neurons and dendrites deteriorate, neuroglia (the connective tissue that fills the spaces between neurons) increase in number. Bodies of *astrocytes,* star-shaped neuroglia, get larger in the aging brain and may prevent neurotransmitters from building up between nerve cells.

One of the most dramatic decreases in basic neurological function related to motor behavior is the deterioration of vestibular awareness (balance). Neurons in one layer of the cerebellum die off fairly rapidly after the age of 60. As previously noted, the cerebellum functions to coordinate voluntary motor behavior and vestibular awareness. Balance as measured by speed-of-return-to-equilibrium and body sway deteriorates significantly during old age. According to Shock (1962), approximately an 80% loss in recovery to equilibrium occurs by age 75. By measuring the postural sway on a balancing platform among three age groups of men 31 to 75 years, Era and Heikkinen (1985) found that as age increased, the sway area increased significantly. At this point, no significant sex differences have been reported in vestibular function or vision among the elderly. In most realistic situations, the maintenance of balance depends on vestibular and visual sensory input. In a study that investigated visual acuity and balance, Brocklehurst, Robertson, and James-Groom (1982) reported that both functions were significantly related to age (i.e., that they decreased with increasing age) and were highly correlated.

Another basic neural process that is crucial to motor behavior is nerve conduction velocity. This function also declines with increasing age. The loss appears to be more prominent in the distal segments of the body and in the lumbosacral regions. Shock (1962) notes that there is a reduction of approximately 27% in the number of nerve trunk fibers with advanced age. The results of such a

Table 2.1   Selected Basic Neurological Changes with Aging

Neurological Structure	Function
Brain weight	Decreases
Number of neurons	Decreases
Nerve conduction velocity	Decreases
Neuron connectivity	Decreases
New axon growth & myelination	Delayed
Neuron excitability	Decreases
Synaptic delay	Increases

loss have been evidenced by several studies showing how reaction and movement time slow with increased age (Pierson & Montoye, 1958; Deupree & Simon, 1963). In general, the decline of both these functions tends to occur most rapidly in the lower parts of the body and in areas of most frequent use, such as the fingers. The general rate of decline appears to be affected by the complexity of the task.

Aging also appears to weaken some of the neurotransmitters associated with autonomic nervous system (ANS) functions. Although the core of the ANS apparently does not undergo any marked deterioration with age, some functional deficits such as blood pressure elevation and a decreased ability to regulate body temperature have been noted (Everitt & Huang, 1980).

A summary list of selected neurological functions that normally show signs of deterioration with aging are presented in table 2.1.

SUMMARY

The blueprint for our heredity is contained in the genes found within the 23 pairs of chromosomes of a human cell nucleus. Genes house DNA, which contains the genetic code determining what is to be transmitted from one generation to the next. In turn, this information is transcribed into molecules of RNA, which provide instructions for human development. Along with such traits as sex and eye color, heredity influences numerous developmental factors such as aging, puberty, and skeletal age. These factors can also be affected by the environment as evidenced by the documented effects of diet and exercise on body weight and aging. There are also strong indications that motor performance is significantly affected by genetic factors.

The nervous system serves three primary functions: (a) sensory, (b) integrative, which includes the memory and thought processes, and (c) motor. The two major parts of the nervous system are the central nervous system (CNS) and the peripheral nervous system (PNS). The primary structures of the CNS are the spinal cord and brain. The basic function of the CNS is to transmit information about the environment and the body to the brain, and to carry information from the brain to muscles and glands, thus producing motor responses and bodily adaptations to environmental demands. The PNS is a branching network of nerves consisting of two distinct systems: the somatic and autonomic. The somatic system controls all the skeletal muscles, whereas the autonomic system is concerned with the automatic regulation of the smooth muscles.

The basic parts of a neuron (or nerve cell) are the cell body, dendrites, and axon. Dendrites are the "receiving" part of the neuron, whereas the function of the axon is to carry information away from the cell body to other cells. The functional connection between the axon and another neuron is called a synapse. There are three basic types of neurons: afferent, efferent, and interneurons. The pyramidal and extrapyramidal motor pathway systems provide the channels through which nerve impulses travel on their way to producing muscular contraction or relaxation. Messages are sent through the nervous system in the form of chemical impulses (waves of electricity). How fast the information travels depends to some degree on the presence of myelin, which forms an insulating sheath around many axons.

Development of the CNS can be described in a sequence of six integrated biological events: cell proliferation, migration, integration, differentiation, myelination, and cell death. The process of proliferation is virtually completed by birth. As proliferation occurs, cells migrate to their final location and elaborate, which is evidenced by the developmental changes that take place in the axon and dendrite structures. Neurons are also involved in an intricate interweaving process (integration) and, in turn, become more specialized (differentiation). During this general process, the critical formation and development of myelin (myelination) also transpires. A normal part of the developmental process is cell death.

The human body's central computer, the brain, undergoes several rapid quantitative and developmental (level of functioning) changes during the first two to three years. During the course of brain development, each hemisphere establishes its own specialized functions; this is referred to as lateralization. Related to this concept is the notion that hand and foot motor control on one side of the body are controlled by areas of the brain in the opposite-side hemisphere.

Several neurological changes occur and become evident during the latter stages of aging. Along with the loss of neurons and brain weight, important functional (process) changes are also evident. Advanced aging may produce abnormal material such as vacuoles, plaques, and tangles. Other materials (e.g., astrocytes) may also develop and fill spaces between neurons, thus inhibiting neural transmission. There are several motor behavior functions that normally show signs of deterioration with aging. Among the most affected by age are vestibular awareness (balance) and nerve conduction velocity (a slowing of movement).

SUGGESTED READINGS

Ganong, W. (1993). *Review of medical physiology.* (16th ed.). Norwalk, CT: Appleton & Lange.

Guyton, A. C. (1991). *Basic neuroscience* (2nd ed.). Philadelphia: Saunders.

Sage, G. H. (1984). *Motor learning and control: A neuropsychological approach.* Dubuque, IA: Wm. C. Brown.

Shepard, G. M. (1988). *Neurobiology* (2nd ed.). New York: Oxford University Press.

Physical Growth Changes

KEY TERMS

physical anthropology
anthropometry
growth curve
puberty
pubescent growth spurt
menarche
physique
somatotype

midgrowth spurt
ossification
osteoporosis
body mass
chronological age
biological age
secular trend

In the previous chapter you learned about the growth and development characteristics of the basic neurological structures. This chapter will focus on changes in the physical growth over the life span. Specific discussions will cover a general overview of physical growth, prenatal development, the adolescent growth spurt, body proportions and physique, height and skeletal growth, and body mass. Methods of estimating maturity level and maturity variations will also be examined.

AN OVERVIEW OF PHYSICAL GROWTH AND DEVELOPMENT

■ *Objective 3.1*

The new perspective on lifelong human development stresses the fact that the developmental process evolves along a continuum from conception to physical death. This perspective is supported through the observation of physical growth changes. Although the first two decades of life are a period of significant growth increases, marked changes also occur during later stages of the aging process. Growth and maturation begin in a cephalocaudal and proximodistal pattern. With the cephalocaudal pattern, growth (especially during the prenatal stage) first occurs in the head and gradually proceeds its way downward to the neck, shoulders, and trunk. At the same time, proximodistal growth proceeds from the center of the body and moves toward the extremities (e.g., the hands and fingers). Individuals grow at a faster rate during gestation and the first year after birth than they will at any other time in their lives. During the early childhood years, physical growth progresses at a relatively uniform and steady rate. In middle and late childhood, growth is a relatively slow, consistent process that may be characterized as the calm before the dramatic biological change that occurs during early adolescence and the onset of puberty. The pubescent growth spurt represents a landmark in physical growth in regard to sex differences and motor performance. Later in early adulthood, physical development usually begins to slow down. And as people age they generally get shorter, lose muscle mass, experience a change in body proportions, and get heavier. Interestingly, in old age there is a tendency for individuals to become lighter than they were during middle adulthood.

■ *Objective 3.2*

We have learned that motor development is tied closely to developmental psychology; the field of **physical anthropology** also provides information and scientific procedures related to the study of biological growth and development. Physical anthropology is basically concerned with the meaningful understanding of the nature, distribution, and significance of biological variation in humans. The branch of science concerned with biological growth and body measurement is referred to as **anthropometry,** which is one of the basic tools used in growth studies. Each physical growth topic discussed in this chapter will include a description of anthropometric measurement techniques.

Anthropometric data used to assess growth take the form of mean value comparisons, percentile rankings, and growth curves. Perhaps a less-known but frequently used procedure in growth studies is to plot the pattern of physical change in individuals or groups on a **growth curve.** One type of curve is called a *distance curve,* which is used to plot growth from one year to the next. The distance curves for males and females exhibit a gradual height increase before a leveling off occurs, as shown in figure 3.1. It depicts typical individual height curves based on data from many individuals. If the values of only one individual were plotted, the graph would be unlikely to show such a smooth curve.

Another technique is to measure the rate of growth using what is known as *velocity curve.* This type of measurement better reflects the individual's growth state at any particular moment than it does the distance or change in growth achieved in years preceding. The velocity of growth refers to increments in growth value from year to year and is represented by points of deceleration or acceleration. Figure 3.2 illustrates a hypothetical velocity curve with different phases of growth rate for height.

The prenatal period (conception to birth) plays a significant role in human development. Events that take place during this period forge the foundation for future development across the life span. In comparison to the other developmental stages of life, the prenatal period presents the greatest variation in human growth and development.

Prenatal Development

■ *Objective 3.3*

Conception   Life begins at *conception* when a single male sperm cell unites with the ovum (egg) in the female's fallopian tube. Fertilization occurs within several days after the egg begins its journey from the ovaries through the fallopian tubes to the uterus.

Germinal Period   The *germinal period* lasts for approximately 14 days after conception. Almost immediately, the 23 chromosomes from the sperm cell nucleus combine with the 23 chromosomes from the ovum to produce 46 chromosomes. The chromosomes then split, yielding 46 pairs, which serve as the template in the development of the individual. Two major developmental events occur during the next 10 to 14 days. One is continued cell differentiation and the other is the firm attachment of the ovum to the wall of the uterus. During this time the divided cells gradually form a spherical mass, which separates into inner and outer segments. The outer segment eventually forms part of the mother-fetus barrier, and the inner mass becomes the fetus.

Embryonic Period   By two weeks after conception, the embryonic disk has folded and formed the distinct layers of cells called an embryo. The *embryonic period* lasts from about two to eight weeks after conception. During this period human form begins to take shape. Although the embryo after eight weeks is only about 1 inch (2.5 cm) and weighs only half an ounce (14 g), the basic parts of the body—the head, trunk, arms, and legs—can be identified. Some of the finer features such as the eyes, ears, fingers, and toes are also discernible. The CNS is relatively developed and the internal organs (e.g., the heart, lungs, reproductive organs, liver, and kidneys) are beginning to function to some degree. All of these developments take place according to a master blueprint that dictates that

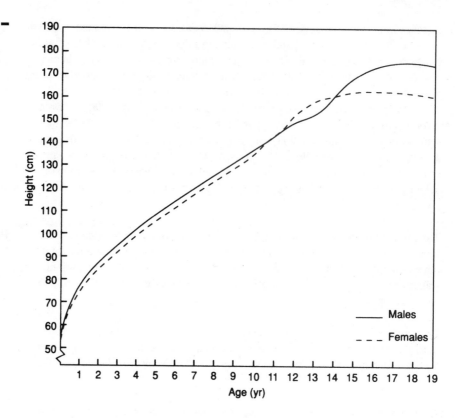

Figure 3.1
Typical distance curves
for change in height.

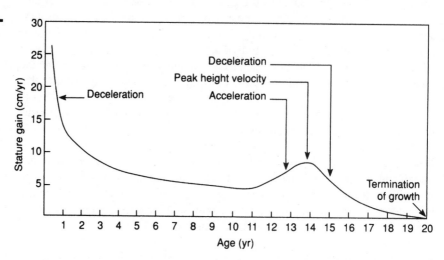

Figure 3.2
An example of a velocity
curve for height
indicating the different
phases of growth.

development starts at the head, then moves to the trunk, and then to the lower extremities, in cephalocaudal order. This developmental blueprint also calls for development to proceed proximodistally from the midline of the body (centering around the spine and heart) outward to the extremities (the shoulders, arms, and hands).

Fetal Period   The *fetal period* of development begins about the eighth week, when the embryo becomes a recognizable human being, and lasts until the time of birth.

By the third month, the fetus is about 3 inches (7.5 cm) long and weighs approximately 1 ounce (28 g). The sex organs are developed to the point where sex can be determined and a number of physical and anatomical features are better defined. The forehead is more prominent and the eyelids, nose, and chin are clearly distinguishable. Although the process may have begun during the embryonic period, it can now be seen that flexible cartilage is being replaced by bone in a process called *ossification*. The fetal period also marks the time at which the first muscle movements in the mouth and jaw probably occur. By the ninth week the muscles of the arms and legs are capable of responding to tactile stimulation.

The fourth month (16 weeks) of fetal development is a period of significant growth. The fetus will increase its length to about 6 inches (15 cm) and weight to approximately 4 ounces (110 g). Whereas a considerable amount of growth has already occurred in the upper body, there is now a growth spurt of the lower parts of the body in the cephalocaudal direction. The toes and fingers have separated, and fingerprint and footprint patterns begin to emerge.

By the end of the fifth month (20 weeks), the fetus is approximately 12 inches (30 cm) in length and weighs about 1 pound (450 g). Features of the skin have become more differentiated and distinctive fingernails and toenails are present.

During the sixth month (24 weeks) the eyes and eyelids are completely formed, the eyelids open for the first time, and a fine layer of hair has begun to form on the head. At this point, the fetus weighs approximately 2 pounds (900 g) and is about 14 inches (36 cm) long.

The seventh month (28 weeks) is sometimes called the "age of viability"—meaning that the fetus has matured enough that it has a fifty-fifty chance of surviving outside of the womb if birth comes prematurely. The internal organs are functioning and the brain now can regulate breathing, body temperature, and swallowing. The fetus is about 16 inches (40 cm) long and weighs close to 3 pounds (1,350 g).

The eighth and ninth months are a period of rapid weight gain. The fetus gains about half a pound per week for a total of 5 pounds (2,250 g) for the two-month period, on the average. Just a few weeks prior to birth (and at delivery), the average fetus is 20 inches (50 cm) long and weighs about 7 pounds (3.2 g). During these months fatty layers are developing that will help to nourish the fetus after birth.

Table 3.1 provides a summary of the major physical growth changes that occur during the prenatal period. Figure 3.3 shows the general body form changes of the developing fetus.

The Pubescent (Adolescent) Growth Spurt

■ *Objective 3.4*

Next to the period of prenatal growth, the pubescent growth spurt, which occurs during early adolescence, represents the life span's most dramatic period of biological change. This period also has landmark significance for the evolution of motor performance and gender differences. These will be discussed in detail in subsequent chapters. At the beginning of adolescence a number of rapid changes occur in physical maturation and growth. This period of biological development is known as **puberty;** hence the term **pubescent growth spurt.** These changes are controlled primarily by genetic factors and hormones. Along with rapid changes in body size and proportions, sexual maturation is one of the most prominent aspects of the pubertal process. In this context, puberty is referred to as the stage of maturation in which the individual becomes physiologically capable of sexual reproduction.

The pubescent growth spurt generally begins at about 10 to 13 years of age in females and 12 to 15 years of age in males, and it lasts for two to three years. Females also go through the pubertal period in a somewhat shorter time than males and so are typically about two years ahead of males in maturity (see fig. 3.16 in regard to height). Add to this the wide range of ages at which individuals enter puberty and the general picture of the possible variation in individual differences becomes quite evident.

In females, the growth spurt usually designates the beginning of sexual maturity, when breasts and pubic hair also first appear. This period of development is highlighted by a universal maturity landmark known as **menarche,** or the time of the female's first menstrual flow. The average age at menarche is about 12.5 years; for many girls the event may occur as early as the age of 10 or as late as 15.5 years and still be considered normal (Hood, 1991; Paikoff, Buchanan, & Brooks-Gunn, 1991). Maturity indicators for males include the appearance of pubic and facial hair, a change in voice, and an increase in the size of the reproductive organs. Sexual maturity, including menarche, will be discussed in greater detail in subsequent sections of the text.

Hormonal change is another prominent feature of the adolescent growth spurt. Puberty begins as increased levels of hormones (originating from the endocrine system) enter the bloodstream in response to signals from the hypothalamus region of the brain. The glands directly responsible for secreting growth and sex hormones are the pituitary gland and the male and female gonads, or sex organs. The sex glands—the ovaries in the female and testes in the male—produce two distinct types of hormones: *androgens* and *estrogens.* Other androgens, called adrenal androgens, are produced in the adrenal cortex as well. Female sex glands also produce *progesterone,* another hormone that plays an important part in reproduction.

Basically androgens and estrogens interact with the growth hormone (GH) to stimulate developmental changes (fig. 3.4). Throughout childhood the

Table 3.1 Major Physical Growth Changes from Conception to Birth

Period	Age (weeks)	Height (cm)	Weight (g)	Notable Characteristics
Germinal	(0–2)			Cell differentiation begins Inner and outer mass formed
Embryonic	(2–8)	2.5	14	Human form takes shape (ears, eyes, arms, legs) Internal organs begin to develop (heart, lungs, reproductive, liver, kidneys) CNS developed
Fetal	12	7.5	28	Sex can be distinguished Movement Bone replacing cartilage Head growth, facial features
	16	15	110	Growth spurt in lower part of body Fingerprints and footprints emerge
	20	30	450	Skin structures form Fingernails and toenails present
	24	36	900	Eyes, eyelids formed Head hair forming
	28	37.5–40	1,110–1,350	Internal organs functioning Age of viability
	32–36	40–50	1,800–3,600	Rapid weight gain Layer of fat forming beneath skin Bones of head are soft

From John W. Santrock, *Life-Span Development*. 2nd edition. Copyright © 1986 Wm. C. Brown Publishers, Dubuque, Iowa. All Rights Reserved. Reprinted by permission.

Figure 3.3
General body form changes of a developing fetus.

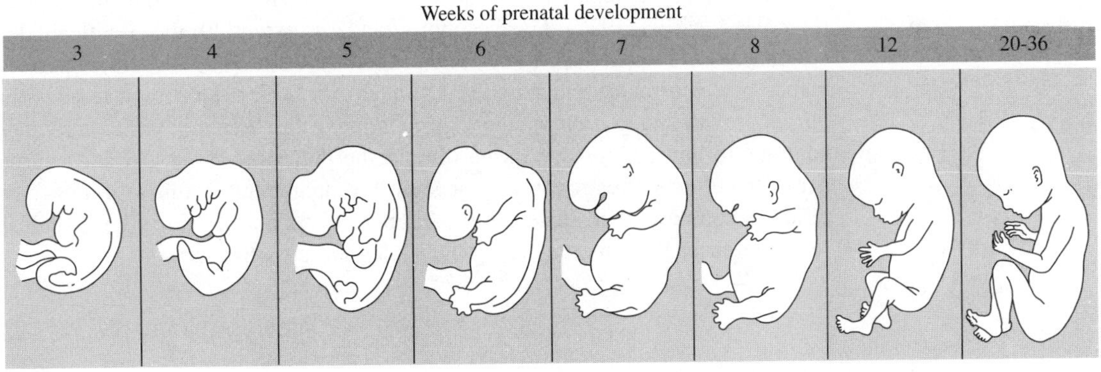

Weeks of prenatal development

| 3 | 4 | 5 | 6 | 7 | 8 | 12 | 20-36 |

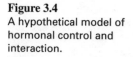

Figure 3.4
A hypothetical model of
hormonal control and
interaction.

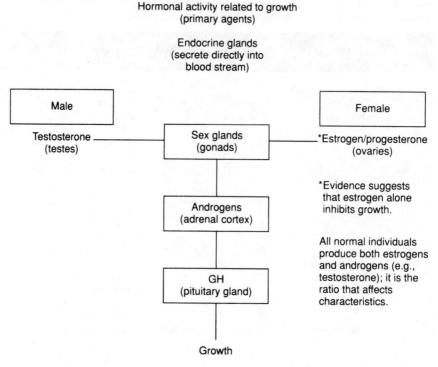

Hormonal activity related to growth
(primary agents)

Endocrine glands
(secrete directly into
blood stream)

male and female body produces approximately equal levels of estrogens and androgens. During puberty, however, the female system begins to produce more estrogens than androgens, and the male system produces more androgens than estrogens.

Androgens play a significant role in hastening the fusion of epiphyseal growth plates in the bones and stimulating the protein synthesis required for the development of muscle tissue. Brooks and Fahey (1985) note that adult males have about 10 times the amount of androgens as prepubertal children and adult women. Following puberty, males secrete about 30 to 200 micrograms of testosterone each day compared with about 5 to 20 micrograms a day in females (Douglas & Miller, 1977). This accounts, to a large extent, for the significant developmental differences in body size, proportions, strength, and motor performance between males and females. Estrogen has even been shown to actually *inhibit* growth and to *promote* the accumulation of fat. So although females may continue to develop muscle tissue (due to the presence of the GH and androgens), the effects of estrogen may account for the greater accumulation of fatty tissue in comparison to males. Table 3.2 provides a list of selected anatomical sex differences that emerge during the adolescent growth spurt.

Table 3.2   Selected Anatomical Sex Differences

Height and weight: males taller and heavier
Shoulder width: males wider, more rotation torque
Forearm length: males longer, more lever torque
Hip shape: insertion of femur more oblique in females
Elbow and knee joints: males parallel; females) (shaped
Leg length: relatively longer in males
Chest girth: males greater thoracic cavity
Center of gravity: males higher, females lower
Fat free weight: males more muscle, bigger bones

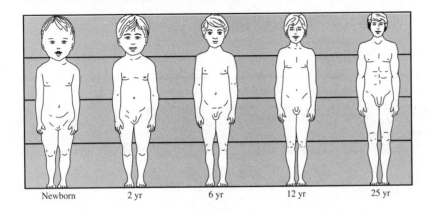

| Newborn | 2 yr | 6 yr | 12 yr | 25 yr |

Figure 3.5
Postnatal body
proportion changes.

Although the development of the human body follows a relatively consistent cephalocaudal/proximodistal pattern of development, specific body parts show varying rates of growth. During the course of development from birth to adulthood, the head size will double, trunk length will triple, the length of the upper limbs will quadruple, and the length of the lower limbs will quintuple. Along with these differential growth rates, observable changes in body proportions and general body build (physique) can be noted.

Body Proportion Changes Relative to the Head   Figure 3.5 shows general postnatal body proportion changes with age. Most notable are the proportional changes of head size in relation to total body length. The head of the newborn accounts for approximately one-fourth of its total height and its legs make up only about three-eighths of its stature. In comparison, the adult head accounts for about one-eighth of total height and leg length for approximately half of stature. The principle of cephalocaudal development is clearly evident here. Interesting to consider also is that in the newborn, head width is very close to that of the shoulders and hips, whereas head width of adults is only about one-third of shoulder width. After birth, shoulder width is generally greater than hip width. During postnatal development, the legs grow at a faster rate than other body parts in moving toward the adult model of body proportions.

CHANGES IN BODY PROPORTIONS AND PHYSIQUE

■ *Objective 3.5*

Body Proportions

Figure 3.6
Measuring head
circumference.

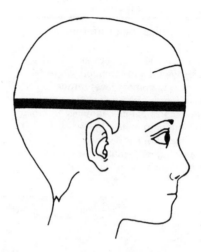

Table 3.3   Average Head Circumference

Age	Mean Inches	Centimeters
Birth	13.8	35
1 month	14.9	37.6
2	15.5	39.7
3	15.9	40.4
6	17.0	43.4
9	17.8	45.0
12	18.3	46.5
18	19.0	48.4
2 years	19.2	49.0
3	19.6	50.0
4	19.8	50.5
5	20.0	50.8
6	20.2	51.2
7	20.5	51.6
8	20.6	52.0
10	20.9	53.0
12	21.0	53.2
14	21.5	54.0
16	21.9	55.0
18	22.1	55.4
20	22.2	55.6

Source: Data from Lowrey, 1978.

Head circumference is a commonly used anthropometric measurement that is used to estimate the rate of brain growth among infants and young children. The head circumference measurement is taken at the level of the plane passing just above the bony ridge over the eyes and the most posterior protrusion on the

- - - - - - - - - - - -

Figure 3.7
Head circumference
growth data for males.

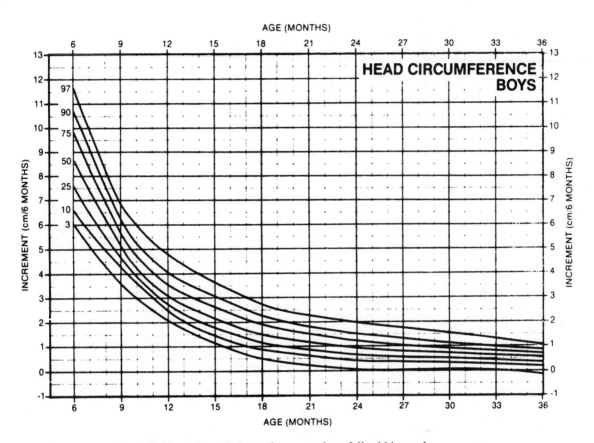

back of the head (fig. 3.6). Normal head circumference values fall within a relatively narrow range for any given age group, with the variance remaining almost constant for the entire growing period. Although there are slight sex differences, with males having a greater circumference, the difference generally does not exceed 1 centimeter at any age. Table 3.3 provides information on average head circumferences for individuals from birth to 20 years of age.

During the early months of postnatal life, growth in head circumference is very rapid and then begins to decelerate. During the first four months there is an increase of about 5 centimeters that doubles by the end of the first year. From that point until age 18, head circumference only increases another 10 centimeters. Head circumference norms, besides being used as an estimate of brain growth, are also used to identify significant abnormal trends (e.g., hydrocephalus). Figures 3.7 and 3.8 provide incremental (velocity) growth charts for males and

Figure 3.8
Head circumference
growth data for females.

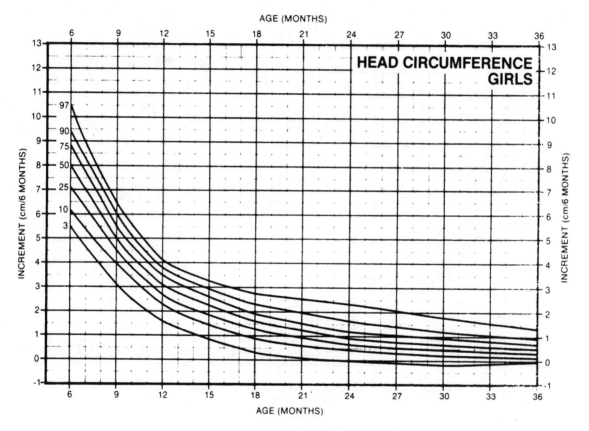

females up to 3 years of age. Charts such as these are frequently used in clinical settings to examine changes in the rate of growth over a specific period of time.

Head size in the very young may have biomechanical implications for motor skill performance. It would seem quite possible that even if the newborn were neurologically ready to attempt walking, a top-heavy body would present problems in maintaining balance. Another commonly seen example is that of a 4- to 5-year-old attempting to perform a backward roll. Due to the larger proportional head size of younger compared to older children, the task of completing a smooth roll over the back of the head often presents difficulty. Few, if any, research findings have shed real light on performance differences attributable to proportional diversity among children.

Changes in Ratio of Sitting Height to Stature   The ratio of sitting height to stature measures the contribution of the legs and trunk to total height. Sitting

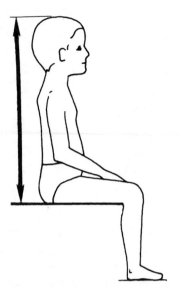

Figure 3.9
Measuring sitting height.

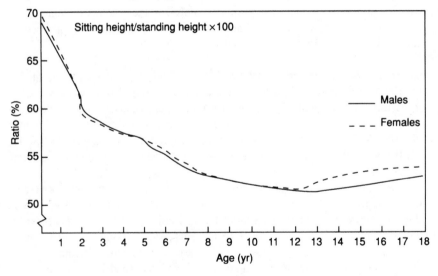

Figure 3.10
Ratios of sitting height to stature.
Source: Data from Hansman, 1970, as reported by Malina, 1984.

height is measured as the height from the seat of the chair in which a person is sitting (with spine erect) to the vertex, or top of the head (see fig. 3.9). Measures of the change in head and trunk height can then be compared with measures of total stature, which includes leg length.

Sitting height is typically 60% to 70% of total body length in the early years and decreases to about 50% when mature height is reached. The most noticeable changes occur in the legs and trunk; the head actually contributes very little to height change after birth. Figure 3.10 depicts the ratio of sitting height to stature for individuals up to 18 years. The rapid growth of the legs is evidenced by the decrease in the ratios over time. The ratio is highest in infancy and

Figure 3.11
(a) Measuring biacromial
(shoulder) width.
(b) Measuring bicristal
breadth (hip width).

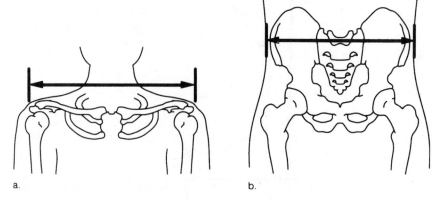

a.   b.

declines throughout childhood into adolescence. Females reach the lowest ratio point around 12 to 13 years, whereas males hit theirs about two years later (at age 13 to 15). Up until approximately 12 years of age, 55% to 60% of overall height gains for both sexes are attributed to leg growth because males and females experience almost identical increases in sitting height (trunk length). Prior to adolescence males and females are proportionately similar in terms of relative leg and trunk length. However, after that point in time, females exhibit, for an equal stature, relatively shorter legs than males.

Changes in Shoulder and Hip Width   The shoulder/hip ratio serves as a basic descriptor of proportional growth in the human body. It also provides information with biomechanical implications for motor performance. The shoulder/hip ratio is determined from two measurements: (a) biacromial breadth (or shoulder width)—the distance between the right and left acromial processes (fig. 3.11); and (b) bicristal breadth (or hip width)—the distance between the right and left iliocristales.

Although hip and shoulder width appear almost equal in the newborn (refer back to figure 3.5), shoulder width is typically greater than hip width for all individuals. Figure 3.12 illustrates shoulder (biacromial) and hip (bicristal) width values for males and females ages 6 to 17 years. Although there are differences between the sexes prior to adolescence, the variances are generally minimal. Beginning with the pubescent growth spurt, the obvious broadening of the shoulders relative to the hips is characteristic in males, whereas females experience greater hip breadth gains relative to the shoulders and waist. The biacromial/bicristal ratio remains relatively constant in males from ages 6 to 11 years, then exhibits a marked increase that continues up to around 16 years. This increase is due primarily to the fact that shoulder breadth is increasing at a more rapid rate than hip breadth. After peaking at approximately 12 years of age, the ratio for females steadily declines as a result of increased hip growth relative to shoulder breadth (fig. 3.13).

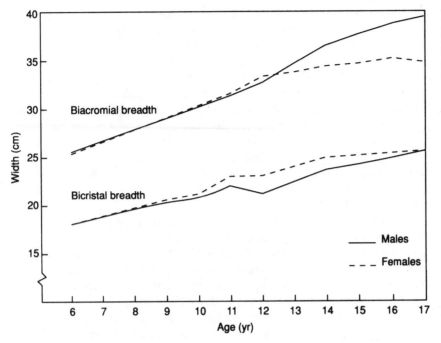

Figure 3.12
Shoulder and hip width values.
Source: Data from National Center for Health Statistics, 1983.

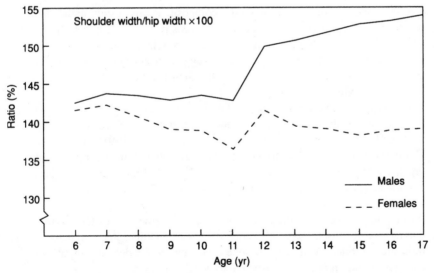

Figure 3.13
Shoulder and hip width ratios.
Source: Data from National Center for Health Statistics, 1983.

Physique

Also referred to as body build, **physique** is a composite of body proportion relationships and body composition characteristics. An informal assessment of body build is made when an individual is identified using terms such as stocky, lean, muscular, or soft-bodied. Various ratios of body mass (weight) to height have been used to provide a simple and convenient estimate of physique; however, several limitations have been noted. Because body proportions change with growth, it is very difficult to develop a scheme of body typing for use over the life span. The best-known and most widely used method of classifying body builds was developed by Sheldon, Dupertuis, and McDermott (1954).

Sheldon's somatotype classifications are a complex rating of standardized photographs used in combination with a height/weight index score. The Sheldon method uses three general body build components to categorize an individual's **somatotype,** or body build: endomorphy, mesomorphy, and ectomorph. *Endomorphy* describes a body type that is soft and round in contour, suggestive of a tendency toward fatness and obesity; *mesomorphy* describes a body with well-defined muscularity and balanced body proportions; *ectomorphy* describes the leanest of the body types, with characteristics typically associated with an extremely thin individual.

Each somatotype is viewed as a continuum for that particular body build rather than a discrete and separate category. Individuals are rated from one to seven (seven being the highest) on each of these body types. For example, a somatotype rating of 2-6-2 (endomorph-mesomorph-ectomorph) describes a mesomorph with only slight endomorphy and ectomorphy characteristics. An important limitation of Sheldon's method is that the ratings were not developed to classify females and children. Furthermore, somatotyping is not a highly scientific method because it is primarily descriptive and subjective. Subsequent investigators have modified Sheldon's method to make the rating process more objective by incorporating data on body composition, by using anthropometric measures, and by including children and females (Carter, 1980; Parnell, 1958).

There appears to be a void in the literature related to changes in physique during the growing years. Some evidence suggests a strong genetic link based on the high percentage of children who acquire their parent's body build. Still, there is no strong evidence indicating consistency in physique from the early years to maturity. There have been reports that males, in general, are more mesomorphic from preschool through young adulthood and that females consistently exhibit more endomorphic characteristics (Malina & Bouchard, 1991). A rather consistent observation is that considerable variation exists within sex, racial, and geographic groups, in children and adults (e.g., Himes, 1988). For an excellent discussion on this topic refer to Malina and Bouchard (1991).

Changes with Aging

As individuals reach middle age, mesomorphic characteristics tend to diminish due to loss of muscle mass; consequently, more endomorphic and ectomorphic traits are exhibited. As a result of bone tissue loss with a decrease in total height, changes in body proportions also tend to occur with aging.

STRUCTURAL DEVELOPMENT

Height

The measurement of body length is one of the simplest and most common measures of structural growth. Height is determined by measuring the individual either while lying down or in a standing position. Until the infant can stand erect without assistance, total body length is measured in the supine position from the

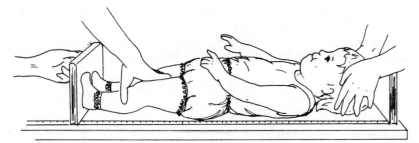

- - - - - - - - - - - - -
Figure 3.14
Measuring recumbent
length in infants.

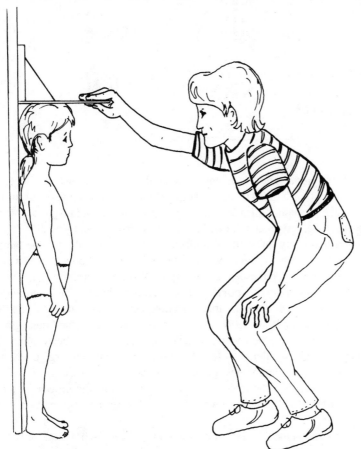

- - - - - - - - - - - - -
Figure 3.15
Measuring standing
height (stature).

vertex (top of head) to the heel. A special slide ruler (fig. 3.14) or anthropometer is used to measure the distance to the nearest one-eighth inch (0.1 cm). The preferred measurement of body length is standing height (stature), which is the distance between the vertex and the floor. To gain greatest accuracy, measurements are taken without shoes, using a sliding headboard (fig. 3.15).

We learned earlier that the embryo at about 8 weeks begins to take human form and has a body length of approximately 1 inch (2.5 cm). During the

■ *Objective 3.6*

Table 3.4   Expected Increase in Height Per Year

Age (yr)	Increase (in.)	
Birth	[Average height = 20 in.]	
0–1	9.8	
1–2	4.8	
2–3	3.1	
3–4	2.9	
4–5	2.7	
5–6	2.7	
6–7	2.4	
7–8	2.2	
8–9	2.2	
9–10	2.0	
10–11	1.9 Male	2.3 Female
11–12	1.9	2.5
12–13	2.2	2.6
13–14	2.8	1.9

Source: Lowrey, 1978.

remainder of prenatal development, a phenomenal rate of growth produces an average length at birth of 20 inches (50 cm). The growth rates that occur during the nine months preceding birth and the first year of life are the fastest that the body will experience. A typical child will increase its birth length by 50% at 1 year and reach approximately one-half of adult height by 2 years of age (see table 3.4). After age 2, the growth rate slows somewhat to average about 2 inches per year until the onset of the pubescent growth spurt. It is not uncommon for children to experience a **midgrowth spurt** in height around the age of 7, although this change is usually less dramatic than the adolescent spurt (Butler, McKie, & Ratcliffe, 1990; Gasser et al., 1991). Growth accelerates before and during puberty, with peak rates of between 3 and 4 inches per year. Until puberty, males on the average are slightly taller than females; however, there is a large overlap, with some females taller than males. Females generally enter the height growth spurt two years earlier than the males and typically complete the process in a shorter period of time (fig. 3.16); however, the age of onset and completion can vary considerably.

Females, on the average, complete their peak growth period by 16½ years of age, whereas males continue to gain in stature for another two years or so. Males are 4 to 6 inches taller than females when they reach mature height. According to Tanner (1978), maximum growth of the long bones may not be reached until about age 25 and of the vertebral column until approximately 30 years, at which time an individual may add one-eighth to one-fourth inch to his or her height. Height remains relatively stable until sometime after the third decade of life when total height begins to regress (fig. 3.17). Figures 3.18 and 3.19 provide percentile charts for comparing individual growth values.

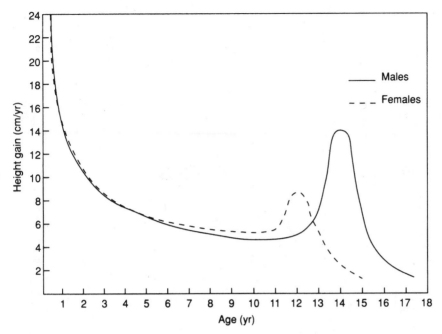

Figure 3.16
Typical velocity curves
for males and females
illustrating height gain
per year.

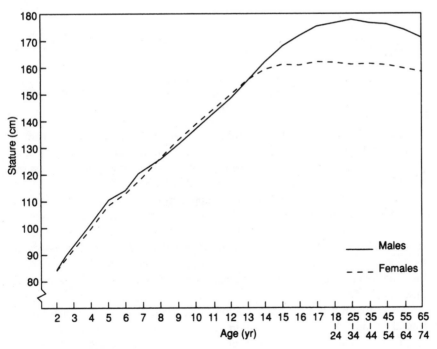

Figure 3.17
Distance curves for
stature.

Source: Data from National
Center for Health Statistics, 1973
and 1979.

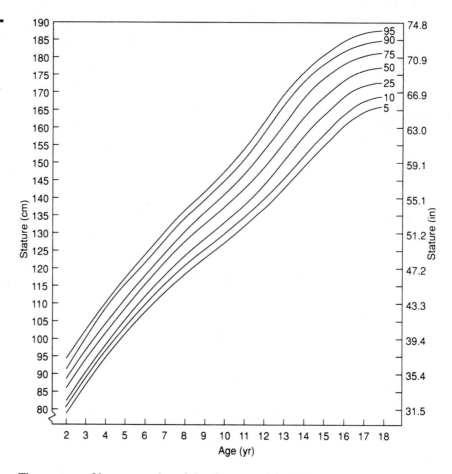

Figure 3.18
Male stature by age
percentiles.

Source: Data from National
Center for Health Statistics, 1976.

Skeletal Growth

The process of bone growth and development that will lead to skeletal maturation begins with the formation of mesoderm cells in the embryo about 16 days after fertilization. Approximately five to six weeks after conception, the first ossification centers appear in the jawbones and collarbones. This is also the approximate time that the long bones of the upper arm and leg appear in a cartilage form. Soon after the cartilage takes the shape of a long bone at about 2 months (fetal age), a ring of true bone begins to form. At the same time, primary ossification centers begin to appear in the midportion (*diaphysis*) of the long bones. In most bones, the primary ossification center appears by the third month of fetal life. The developmental status of a human fetus skeleton at 18 weeks is shown in figure 3.20.

Ossification is the transformation of cartilage to true bone. It proceeds from the center of the shaft outward in both directions until the entire shaft is ossified. More than 800 ossification centers appear in the body after birth. Shortly after birth, secondary ossification centers appear at the ends of the shaft (*epiphyseal areas*), forming the epiphyseal plate, or "growth plate." Basically, cartilage

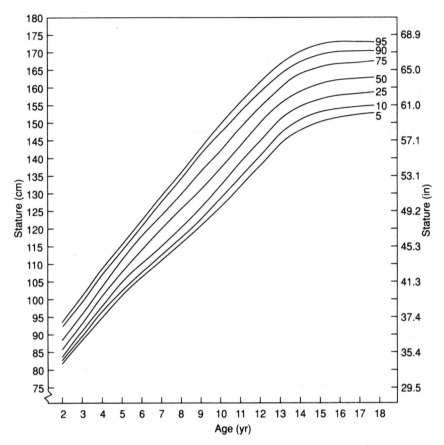

Figure 3.19
Female stature by age
percentiles.
Source: Data from National
Center for Health Statistics, 1976.

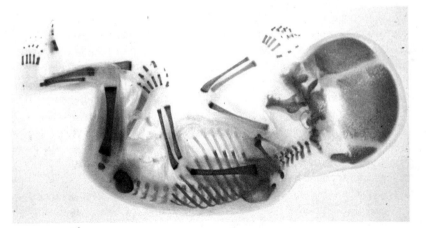

Figure 3.20
A fetal skeleton at 18
weeks. Dark areas denote
ossified portions and
spaces between cartilage
models.
© Carolina Biological Supply
Company.

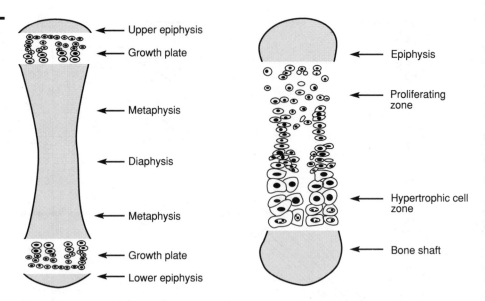

Figure 3.21
(Left) Long bone with upper and lower epiphyses and (right) general process of bone formation.
Source: From Tanner, 1978.

cells are nourished by the blood in a reservoir next to the epiphysis. They begin to increase in size in the *proliferating zone* before moving on to the *hypertrophic zone* where the cells arrange themselves in vertical columns. At that point, the cartilage cells erode and true bone begins to grow as collagen is secreted by *osteoblasts* (the bone cells engaged in producing true bone). *Osteocytes,* the principal cells of mature bone, become trapped in the tissue matrix and collect calcium and phosphorus. Essentially bone is a matrix composed primarily of collagen fibers and noncollagenous proteins. When full maturity is completed, the epiphysis fuses to the *metaphysis* (the area between diaphysis and epiphysis) and the growth plate is erased (fig. 3.21).

The mechanism controlling bone girth (appositional growth) is not as complex as the cartilage replacement processes responsible for growth in length. New rings of bone are added in circular layers on top of the old layers formed earlier. The bone has a compensating mechanism that controls thickness. As new rings are added old layers are removed from the inner circumference of the bone sleeve, permitting the shell to become thicker. This process continues until each of the long bones reaches full maturity in late childhood to early adulthood.

Another important mechanism of long bone growth is modeling resorption. At birth the bone shaft tends to flare out at its ends, hence creating a thinner middle. As the bone grows in length, the end becomes wider and the shaft heavier. The mechanism for maintaining the linear form of the bone is called *modeling resorption.* As the ends become increasingly thick, the excess bone tissue is removed by a process of resorption or breakdown, thus permitting the bone to maintain its linear shape while it is growing longer.

During the growth of short bones of the wrist and fingers, the core of the bone in the center of the cartilage steadily enlarges. At the same time, the original cartilage itself is growing appositionally as more cartilage is deposited on its surface. Growth of new bone tissue proceeds faster than growth of cartilage; thus more and more bone is added to the developing structure. This continues until the bone matures fully.

Epiphyseal plate closure, indicative of long bone maturity, generally starts somewhere between 13½ and 18 years of age, beginning at the distal humerus. Most of the long bones reach maturity at about 16 to 18 years of age in females and 18 to 21 in males. However, there is some evidence that maximal growth of the long bones may not be achieved until approximately age 25 (Tanner, 1978). Cranial development follows a different postnatal growth curve. The majority of growth is completed by 3 years, with 50% attained by the ninth month. There are areas of the braincase, however, that do not reach maturity until well into adulthood.

Females, from birth, typically have a more developed skeleton than males. Females are approximately 20% more advanced in skeletal maturity than boys, which in practical terms is a difference of about two months at age 1 and two years at age 10. There are also indications that there is less variability in rate of ossification for females. Greater size in the male after puberty is due to the longer growing period (approximately two years) and the increased production of adrenal androgens and testosterone. As will be noted later in the discussion of maturity estimates, skeletal age may precede or lag behind chronological age by as much as a year.

Normal aging causes the bones to lose mass and the total height to decrease. Women begin to lose bone mineral at about age 30 and men at approximately 50 years of age. Stature decreases with age because of an increase in postural kyphosis (rounding of the back), compression of intravertebral disks, and deterioration of vertebrae (Brooks & Fahey, 1985). Estimates of height decreases from 35 to 75 years of age are about 1 inch for males and 2 inches for females.

Changes with Aging

The loss of bone mass presents a more complex problem, especially in women. Bone loss in women begins slowly during the third decade (0.75% to 1% per year) and increases to a higher rate (2% to 3% per year) shortly before and after menopause. The total bone mineral mass loss by age 70 is approximately 25% to 30%. Bone loss estimates for men at age 70 are about half what women experience (12% to 15% by age 70). Age- and sex-related changes in the width of cortical bone are illustrated in figure 3.22.

Osteoporosis has become a familiar term to the general public; it refers to the loss of total bone mass to such an extent that the skeleton is unable to maintain its mechanical integrity. The development of this condition is a complex process influenced by a variety of nutritional (related primarily to calcium), physical, hormonal, and genetic factors. In the scientific community there is some debate concerning the dividing line between normal aging of bones and the more-pronounced brittleness of bones generally associated with osteoporosis. As a result of osteoporosis bones tend to break more easily, which increases the risk of fracture. Studies have reported that postmenopausal women are four times more likely to break bones than men and that 25% of these women have suffered

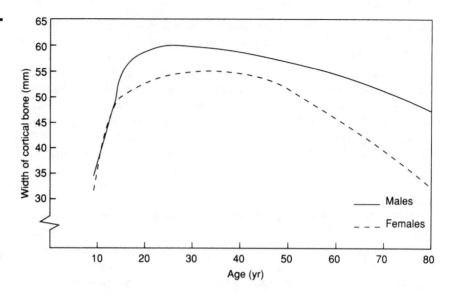

Figure 3.22
Age changes in width of
cortical bone (second
metacarpal).

at least one broken bone by age 65 (Hogue, 1982; Sterns, Barrett, & Alexander, 1985). Short-term mortality rates have also been shown to increase because of certain types of fractures.

BODY MASS

■ *Objective 3.7*

Body weight and body composition are terms frequently used to describe an individual's **body mass.** Body weight is a general descriptor of total body mass that is a composite of all tissue components. Body composition, in contrast, is a description of the various independent tissue components—namely, lean body mass (fat-free weight), and body fat. Thus, body weight equals lean body mass plus fat. The following discussion of body composition will focus on skeletal muscle tissue (lean body mass) and body fat. It should be noted, however, that lean body mass (for body weight measures) encompasses all of the body's nonfat tissues including skeleton, muscle, water, connective tissue, organ tissues, and teeth. The growth and development of cardiac (heart) muscle tissue will be discussed in chapter 4.

Body Weight

The assessment of body weight is the most common anthropometric measure used to estimate body mass. Instrumentation is relatively simple in that the individual stands on clinical scales (preferably beam-type) without shoes and dressed in as little clothing as possible (preferably briefs or bathing suit). Infant scales and bed scales are typically used for the disabled, elderly, or other individuals not capable of standing without aid. Weight is measured to the nearest tenth of a kilogram (or quarter of a pound).

The embryo at 8 weeks weighs about half an ounce (14 g) and by the end of the fifth month the fetus weighs about 1 pound (450 g). The last two months of

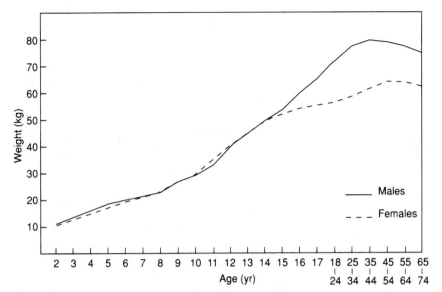

Figure 3.23
Distance curves for body weight.
Source: Data from National Center for Health Statistics, 1973 and 1979.

prenatal life is a period of rapid weight gain. The fetus gains about half a pound (225 g) a week for a total of 5 pounds (2,250 g) on the average. Just a few weeks before birth (and at delivery), the average fetus weighs about 7 pounds (3.2 g). Females, on the average, are about half a pound smaller than males at birth.

Total body mass generally decreases during the first few days after birth due to "transition" factors such as maternal milk supply and level of physiological stress. However, birth weight is usually regained within two weeks. Body weight doubles after three to four months, triples at one year, quadruples by the second year, and by age 7 increases seven times the birth weight (Lowrey, 1986). Most children will reach one-half of their mature body mass by age 10. Figure 3.23 illustrates a distance curve for body weight among individuals 2 to 74 years of age.

The pubescent growth spurt for body weight is similar to the height curve except that it generally continues for a longer time. Female growth patterns reveal a particularly rapid weight gain between the ages of 12 and 13. As in the growth pattern for height, males begin the growth spurt later than females and add more body size. Males increase their body mass by about 45 pounds (20 kg) and females by approximately 35 pounds (16 kg) by early adulthood. Mature body weight is about 20 times that of birth weight. Total body mass begins to decline with aging as a reflection of bone and muscle tissue losses. Figures 3.24 and 3.25 compare individual growth values using percentile charts.

Body weight is a general measure of body mass and therefore does not break down the contribution of different tissues making up this collective measurement. A measure of body composition, however, independently assesses two types of tissue: muscle tissue (lean body mass) and body fat. A variety of

Body Composition

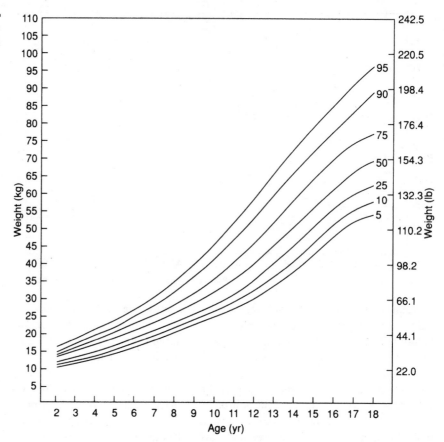

Figure 3.24
Male body weight by age
percentiles.

Source: Data from National
Center for Health Statistics, June
1976.

measurement techniques are available to estimate total body fatness, lean body
mass, or both. All of the measurement descriptions that follow are only *estimates*
because they are indirectly determined. The only direct way of taking the measurements is by making a biochemical analysis of cadavers.

Some of the more common techniques for measuring body composition are
underwater weighing, radiographs (X rays), anthropometric dimensions (skinfolds), and body mass index. Although potassium-40 (a naturally occurring iso-
tope) is also used to measure and predict body composition (i.e., lean body
mass), this method is almost completely restricted to research because the necessary equipment is so expensive.

Underwater (hydrostatic) weighing is considered the most accurate indirect
means of measurement and serves as a standard for other methods such as skinfold measurement. In underwater weighing, the percentage of body fat is com-
puted by finding the ratio of body weight to body volume, called body density.
Body volume is equal to the loss of weight in water with the appropriate temperature correction for the density of water. Using other equations, total body fat and
lean body weight can also be measured. The equations used to predict body

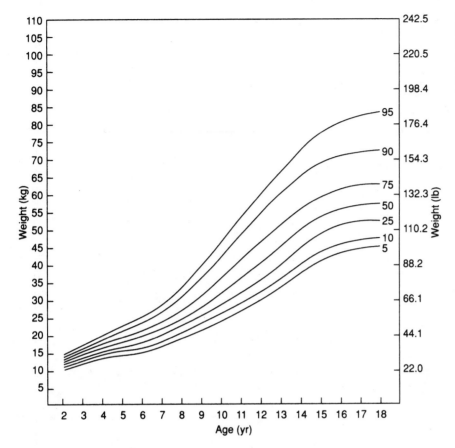

Figure 3.25
Female body weight by
age percentiles.
Source: Data from National
Center for Health Statistics, June
1976.

density with underwater weighing were formulated from the chemical analyses
of human adult cadavers. The use of the same equations for younger populations
has been known to give rise to problems of developmental interpretation
(Lohman, 1989).

In the not too distant past, the body fat content of children was estimated
from an adult model on the assumption that the child's body composition was
chemically equivalent to that of a mature individual. Evidence suggests, however, that pre- and postadolescent children are not chemically mature and may
have a fat-free body density that is lower than that of an adult. Body density, in
this context, refers to water content and bone mineral content of the fat-free body.
The water content of the adult fat-free body is typically 71% to 73%. During the
pre- and postadolescent period, water content of the fat-free body appears to decrease from slightly over 80% to 72% (fig. 3.26). The adult level appears to be
reached around 15 years of age. Related research suggests that these changes are
a function of age (Lohman, Boileau, & Slaughter, 1984). Bone mineral content
also varies between the adult and child models (fig. 3.27). In general, bone density increases with age from the early fetal period to adulthood, with male bones

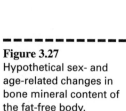

Figure 3.26

Hypothetical age-related changes in water content of the fat-free body.

From: "Body Composition in Children and Youth" by Timothy G. Lohman, Richard A. Boileau, and Mary H. Slaughter. In *Advances in Pediatric Sport Sciences I,* page 38 by Richard A. Boileau (Ed.), 1984. Champaign, IL: Human Kinetics. Copyright 1984 by Human Kinetics Publishers. Reprinted by permission.

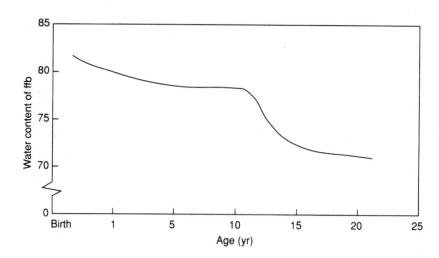

Figure 3.27

Hypothetical sex- and age-related changes in bone mineral content of the fat-free body.

From: "Body Composition in Children and Youth" by Timothy G. Lohman, Richard A. Boileau, and Mary H. Slaughter. In *Advances in Pediatric Sport Sciences I,* page 43 by Richard A. Boileau (Ed.), 1984. Champaign, IL: Human Kinetics. Copyright 1984 by Human Kinetics Publishers. Reprinted by permission.

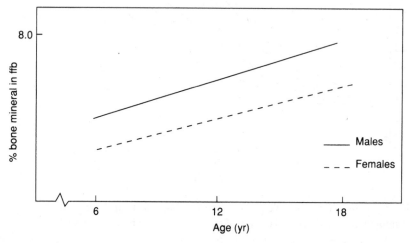

being heavier than female bones. In addition, there is some evidence that the bones of blacks are heavier than those of whites.

In summary, children and youth are likely to have higher body water content and lower bone mineral than adults, resulting in a lower fat-free body density. Therefore, body composition estimates for children from adult equations may overestimate values. Updated equations based on better estimates of children's fat-free body composition are available (see Lohman, 1989). These estimates may provide more accurate information about children's body composition, in particular, and about patterns of growth and development, in general.

Radiographs (X rays) are used to assess regional changes in fat, muscle, and bone during growth. When proper laboratory facilities are not available, subcutaneous skinfolds can also provide accurate regional information about growth changes. Skinfold measurements are probably the most popular means of assessing body composition. The rationale for using skinfold measurements is based on

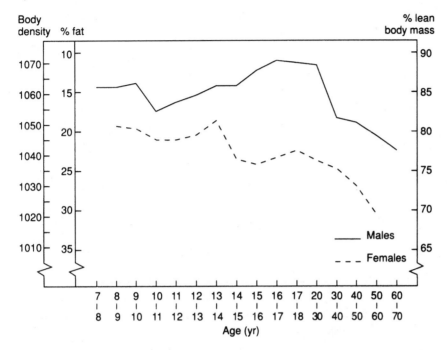

Figure 3.28
Body composition
change as a function of
age.
Source: Data from G. A. Brooks
and T. D. Fahey, *Exercise
Physiology: Human Bioenergetics
and Its Applications.* Copyright ©
1985 Macmillan Publishing
Company, New York, NY.

the fact that a relationship exists between the amount of fat located directly beneath the surface of the skin and body composition (internal body fat).

Skinfold equations are derived using statistical techniques of multiple regression that predict the results of the underwater weighing procedure. It is essential that individuals be measured using an equation derived from a compatible population; this has implications for measuring children and adults. The triceps skinfold and the more recently introduced medial calf skinfold correlate more closely with body density data derived from underwater weighing procedures. Limitations of this method include its use for measuring very thin or obese individuals, skinfold caliper accuracy, and tester reliability.

Body mass index is a calculation of body fat based on height and body weight:

BMI = body weight [kg]/height^2 [m]

This method has been used more in recent years as an alternative to skinfolds because it can be easily administered. The predictive accuracy of this as an estimate of body fat has not been extensively studied. The data that do exist indicate that it warrants only a fair rating, less accurate than the triceps skinfold method.

On the average, males at all ages have greater body densities and greater lean body mass than females (e.g., Malina, 1984; Parizkova, 1968). While the differences are closer during the childhood years, females have greater amounts of body fat across the life span. This difference is magnified as adolescence approaches and continues on through development. Figure 3.28 shows selected body composition values for ages 7 to 70. During the pubescent growth spurt,

Figure 3.29
Median estimated
midarm muscle
circumference and
triceps skinfold.
Source: Data from Malina, 1984.

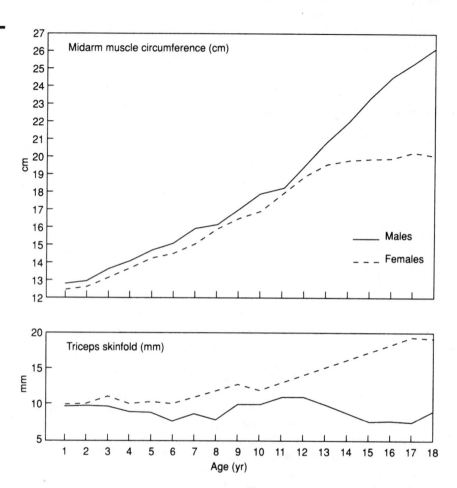

males experience a rapid increase in lean body mass and a decrease in percentage of fat. Females also experience an increase in lean mass, although less than males, and an increase in body fat. From puberty to 18 years of age, females gain about 4.5 kilograms of fat. Data from Malina (1984) indicate that estimated midarm muscle circumference reveals growth patterns similar to lean body mass (fig. 3.29). During the years 1 to 18, males have larger estimated midarm musculature and these differences become greater with the pubescent growth spurt.

Body Fat Growth Body fat layers appear in the fetus between the seventh and eighth prenatal months. Between birth and 6 to 9 months, body fat mass will increase 10% to 20% before tapering off to a plateau that may last from five to seven years. On the average, percentage of body fat increases from about 16% at birth to between 24% and 30% at 1 year. By age 6 body fat decreases to approximately 14% of body weight. Some individuals will experience an increase during the midgrowth spurt of 5 ½ to 7 years, which generally ceases after that period in males but continues in females through the adolescent years. As noted earlier, females have more body fat (absolute and relative) than males at all ages, with differences being more apparent during and after puberty.

Fat and muscle cells grow both in number (*hyperplasia*) and in size (*hypertrophy*). The adult of average body size has between 25 to 50 billion fat cells. Obese individuals may have 75 billion or more (McArdle, Katch, & Katch, 1991). The size of fat cells in children up to 1 year is about one-fourth that of adult fat cells. Fat cell size triples during the first six years, then remains relatively stable until the onset of puberty. At this time, very little information is available on fat cell development during adolescence. What exists suggests that cell size increases during the growth spurt because the cells of adults are significantly larger than cells at age 13 or in late adolescence.

By the end of the first year of life, the number of fat cells is about three times what it was at birth. It is believed that most of the fat cell production takes place during the last trimester before birth. After the first year, the number of cells increases gradually until the onset of the pubescent growth spurt, at which time the number increases significantly. Although the evidence is not decisive, there is speculation that fat cells are particularly prone to overmultiply if overfeeding occurs during the early stages of life (Hager, Sjostrom, & Arvidsson, 1977; Oscai et al., 1972). Relationships have been drawn between overfeeding, excessive fat cell multiplication, and the weight control problems that individuals may experience in later life. Although there is some disagreement, it is generally believed that by adulthood (or perhaps before) the number of cells becomes fixed and increases in size only, except in cases of extreme obesity (Brooks & Fahey, 1985). Obese individuals can have as many as twice the normal number of fat cells and these can be up to 40% larger than those of lean individuals. When individuals lose weight, their existing fat cells shrink in structural size rather than change in cell number. When individuals gain fat tissue as a result of overeating, their fat cells enlarge but no new cells are created.

Skeletal Muscle Tissue Growth Muscle fibers develop in the premuscular mesodermic tissue. After the 16th week of fetal development, the first muscle fibers can be identified. Growth of voluntary muscle during the prenatal period is both hyperplastic and hypertrophic. The hyperplastic phase continues until shortly after birth, after which cell growth is predominantly achieved by increasing cell size hypertrophy. The increase in mass of a skeletal muscle of up to 50 times during postnatal development is due almost exclusively to a massive increase in the size of individual fibers (Lang & Luff, 1990). Thickness of the muscle fibers at birth is approximately one-fifth that in an adult. The essential element in muscle composition, protein, increases gradually during the fetal period and, by birth, reaches a level very near that found in adults. Approximately 15% of total body mass of the fetus halfway through its development consists of skeletal muscle tissue. This percentage increases progressively to about 25% of birth weight for both sexes. Approximately 40% to 50% of body mass in the adult male is skeletal muscle. Females generally do not show a significant increase in lean body mass during puberty and at maturity have only about 60% of the muscle mass and force characteristics of the male.

The differentiation of fiber types and sizes is frequently used to describe muscle growth and development. Basically, in the adult, *type I fibers* are described as "slow twitch," are relatively fatigue resistant, and are associated with the performance of aerobic endurance activities (e.g., long-distance running and cycling). *Type II fibers,* in contrast, are "fast twitch" and have the capacity to contract more rapidly and forcefully, but they also fatigue rapidly. Fast-twitch fibers are associated with the performance of short-duration (anaerobic) activities (e.g., sprinting, jumping, and weight lifting). At birth, type I fibers constitute approximately 50% to 60% of all muscle fibers. During this same period, only about 25% of type II fibers are found (Colling-Saltin, 1980). The amount of undifferentiated fibers ranges from 15% to 20%. In adults there are very few, if any, undifferentiated fibers.

Studies that have investigated muscle fiber changes during adolescence have noted the possible significant effects of pubertal hormones on growth and differentiation. Researchers have found a relatively high percentage (7% to 13%) of "transitional" fibers in adolescents (du Plessis et al., 1985). *Transitional fibers* are ones in which the functional features of the enzyme system may develop as either aerobic (type I) or anaerobic (type II), depending on the physical training to which the fibers are treated. These fibers supposedly develop after infancy due to the influence of pubertal hormones. Such factors may have implications for motor programming and training with adolescents.

In regard to muscle fiber size, the typical adult has a fiber diameter of 50 to 60 micrometers. At about the 14th or 15th week of fetal development the fiber is approximately 6 micrometers, and by birth the diameter has doubled. By the end of the first year the fiber size will be about 30% of adult size, and at 5 years it will reach approximately 50%. It should be noted, however, that size (and type) of muscle fiber can vary considerably in fetuses of the same age. There does not appear to be any major differences between upper and lower extremities or proximal and distal muscles, nor any significant variance between the sexes. Although several hypotheses have been proposed concerning how muscle growth and differentiation take place in both the prenatal and neonatal period, the strongest is that the pattern is inherited and genetically coded in the nucleus of the muscle cell.

Changes with Aging

Significant changes in body fat and lean muscle mass take place with aging. Whereas height decreases with age, weight increases steadily beginning in the third decade and continues until age 55 to 60 years, when it declines. An increase in body weight is usually accompanied by gains in body fat and a decrease in lean body mass. After age 60 or so, total body weight decreases despite the increase in body fat. At 17 years of age, the average male is 15% fat and by 60 years increases to about 28%. Females change from about 25% body fat at 17 years to approximately 39% at 60 years. These values can vary to quite an extent

Table 3.5 Average Estimated Changes in Body Composition for Males 20–55 Years (Standard Weights Refer to Men 176 cm Tall)

Age	Standard Weight (kg)	Standard Fat (%)	Fat (kg)	Fat-free Weight (kg)
20	67.6	10.30	6.96	60.6
25	69.9	13.42	9.38	60.5
30	71.3	16.20	11.55	59.8
35	72.9	18.64	13.59	59.3
40	74.1	20.74	15.37	58.7
45	75.5	22.50	16.99	58.5
50	76.0	23.92	18.18	57.8
55	76.8	25.01	19.20	57.6

From J. Brozek, "Changes of Body Composition in Man During Maturity and Their Nutritional Implications" in *Federation Proceedings,* 11:787, 1952. Copyright © 1952 Federation of American Societies for Experimental Biology. Reprinted by permission.

Table 3.6 Summary of Physical Growth Changes with Aging

Physical Growth Characteristic	
Stature	Decreases
Bone mass	Decreases
Body weight	Increases
Percent body fat	Increases
Lean body (muscle) mass	Decreases

depending on lifestyle and tend to be less in individuals who are physically active. Table 3.5 shows typical estimated changes with increasing age in various body composition characteristics for adult males (see also figure 3.28).

As individuals age, the distribution of body fat also tends to change. With aging there is an increased tendency for a large proportion of the total body fat to be located internally, rather than subcutaneously (Schwartz et al., 1990). Abdominal girth in males increases 6% to 16%, while females show an age-related gain of 25% to 35%. Sex differences in distribution are also apparent, with males accumulating more body fat on the back and females around the waist and upper arms.

Lean body mass tends to decrease with age. This overall effect is due primarily to the loss of bone mass and the reduction of muscle mass. Most individuals experience a marked deterioration in muscle mass with aging. For people over 60 years of age, the loss of muscle mass can be as high as 30% to 50% of that found in young adults. Table 3.6 presents a summary of physical growth changes with advancing age.

MATURITY ESTIMATES

■ *Objective 3.8*

Because maturation is a qualitative process, it can be a challenge to determine in quantitative terms how far an individual has progressed along his or her road to full maturity. Level of maturity is generally estimated from chronological age and biological (developmental) age characteristics.

Chronological age refers to the age of an individual in relationship to standard calendar days. To estimate level of maturity using this method alone is of limited value because of the wide variance in growth and development that is possible at any single age. An accurate calculation of chronological age does provide the basic factor with which measures of growth and development are compared when estimating level of maturity. The measurement of **biological age** generally provides a better method for estimating maturity by providing information that may be compared with the individual's age. Characteristics commonly used in a measurement of biological age include estimates of morphological age, dental age, sexual age, and skeletal age.

Morphological Age

The term *morphology* refers to the form or structure of an individual. Morphological age is usually estimated from height. A measure of height is easy to attain and there are several height percentile charts that indicate where an individual ranks for a given age and sex. A frequently used general assessment of a person's maturity might indicate that he or she is above or below average for age and sex. A measure of height alone, however, is not a good estimate of maturity because individuals differ in mature height. To be assessed as above average in height at a specific time in a child's life may signify either a rapid pace of growth or may be an average rate of growth for a child who is going to be above average in height as an adult.

Dental Age

The age at which teeth erupt may also provide information on approximate level of maturity. Eruption of both deciduous (temporary) and permanent teeth in most individuals occurs fairly predictably. The deciduous dentition erupts from 6 months to 2 years; permanent teeth generally appear from about 6 to 13 years of age. One of the limitations of attempting to use dental age to gain a complete perspective of full maturity is the lack of available information on dentition growth between the ages of 2 to 6 years and after age 13. Recent research has focused on dental calcification seen in X rays as a promising indicator of dental age.

Sexual Age

This biological estimate of sexual age refers to the assessed maturity of primary and secondary sexual qualities. Rather subjective maturity rating scales are often used to evaluate the status of sexual development in the reproduction organs, pubic hair, and breast development and menarche in females. Of all the sexual characteristics of females, the age of menarche (first menstrual flow) is the most common standard of maturity.

Table 3.7 Age at Menarche

Menarcheal Age (yr)	Cumulative % of Females
<10	2.0
11	12.8
12	43.3
13	73.2
14	91.7
15	98.3
16–17	99.7

Source: Data from U.S. Department of Health, 1973.

In most cases, menarche occurs relatively late in the adolescent growth period, after growth in height has peaked (table 3.7). Approximately 96% of females experience menarche between the ages of 11 and 15. It can also be estimated from this data that most females will have their first menstrual flow between 12 and 13 years of age. Marshall and Tanner (1969) note that menarche does not typically occur more than five years following the onset of puberty. Although menarche indicates maturity in uterine development, it does not always denote full maturity of the female reproductive system. Initially menstrual cycles may be very irregular, and it may be several years after periods begin that a female becomes fertile. Figures 3.30 and 3.31 illustrate the normal range of the development of selected sexual characteristics in males and females.

Skeletal Age

Perhaps the best method of maturity estimate is derived from evaluating the successive stages of skeletal growth as seen in radiographs (X rays). Also referred to as bone age, skeletal age provides a good estimate of maturity in that its development spans virtually the entire period of active growth and maturation. All normally developing people, regardless of whether they are early, average, or late maturing, will eventually reach a stage of complete skeletal maturity. Skeletal maturity is usually measured by x-raying the bones of the left wrist and hand complex, which consists of some 29 separate ossification centers. The hand and wrist, however, only provide useful information from about 18 months of age and older. Other parts, such as the leg, knee, and foot, have also been recommended for measuring the skeletal age of younger and older populations. The progressive enlargement and change in shape of ossification centers can be detected on X ray and compared with other X rays standardized to represent skeletal ages at six-month and one-year intervals (figs. 3.32 and 3.33).

Along with the differences in skeletal growth, these X rays can differentiate the male and female maturity rates. Females, in general, are more biologically

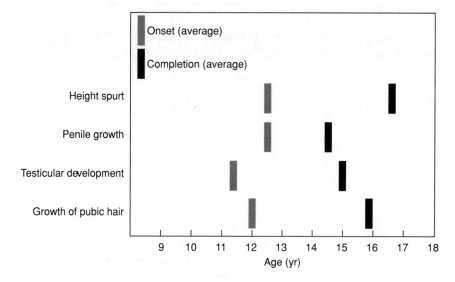

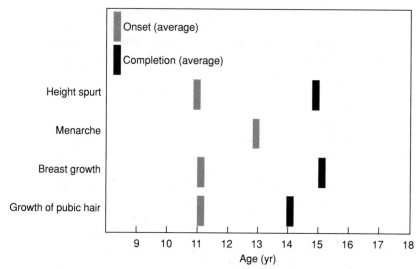

mature than males. Areas that appear as spaces between the wrist and fingers (black or grayish in appearance) are cartilaginous bone and not yet calcified. As the bone matures, or ossifies, it becomes more dense and thus is more visible in the X ray. Females are approximately 20% more advanced in skeletal maturity than males. In practical terms, the difference is about two months at age 1, and two years at age 10. Popular atlases of skeletal maturity include those developed by Greulich and Pyle (1959) and Tanner and associates (1975).

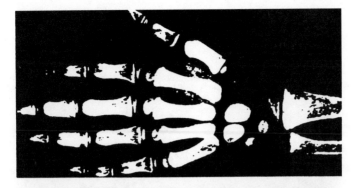

Figure 3.32
X ray of the hand showing the skeletal maturity of a male at 48 months or a female at 37 months.

S. I. Pyle, *A Radiographic Standard of Reference for the Growing Hand and Wrist.* © Year Book Medical Publishers, Chicago, IL, 1971.

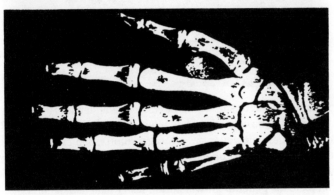

Figure 3.33
X ray of the hand showing the skeletal maturity of a male at 156 months or a female at 128 months.

S. I. Pyle, *A Radiographic Standard of Reference for the Growing Hand and Wrist.* © Year Book Medical Publishers, Chicago, IL, 1971.

MATURITY VARIATIONS

■ *Objective 3.9*

Although diet and other health factors may influence the rate of maturity, it is the biological clock within the individual that sets the pace of maturation. Individuals are generally grouped into categories of early, average, and late maturers, according to where they fit on various maturational scales. We are often concerned with the development of individuals who appear to be late maturers or early maturers. As noted earlier, it can be difficult to quantify a rather qualitative aspect of growth and development. However, the extremes can be established with more accuracy than levels in the midrange of maturity.

Early maturers are those in whom the maturity characteristics are in advance of their chronological age, whereas late-maturing individuals would be behind in relation to the standard. For example, an individual with a chronological age of 10 and skeletal age of 13 would be categorized as being ahead (early maturing) of the normal maturation pace. A female who experiences menarche at 10 years (when the average is 12.5 years) would also be an early maturer. In contrast, a female who does not experience menarche until her 15th birthday would be considered late maturing, as would the female who is 8 years old with a skeletal age of 6 years.

What is early or late? Keogh and Sugden (1985) suggest that when comparing chronological age with skeletal age, a 20% difference from mean skeletal age should be considered early or late, respectively. For example, if a 10-year-old

Figure 3.34
Secular trend for
menarche.

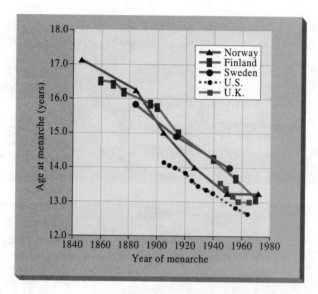

is average maturing, he or she would have a skeletal age of 10 years; a skeletal age of 8 years would be considered late maturing, and a skeletal age of 12 years, early maturing. These, however, are arbitrary cutoffs, and clinical practice often uses up to one to three years on either side of chronological age. Early maturers, on the average, take four to five fewer years to achieve full skeletal maturity and tend to complete the pubertal growth period faster than late maturers.

Although differences in maturity levels are apparent prior to the pubescent growth spurt, they appear to be most pronounced during adolescence. Malina (1984) notes that early maturers are generally taller and heavier for their age than late-maturing peers. Late maturers do, however, catch up to early maturers in stature during late adolescence or early adulthood. Various maturational rates may also affect body build. Early-maturing males are generally more mesomorphic, and early-maturing females more endomorphic. Later-maturing males and females tend to have less weight for stature, whereas early maturers of both sexes generally have more body fat during adolescence.

In terms of interaction with peers, early maturation appears to be advantageous. Early maturers are more likely to be elected to class offices, be rated as popular, and excel in athletics. The superior size and advanced physiological characteristics (e.g., strength, speed) that are associated with early maturing are in accord with male sex-role expectations.

■ *Objective 3.10*   These maturational differences are *individual* differences. As a group, today's population is taller and heavier and matures earlier than the population of generations past. These observations are referred to as **secular trend.** Today the average female reaches adult height two years earlier than did females at the turn of the century. Figure 3.34 illustrates the secular trend for menarche. Note that for Norwegian girls in 1840 menarche occurred at around age 17, compared with just over 13 years of age in 1969. For American girls, the mean age of menarche has been declining an average of about four months per decade. Thus, due primarily to improved health conditions and nutrition, females are maturing earlier (sexually) (Brooks-Gunn, 1991; Eveleth & Tanner, 1990).

Most males today reach adult stature by age 18 or 19, whereas about 75 years ago maximal height was not reached until age 25 or 26. Since 1900 children from average households have increased in height between the ages of 5 and 7 years by approximately 1 to 2 centimeters each decade. Secular differences in stature appear to be most evident during the adolescent growth period due to the faster rate of maturation in more recent generations. The difference between a 15-year-old male of 1880 and 1960 was approximately 5 1/2 inches (14 cm). The difference in maximal adult height during that same period was of a lesser magnitude—about 3 inches (8 cm).

It has been suggested that secular trend in size and maturation in individuals from developed countries has leveled off in recent years (Malina, 1978). Speculation is that the factors associated with secular trend, in general, such as improvement in nutrition, reduction in the incidence of disease, better health practices, and greater genetic hybridization, have been maximized. A negative secular trend that has been noted in recent years is that children of the 1980s carried more body fat than their counterparts of the 1960s (Gortmaker et al., 1987; NCYFS II, 1987). In addition, there is evidence that teenagers are less physically active than they were a decade ago (Centers for Disease Control, 1992) and considerable speculation that the same general trend (more TV, etc.) applies to children.

All of the physical growth characteristics described in this chapter are related in some manner to motor performance. Although research has produced few conclusive findings, especially with young populations, some generalizations can be offered.

One of the original lines of study that has been followed up in recent years is that of the relationship between body length and weight at birth to the onset of independent walking (Jaffe & Kosakov, 1982; Norval, 1947). A common finding of all investigations has been that infants who are longer for their weight tend to walk at an earlier age in comparison to shorter and fatter babies. However, follow-up studies have revealed that delayed groups usually catch up quickly and that initial walking patterns of the early group are quite immature. Similar inconclusive findings concerning young populations have also been reported concerning the relationships between body size, body mass, body fat, selected anthropometric measures, and strength-related tasks (e.g., Gabbard, 1982, 1985).

With school-age and older populations, the effects of excess body weight (and body fat) on tasks that require the body to be propelled (e.g., running, jumping) generally are negative. In tasks requiring that an object be propelled or force be applied, increasing body size usually has a positive effect on performance (Malina, 1984). It should be noted, however, that the overall contribution of age, stature, and body mass to variation in motor performance is generally low to moderate at best.

IMPLICATIONS FOR MOTOR PERFORMANCE

■ *Objective 3.11*

Similar relationship levels have also been found in regard to strength performance and somatotype. In general, mesomorphy and endomorphy are related positively to strength, whereas ectomorphic characteristics are not. Relationships between body build and motor performance (especially with activities involving body propulsion and agility) are generally low between endomorphy and task performance. The inconsistency here is that correlations between individuals having mesomorphic qualities and motor performance are generally positive but low.

As noted earlier, the proportion of the infant's head in relationship to the remainder of the body may influence motor performance. Because infants are relatively "top heavy," their ability to balance and perform such tasks as body rolling may be hindered. As body proportions change during childhood an individual's center of gravity may also vary considerably. Although the center of gravity remains relatively constant in proportion to total height with age, the center of gravity is higher in children because they carry a greater proportion of weight in the upper body. In general, females have a lower center of gravity than males, which contributes to the better performance of females in tasks requiring balance. From the age of 6 fetal months to 70 years the center of gravity descends gradually from the level of the seventh thoracic vertebra to the level of the first sacral (pelvis) segment (Swearingen et al., 1969).

The structural and anatomical changes that occur during adolescence may also contribute to motor performance differences between the sexes. As a result of the pubescent growth spurt, males generally become taller and heavier, develop wider shoulders and longer legs and forearms, and gain in lean body weight. Females during this period will develop a lower center of gravity.

The implications of these characteristics for performing motor tasks are not decisive but from a biomechanical perspective warrant consideration. The female's lower center of gravity and shorter legs give her better balance. Shorter leg length and wider hips, however, are disadvantageous in running (especially sprinting) and jumping events, when compared to the performance of males whose legs are proportionately longer and their hips narrower. The longer arms and wider shoulders of the male may also provide a mechanical advantage for motor tasks involving throwing and striking (Haubenstricker & Sapp, 1985; Oxedine, 1984). Research findings have estimated that body size and structure may account for 10% to 25% of the variance in motor performance scores (Espenschade, 1963; Haubenstricker & Sapp, 1985).

SUMMARY

Characteristic of the changes in lifelong physical growth are the cephalocaudal and proximodistal patterns of development. The study of these changes can be measured using a variety of instruments and techniques.

Individuals grow faster during gestation and the first year after birth than at any other time in the life span. After conception, three major periods of physical growth are associated with the rapid changes that occur before birth: the germinal period, the embryonic period, and the fetal period.

Next to the rapid changes in growth that appear during the prenatal stage, the growth spurt that takes place with puberty represents the life span's most dramatic period of biological alteration. In males puberty is characterized by significant sexual, anatomical, and physiological changes. For females the changes are less dramatic, but a major milestone reached during this period is menarche—the first menstrual flow and the universal indicator of sexual maturity.

Along with the rapid changes in physical maturation and growth are the emergence of several other landmarks in motor performance and the evolution of gender differences. One of the primary mechanisms at work in the changes associated with puberty is the endocrine system and the increased levels of hormones that it releases. The hormones having the greatest impact at this time are androgens, estrogens, and the growth hormone. Of those hormones, the greater quantities of androgens produced in males account, to a large extent, for the significant differences that are found between the sexes.

Specific body parts undergo dramatic changes during the course of physical growth and development. Along with a doubling of head size, there are also comparable changes in limb length (especially the legs), trunk length, and shoulder and hip width. Prior to puberty, males and females appear to have similar body builds; however, noticeable differences occur thereafter, as evidenced by higher mesomorphic rating in males. With advancing age, mesomorphic characteristics tend to diminish and more endomorphic and ectomorphic traits appear. Changes may also occur in body proportions due primarily to a loss of bone tissue.

A typical child will increase birth length by 50% at 1 year of age and reach approximately one-half of mature height at 2 years of age. Most females complete their peak growth period by age 16½, while males continue to gain in stature for at least another two years.

The elaborate process of bone ossification begins shortly after conception and continues until maturity when epiphyseal plates and fusion areas are closed. Although growth of the long bones is generally complete by the early twenties, portions of the braincase do not reach maturity (or become fused) until well into adulthood.

With normal aging, stature decreases slightly and the body has a tendency to develop kyphosis. More dramatic, however, is the loss of bone mass that occurs, especially in females. By the age of 70, some females may lose as much as 30% of total bone mineral mass. Related to bone loss is the condition of osteoporosis, which causes bones to break more easily and fractures to occur in old age. Bone loss estimates for males are about one-half the estimates for females.

Body mass characteristics (body weight and body composition) also undergo numerous changes over the life span. The growth spurt for body weight is similar to the height curve, except that it generally continues for a longer time. Males usually begin the spurt later and add more body size. Due primarily to the normal loss of bone and muscle tissue with advancing age, total body mass declines in later years.

Males at all ages have greater body densities and greater lean body mass than females. During the pubescent growth spurt, males experience rapid gains in lean body mass and decreases in the percentage of body fat. Females, on the other hand, characteristically have greater amounts of body fat than males throughout the life span, with the difference being more evident after puberty. With advancing age, body weight usually increases as body fat increases and lean body mass decreases. The loss of muscle mass in individuals over the age of 60 may be as high as 30% to 50%.

Estimating the level of physical maturity is best accomplished using measures of biological age rather than chronological age. Commonly used measurements include morphological age (usually height), dental age (age of tooth eruption), sexual age (menarche and appearance of sexual characteristics), and skeletal age. Of these methods skeletal age, which is derived from comparisons of radiographs, is perhaps the most accurate estimate of physical maturity.

Individuals are often categorized as early, average, or late maturers, according to where they fit on various maturational scales. The most common type of evaluation, though subjective, is to assess level of maturity by making comparisons to others of the same age. Differences in maturity levels are observable prior to adolescence but become more pronounced during or following puberty. In general, early maturers are taller and heavier for their age than their late-maturing peers. Late maturers often "catch up" to others during late adolescence or early adulthood, however.

Observations of secular trend indicate that today's population is maturing earlier and is taller than that of about 75 years ago. The explanation for this is complex but includes the factors of improved nutrition, reduced incidence of disease, and better health practices.

Though few conclusive findings have been reported, several generalizations can be made about the implications of change in physical growth for motor performance. One of the most evident is the relationship between the specific structural and anatomical changes that occur during adolescence and selected motor performance differences between the sexes.

SUGGESTED READINGS

Cooper, C. (1990). Bone mass throughout life: Bone growth and involution. In R. M. Francis (Ed.), *Osteoporosis: Pathogenesis and management.* Lancaster, UK: Kluwer Academic Publishers.

Lowrey, G. H. (1986). *Growth and development of children* (8th ed.). Chicago: Yearbook Medical Publishers.

Malina, R. M., & Bouchard, C. (1991). *Growth, maturation, and physical activity.* Champaign, IL: Human Kinetics.

Sinclair, D. (1989). *Human growth after birth* (5th ed.). London: Oxford Medical Publications.

Tanner, J. M. (1989). *Fetus into man* (2nd ed.). Cambridge, MA: Harvard University Press.

Physiological Changes

KEY TERMS

maximal oxygen uptake (VO_2max)
aerobic power
anaerobic power
heart rate
cardiac output

vital capacity
blood pressure
basal metabolic rate (BMR)
muscular strength
flexibility

Important to any comprehensive discussion of lifelong motor development is an understanding of the basic physiological changes associated with exercise and motor skill performance. In the context of this aspect of developmental exercise physiology and motor development, this chapter will focus on the cardiovascular and respiratory systems, basal metabolic rate, muscular strength, and flexibility. Along with the functions of the central nervous system, these components provide the basis for motor skill performance and lifelong physical fitness. In recent years, the term *health-related fitness* has been used to describe those aspects of physical fitness related to functional health and preventive medicine (aerobic fitness, muscular strength and endurance, flexibility, and optimal body composition), rather than motor performance (speed, power, agility, and coordination).

The next chapter will provide a discussion of the particular effects of physical activity and exercise on the basic physiological components of change outlined in the following pages. For a more detailed discussion of physiological change as a result of various intensities of exercise, refer to the Suggested Readings.

CARDIO-RESPIRATORY DEVELOPMENT

The term *cardiorespiratory* will be used in this discussion to refer to selected aspects and functions of the cardiovascular and respiratory (pulmonary) systems. Early system growth and development, as well as the developmental aspects of selected functions and capabilities of these systems, will be covered.

Basic Structure and Function

■ *Objective 4.1*

The *cardiorespiratory system* consists of the heart, its cardiovascular network of arteries, capillaries, and veins, and the respiratory (pulmonary) system. The heart is a four-chambered muscular organ that weighs less than 1 pound, yet has the power to beat approximately 40 million times per year. The heart may be viewed as two separate pumping units. The chambers that comprise the right side of the heart receive blood returning from all parts of the body and then pump blood to the lungs. The left side of the heart receives oxygenated blood from the lungs and pumps it into the aorta for distribution throughout the body by way of the circulatory system (fig. 4.1).

The heart is composed of three layers: the pericardium, the outer layer composed of fibrous and adipose tissues; the myocardium, the middle layer composed of cardiac muscle tissue; and the endocardium, or inner layer. Cardiac muscle tissue differs from skeletal muscle tissue in that cardiac muscle is involuntary. The individual fibers of cardiac tissue are multinucleated cells interconnected in a close network. Consequently, when an individual cell is stimulated, the action potential, or impulse, speeds through the entire heart, causing it to function as a unit.

The process by which air is brought into and exchanged with the air in the lungs is referred to as pulmonary (respiratory) ventilation. Air enters through the nose and mouth and flows into the ventilation system where it is adjusted to accommodate body temperature. The air is then filtered and humidified as it passes through the trachea. The inspired air then passes into two bronchi, the large tubes

Figure 4.1
The heart and direction
of blood flow.

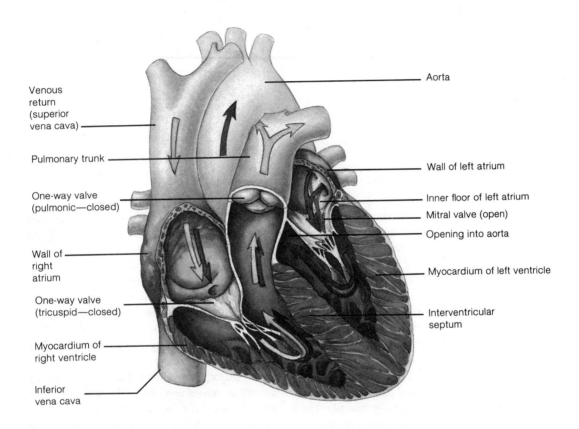

Aorta

Venous
return
(superior
vena cava)

Pulmonary trunk

Wall of left atrium

One-way valve
(pulmonic—closed)

Inner floor of left atrium

Mitral valve (open)

Opening into aorta

Wall of
right
atrium

Myocardium of left ventricle

One-way valve
(tricuspid—closed)

Interventricular
septum

Myocardium of
right ventricle

Inferior
vena cava

serving as primary conduits in each of the lungs. The bronchi further subdivide into numerous bronchioles that conduct the air to be mixed with existing air in the alveoli, or terminal branches of the respiratory tract (fig. 4.2).

Heart Growth   The heart begins its development as a single tube. Its first pulsations may occur as early as the third week of conception, even though there is no blood for it to circulate because the fetus is getting oxygen through the umbilical cord. By 12 weeks the circulatory system is operating and by sometime between the fourth and fifth fetal months, the heartbeat is regular and strong enough to be heard using a stethoscope. The greatest increases in heart weight occur after the first month of postnatal life. The weight doubles during the first year, quadruples by age 5, and by age 9, has increased six times. From 9 to 16 years of age the

**Early Growth and
Development**

■ *Objective 4.2*

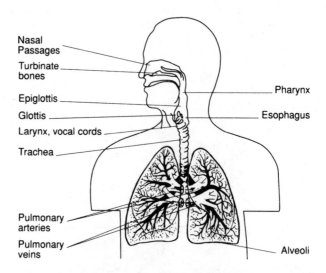

Figure 4.2
Basic structure of the respiratory (ventilatory) system.

heart undergoes a second phase of rapid growth that parallels the general growth rate during that period (Lowrey, 1986). All of the functional mechanisms found in the adult electrocardiogram (ECG) are also evident in the fetus after about five weeks.

In the fetus, the vessels that develop from the main trunk of the heart grow in direct proportion to the parts of the body supplied, indicating that a relationship exists between caliber of growth and the volumes or weights of the regions supplied. At birth the right ventricular muscle outweighs the left by about 13%, but by the fifth month they become equal in weight. By 7 years of age the left wall is two to three times thicker than the right, as it is in the adult.

The thickness of the walls of the major veins doubles between the time of birth and puberty, and their cavity size increases approximately two and a half times by early maturity (Hurst, 1982). In newborns, the total blood volume of the heart, one indicator of size, is approximately 200 milliliters. By the end of the first year, this value doubles, and at the time of maturity, blood volume for males is 5,500 milliliters and 4,200 milliliters for females. There is very little difference between the sexes until puberty; after that, the male heart is about 15% larger, even with adjustment for body weight.

Respiratory Component Growth   By the sixth fetal week, the trachea, bronchi, and lung buds are evident and migration toward the thoracic region begins. By the third fetal month the journey is almost complete, and thereafter the lungs begin to gradually descend into their destined location. At birth the larynx is approximately one-third of its adult size but is about the same size in relation to the rest of the body as it is in the adult. From about the third year on, the larynx is longer and wider in males than in females. At the time of delivery the trachea is roughly one-third (about 4 cm) of its adult length and by puberty will increase both its anteroposterior and lateral diameters nearly 300%.

Anatomical growth of the lungs follows individual patterns for each of the major ventilatory components. Development of the lungs begins as a lung bud protrusion appearing between the third and fourth fetal week. Growth proceeds when this protrusion branches, and by about the 16th fetal week the bronchial tree is completely formed. The airways canalize around 20 weeks and a little later alveoli develop from the distal branches of the airway. Bronchial cartilage begins to form at 10 weeks, and by 25 weeks its distribution resembles the adult model. At about 13 weeks, mucous glands appear and are functional by 24 weeks. Cilia make their appearance around the 13th fetal week and extend to the peripheral bronchi. By 15 weeks bronchi sensory and motor nerves can be detected along the pathways to the most distal passages. Capillaries appear about the 20th week and multiply rapidly near the developing airways when the alveoli begin to grow. Development of the alveolar-capillary interface for the exchange of oxygen and carbon dioxide is essential for independent existence after birth; this process is seldom complete before the 26th fetal week.

The weight and capacity of the lungs doubles in the first six months, triples by age 1, and by maturity increases approximately 20-fold. There will also be an increase of two to three times in the diameter of the airway. The diameter of the bronchi doubles by age 6 and that of the bronchioles and trachea by approximately age 15. About 20 million primitive alveoli are present at the time of birth. During the first three years after birth, the increase in lung size is primarily due to increases in the number of alveoli. By age 8 there are about 300 million alveoli—about the same number found in the adult. After age 8 the alveoli begin to increase in size until the chest wall ceases growing (Forfar & Arneil, 1992).

Aerobic Power

■ Objective 4.3

Sometimes referred to as maximal oxygen consumption and maximal aerobic power, **maximal oxygen uptake (VO_2max)** is the maximum amount of oxygen that an individual can use per unit of time. This amount is usually expressed in liters or milliliters of oxygen per minute. Maximal oxygen uptake is considered the best measure of cardiorespiratory fitness as indicated by graded maximal working capacity tests using either a cycle ergometer or a treadmill.

The association between VO_2max and aerobic power lies with the energy sources of the body. For the body to do mechanical work, it depends on the splitting of adenosine triphosphate (ATP) to release energy for muscle contraction. This can be supplied from: (a) limited stores of phosphagens, (b) glycolysis, which produces lactic acid, or (c) the conversion of substrates to carbon dioxide and water in a phase of carbohydrate breakdown called the Kreb's cycle and the electron transport system. The first two sources are considered anaerobic processes because additional oxygen is not required. Muscle contractions that result from anaerobic sources generally cannot be sustained longer than 40 or 50 seconds. Motor performance activities that are considered more anaerobic are sprints, jumping, weight lifting, and other movements where the required power intensity is high and the duration relatively short. In contrast, the third conversion method requires oxygen, meaning that it uses **aerobic power.** Muscle contractions utilizing oxygen as an energy source can last several minutes and, in trained

Figure 4.3

Maximal oxygen uptake
as a function of age.

Reprinted with the permission of
Simon & Schuster from the
Macmillan college text: *Fitness,
Health, and Work Capacity* by
Leonard A. Larson. Copyright ©
1974 by Macmillan Publishing
Company, Inc.

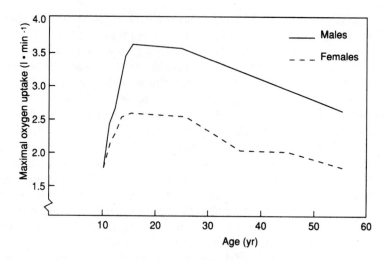

individuals, for hours. Activities primarily, though not exclusively, associated with aerobic power include long-distance running, cycling, cross-country skiing, and swimming. These activities also involve anaerobic processes to a degree.

Figure 4.3 illustrates the relationship between maximal oxygen uptake and age. As a general trend, maximal oxygen uptake increases with growth through childhood, with both sexes having similar values until approximately 12 years of age (Krahenbuhl, Skinner, & Kohrt, 1985). Maximal oxygen uptake in males continues to increase until about age 18, while females show little improvement beyond the age of 14. The increase in VO_2max for males with puberty appears to correspond with the peak increase in height and with the time when androgen secretion increases dramatically. After age 25, capacity declines steadily. Both children and adults have shown improvements in aerobic capacity through training, with average increases ranging from 6% to 20%.

Maximal oxygen uptake is strongly related to lean body mass, which accounts to a great extent for the difference between the sexes after puberty. As noted earlier, lean body mass following puberty tends to level off in females as additional mass, primarily in the form of body fat, is added. Males, in contrast, continue to gain lean body mass. Approximately 69% of the individual differences in VO_2max can be explained simply by differences in body weight (McArdle, Katch, & Katch, 1991). Values for young adult females are typically 15% to 30% lower than males. Among trained athletes, the difference ranges from 12% to 20%. Although aerobic capacity has been shown to increase with growth through childhood, Zwiren (1989) concluded that, when expressed relative to body mass, aerobic capacity remains the same or decreases with age.

Along with training and physical characteristics, maximal oxygen uptake is also influenced by heredity. Based on the findings that genetic factors appear to influence the growth of body size and mass as well as muscle fiber type, the heritability estimate for maximal oxygen uptake is believed to be over 90% (Cunningham, Paterson, & Blimke, 1984; McArdle, Katch, & Katch, 1991).

Body weight is the major physical determinant in maximal oxygen uptake capacity; however, several other factors are also related. These factors include cardiac output, vital capacity, heart rate, stroke volume, and the oxygen-carrying properties of the blood, including blood volume and hemoglobin. Differences between the sexes are primarily related to differences in the values associated with these factors and differences in heart size. Table 4.1 indicates adult sex differences in maximal oxygen uptake.

In contrast to the characteristics of aerobic power, **anaerobic power** is the maximum rate at which metabolic processes can occur without additional oxygen. Sustained aerobic activity depends on a continuous transport of oxygen to the working tissues, as measured by VO_2max. However, when exercise continues longer than a few seconds, more energy for ATP resynthesis must be generated through the anaerobic reactions of glycolysis. More intense, all-out activity requires levels of energy that far exceed what can be generated by the respiratory processes alone. Consequently, the anaerobic reactions of glycolysis predominate and large amounts of lactic acid accumulate within active muscle and in the bloodstream. The level of lactic acid in the blood is the most common indicator of anaerobic activity. As the intensity of exercise increases and VO_2max is reached, the amount of lactic acid in the blood increases quickly. Under these conditions, work can only be maintained for a few minutes. Anaerobic power is generally measured in tests requiring that a supramaximal exercise be performed for a minute or less. Popular tests include sprinting up a flight of stairs (Margaria Power Test), arm or leg cranking on a cycle ergometer (Wingate Anaerobic Test), jumping-power tests, and short sprints.

Anaerobic Power

Anaerobic power increases with age up to early adulthood no matter whether the values are expressed absolutely or relative to body weight (Bar-Or, 1983; Zwiren, 1989). In anaerobic tests that last one minute or less, children exhibit significantly lower scores than adolescents and adults. Whereas maximal aerobic power generally does not change and may even decrease relative to body weight, anaerobic power increases progressively with growth. The differences in anaerobic power cannot be accounted for by muscle tissue mass. The markedly lower capacity of young children reflects, to a great extent, a qualitative deficiency in the muscle. The major age-related difference appears to be in glycolytic capacity—the resting concentration of glycogen and the rate at which lactate is produced with its utilization. Maximal lactate levels are lower in children than in older populations. The level of acidosis at which the muscle can still contract is also used as an indicator of anaerobic capacity. Following a pattern similar to glycolytic capacity, children do not reach the levels of acidosis of which adolescents and young adults are capable. Irrespective of measured values, maximal acidosis also increases with age.

Females, in general, do not have the anaerobic power of males, especially after puberty. Preliminary studies with animals suggest that lactate production is related to the level of circulating androgens, thus explaining the male advantage after puberty (Krotkiewski, Kral, & Karlsson, 1980).

Table 4.1 Sex Differences in Maximal Oxygen Consumption (50th Percentile)

Age (yr)	Males	Females
20–29	40.0	31.1
30–39	37.5	30.3
40–49	36.0	28.0
50–59	33.6	25.7
Over 60	30.0	22.9

Source: Data from Pollock, J. H. Wilmore, and S. M. Fox, *Health and Fitness Through Physical Activity*. Copyright © 1978 John Wiley & Sons, Inc., New York, NY.

Heart Rate

■ *Objective 4.4*

Due to the relative ease of monitoring, **heart rate** has been one of the most commonly utilized measures of cardiovascular response to exercise and of energy expenditure. Because heart rate is the major determinant in cardiac output (the amount of blood pumped per unit of time), it is also an important factor in maximal oxygen consumption. Heart rate measures can be taken at rest, at a given rate of exercise (submaximal), and at maximal level of exercise (for maximal heart rate). A rough estimate of maximal heart rate can be obtained for individuals by subtracting their age from 220. A more direct and accurate measure is obtained by measuring the highest heart rate during a treadmill test that pushes the individual to maximal oxygen uptake.

Resting heart rate levels decrease with age; females average 5 beats per minute (bpm) higher than males (table 4.2). Bar-Or (1983) reports a similar pattern for heart rate at a submaximal exercise level; namely, that submaximal heart rate also declines with age and that females exhibit higher rates. Under these exercise conditions, it is not unusual for a preadolescent child to have a heart rate 30 to 40 beats per minute higher than a young adult performing the same task. The difference is due in part to the greater relative exercise intensity performed by the younger individual and the physiological compensation for a lower stroke volume.

The reported reasons for the higher rates for females are not conclusive. A possible reason, at least among adults, is the lower blood hemoglobin levels of women. However, the hemoglobin levels of preadolescents are similar and females still exhibit greater heart rates. Another explanation has been related to stroke volume differences between the sexes. The lower stroke volume and cardiac output of females may be compensated for by increased heart rate.

Londeree and Moeschberger (1982) conducted a comprehensive review of existing information on maximal exercise heart rate among sampled individuals 5 to 81 years of age (fig. 4.4). The review showed that maximal exercise heart rate declines with age and that as age increases the decline accelerates slightly. It was also reported that no significant differences between the sexes or categories of race for maximal exercise heart rate are apparent.

Table 4.2   Average Heart Rate at Rest

Age	Average Rate *
Birth	140
1 mo	130
1–6	130
6–12	115
1–2 yr	110
2–4	105
6–10	95
10–14	85
14–18	82
20–29	66
30–39	65
Over 40	64

Source: Data recalculated from Lowrey, 1978 and Pollock et al., 1978
*Female values are an average 5 bpm higher than males.

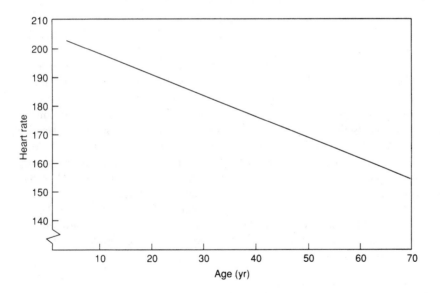

Figure 4.4
Maximal exercise heart
rate as a function of age.
This figure is reprinted with
permission from the *Research
Quarterly for Exercise and Sport,*
vol. 53, no. 297–304, 1983. The
*Research Quarterly for Exercise
and Sport* is a publication of the
American Alliance for Health,
Physical Education, Recreation
and Dance, 1900 Association
Drive, Reston, VA 22091.

Cardiac Output and Stroke Volume

Cardiac output is the major indicator of the functional capacity of the circulatory system. Output refers to the rate that blood is pumped by the heart (heart rate) and the quantity of blood ejected with each stroke (stroke volume). As previously noted, cardiac output is the primary determinant of maximal oxygen uptake.

In general, as heart volume increases with growth and age, maximal cardiac output gradually increases much like the pattern of development in maximal oxygen uptake (fig. 4.3). Children have a significantly lower stroke volume than adults at all exercise levels and thus a lower cardiac output. This is compensated for by higher heart rates. Although not dramatic, males have a higher stroke volume than females at all levels of exercise (Bar-Or, 1983) partly due to the biological advantage of a larger heart, thus greater volume and pump capabilities, especially after puberty.

Vital Capacity and Pulmonary Ventilation

Vital capacity refers to the total volume of air that can be voluntarily expired following maximal inspiration. Vital capacity consists of the tidal volume, the air moved during either inspiration or expiration, plus the available reserve volumes. The movement of air in and out of the pulmonary system is called ventilation (breathing).

The growth pattern of vital capacity indicates that as males and females age and grow larger in body size, their vital capacity increases (fig. 4.5). Vital capacity is strongly linked to body size. Males and females experience similar increases until the onset of puberty, when the larger body size of the male allows for greater absolute values. At 20 years of age, males have approximately 20% greater vital capacity than females. Because females enter puberty earlier than males (at 10 to 12 years of age), body sizes are relatively equal at that time and their vital capacities are much closer to equal.

As with vital capacity, the absolute volume for pulmonary ventilation increases with age and body size until young adulthood (Fahey, 1990). An average 6-year-old may have a maximal ventilation of 30 to 40 liters per minute whereas a young adult can reach 100 liters per minute quite easily. However, children's maximal ventilation values are not significantly different than adolescent and adult values when expressed relative to body weight. Once again, due to body size differences, maximal ventilation is normally higher in males. Interestingly, however, due to the smaller lung capacity in children, their respiratory rates are markedly higher than adults when performing a submaximal exercise task. Ventilation at any given oxygen uptake is higher in children, and the ventilatory "breaking point," which signifies the onset of blood lactate, appears earlier than in adolescents and adults. During a submaximal test, a child of 6 years can have a ventilation value 50% higher than an individual of 17 years (Bar-Or, 1983). Research related to ventilatory equivalent, a numerical expression of ventilatory efficiency, though not conclusive, suggests that it decreases with age, indicating that children are less efficient at ventilating.

BLOOD CHARACTERISTICS

Blood Volume

■ *Objective 4.5*

The increase in total blood volume maintains a relatively constant relationship to growth in body weight. The average newborn infant has a blood volume of 85 milliliters per kilogram relative to body weight, increasing to about 105 milliliters shortly after birth, then decreasing during the first few months. Thereafter, the average blood volume is 75 to 77 milliliters per kilogram, which is maintained through adolescence and adulthood. Males' larger body mass, especially after puberty, results in a greater total and relative blood volume than that of females. The relative difference is approximately 10 milliliters per kilogram.

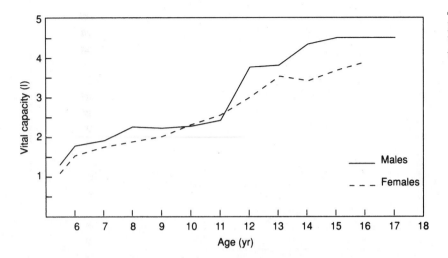

Figure 4.5
Vital capacity in relation
to age.

The red blood cell is one of the most specialized cells in the body. Its major compound, hemoglobin, accounts for more than 95% of the total protein and for about 90% of the dry weight of the cell. The function of hemoglobin is to combine reversibly with oxygen, allowing red blood cells to collect oxygen from the lungs and deliver it to the tissues.

Red Blood Cells Content

By the second week of gestation the developing embryo begins to produce red blood cells. The production of these structures outside of the skeletal system reaches a peak at about the fifth fetal month. Thereafter production takes place in the bone marrow. The red blood cell count increases from approximately 1.5 million per cubic millimeter at 12 weeks of gestation, to about 4.7 million at birth (Rudolph, 1982). A marked decrease in the rate of red blood cell production continues for six to eight weeks after birth. This period marks the peak of red cell destruction and decreased hemoglobin values. The life span of a red cell is about 90 days shortly after birth, compared to 120 days in adults. After this period, cell production increases gradually throughout childhood and adolescence.

The number of red blood cells and, consequently, the amount of hemoglobin (excepting the two-month period in which both decrease) are closely related to growth in body mass and blood volume (fig. 4.6). Values in males and females first begin to diverge during adolescence. In the female, the gradual increase starts in preadolescence, continues into early puberty, and then levels off. In contrast, increases in the male appear to be related to sexual and physical maturity. The higher values of androgens in males after puberty have been associated with higher concentrations of hemoglobin. During adolescence in males, blood levels vary with sexual maturity (Rudolph, 1982). Higher hemoglobin levels in males help to explain higher anaerobic capacity.

Blood pressure is the product of cardiac output and peripheral vascular resistance; an increase in either cardiac output or peripheral resistance will raise blood pressure. Systolic pressure refers to how hard the heart is working and the

Blood Pressure

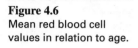

Figure 4.6
Mean red blood cell
values in relation to age.

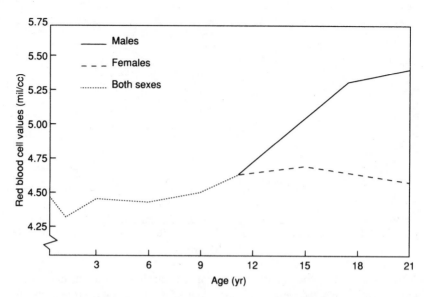

strain against the arterial walls during ventricular contraction. Diastolic pressure is an indication of peripheral resistance or of the ease with which blood flows from the arterioles into the capillaries. Due to the higher contractile forces of the heart during exercise, there is a marked increase in systolic pressure compared to diastolic pressure. Diastolic blood pressure changes only slightly with physical exertion.

Systolic blood pressure rises steadily during childhood and increases at a much faster rate during adolescence until approximate adult levels are reached. The average systolic and diastolic blood pressure for young adults at rest is about 120/80 (mmHg), though this varies related to age and size. During submaximal exercise, children respond with lower systolic and diastolic pressures than individuals of adolescent age. Bar-Or (1983) notes that this is associated with the child's lower cardiac output and stroke volume and may also reflect a lower peripheral vascular resistance due to shorter blood vessels. Males have a higher systolic pressure than females after the age of 12 (fig. 4.7). As with measures of other blood characteristics, this change first appears at about age 12 in females and rises markedly during puberty in males. Again, this difference between the sexes has been attributed primarily to the greater heart size and increased blood volume in males. Diastolic pressure changes very little during adolescence with no apparent difference between the sexes.

BASAL METABOLIC RATE

■ *Objective 4.6*

Basal metabolic rate (BMR) refers to the amount of heat produced by the body during resting conditions. This physiological process reflects the minimum level of energy required to sustain the body's vital functions in the waking state. BMR is usually determined indirectly by measuring submaximal oxygen consumption and is recorded in terms of the heat produced per square meter of body surface.

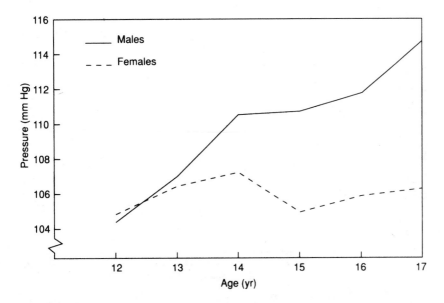

Individuals are measured after having fasted for at least 12 hours so there is no increase in metabolism due to the energy required to digest food. The BMR is frequently used to establish an energy baseline for constructing programs of weight control.

Basically, BMR declines continuously from birth to old age (fig. 4.8). Males have slightly higher values at all ages; the values for adult females are 5% to 10% lower than those for adult males. This difference is largely because females generally have more fat tissue, and fat is metabolically less active than lean body tissue. This difference between sexes is almost eliminated when BMR is expressed relative to fat-free or lean body mass. The hormonal effects of increased androgens in males has also been mentioned as a possible contributing factor in sex differences.

MUSCULAR STRENGTH

■ *Objective 4.7*

Previous discussions regarding muscles have focused on their neurological structure and tissue development. This section will provide information concerning the force characteristics of a muscle and their relationship to motor performance. **Muscular strength** refers to the maximum force or tension generated by a muscle or muscle groups. Tension of the muscle in partial or complete contraction without any appreciable change in length is identified as *isometric* contraction. This type of muscular force is referred to as *static strength* and is generally measured using a dynamometer and tensiometer. If the muscle varies its length when activated to produce force, the contraction is *isotonic*. Isotonic force is also known as *dynamic strength* and is usually determined by a one-repetition maximum, which refers to the maximum amount of weight lifted one time. Some exercise physiologists also refer to *isokinetic* contraction. This is a muscular

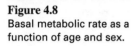

Figure 4.8
Basal metabolic rate as a
function of age and sex.

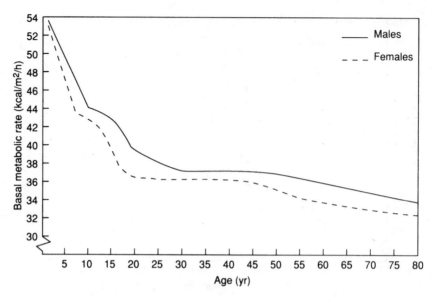

Figure 4.8
Basal metabolic rate as a
function of age and sex.

contraction in which the muscle shortens at a constant velocity. For training purposes, isokinetic machines can be set to move at various specified velocity rates and resistance levels.

Most past descriptive studies of children have documented the development of static strength using either a hand-grip dynamometer or, for other parts of the body, a cable tensiometer. With more recent advances in scientific instrumentation and the emergence of microprocessor technology, it is now possible to update this body of information (especially with younger populations) by quantifying more accurately the muscular forces generated during a variety of movements.

Figures 4.9 and 4.10 show means of strength characteristics (absolute and relative to body weight) for individuals 10 to 59 years of age. Basically, muscular strength increases with age in both sexes and reaches a peak between 25 and 29 years of age (Eckert, 1987). There is, however, considerable variability within and between the sexes. The development of strength is basically symmetrical on both sides of the body, although the dominant side is slightly stronger. The legs contribute approximately 60% to total body strength.

For young children, very little information is available below the age of 6 years. Metheny (1941) reported that grip strength in males increased 359% between the ages of 6 and 18 years, whereas the values for females improved 260% during the same period. Other studies have reported similar age-related increases for individuals over this age span. The significant physiological and structural changes associated with the adolescent growth spurt are reflected in both capacity and differences between the sexes (figs. 4.9 and 4.10). Up to the onset of puberty (around 13 to 14 years), the sex differences are much smaller, although males exhibit greater strength values at all age levels. Much of the difference is attributed to greater muscle mass and androgen production in males.

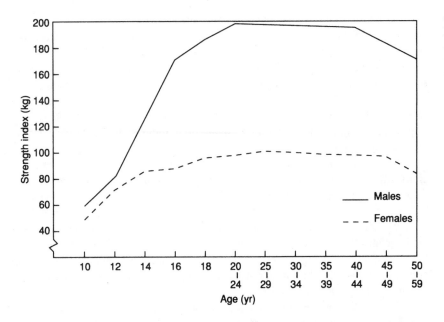

- - - - - - - - - - -
Figure 4.9
Strength index (means)
of right and left grip of
arm.

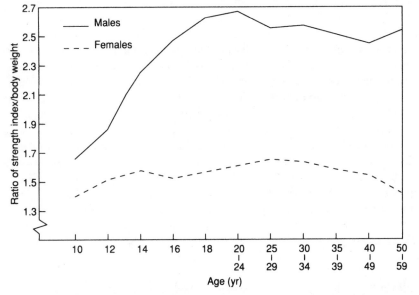

- - - - - - - - - - -
Figure 4.10
Ratio of strength index to
body weight.

When evaluating the differences between the sexes, three approaches may be taken: (a) relative to muscle cross-sectional area, (b) on an absolute basis as total force, and (c) relative to body weight or lean body mass. Regardless of sex, human skeletal muscle, based on cross-sectional analysis, generates approximately 3 to 4 kilograms of force per square centimeter (McArdle, Katch, & Katch, 1991). Here the difference is related to the size of the muscle cross section; for the same-size muscle section, there is little difference.

When strength is compared on an absolute basis, males are usually considerably stronger than females (fig. 4.9). Males are typically 30% to 50% stronger than females in most muscle groups. The difference is generally less when comparing the strength of the trunk and legs, as opposed to the shoulders and arms. As noted earlier, much of this difference, especially after puberty, is related to higher androgen levels and larger lean body mass in males. Researchers have also noted that the greater levels of androgens are an advantage in muscle hypertrophy, an increase in muscle size in response to training. Adult males have about 10 times the amount of androgens as prepubescent children and women.

When strength values are expressed relative to body weight, males are generally still stronger (fig. 4.10). This difference is due in part to the fact that the measure of body weight reflects total body fat and lean body mass. Since females have more body fat and less lean body mass than males, their strength per unit of body weight is lower. When strength is calculated relative to lean body weight, the large absolute differences are reduced considerably. This information has led researchers to conclude that sex differences are a function of differences in lean body weight and distribution of muscle and fat in the body segments (Heyward, Johannes-Ellis, & Romer, 1986).

The decline in muscular force production after it peaks (at 25 to 29 years of age) is closely related to the loss of muscle mass. A decrease in mass of approximately 25% to 30% takes place with aging.

Closely associated with strength and the ability of the muscle to produce force is *muscular endurance.* In the purest form, muscular strength refers to maximum force or tension generated by a muscle or muscle group in a single maximum contraction. Muscular endurance, on the other hand, is the ability of a muscle or muscle group to perform repeated contractions. That is, endurance is the ability to sustain strength performance. Commonly used tests of muscular endurance include push-ups, sit-ups, and pull-ups. According to two nationwide physical fitness surveys, children gain steadily in sit-up and pull-up performance from 6 to 9 years of age, with males generally continuing to increase through adolescence and females showing indications of leveling off around the time of puberty (NCYFSI, 1985; NCYFSII, 1987). Other gross motor activities such as running, swimming, bicycling, and gymnastics require varying levels of muscular endurance. The primary component in the initiation of any muscular endurance task, however, is muscular strength.

FLEXIBILITY

■ *Objective 4.8*

Flexibility is the degree of ability to move body parts through a range of motion without undue strain. This is an essential component of health-related fitness for optimal performance of motor skills and for preventing muscle injury. Research on the basic tenets of flexibility has produced the following conclusions: (a) flexibility is specific to each joint, (b) flexibility is not related to the length of limbs, (c) strength development need not hinder range of motion, (d) activity levels are a better indicator of flexibility than age, and (e) females tend to be more flexible than males.

Most flexibility tests are classified as relative or absolute measures. Relative tests measure flexibility relative to the length or width of a specific body part; thus the movement and influencing body parts are measured. Tests that measure only the movement are referred to as absolute. The most commonly used instruments for testing flexibility are various types of flexometers and goniometers (linear or rotary measuring devices) and the sit-and-reach box. In recent years the sit-and-reach test, which is an absolute measure of hamstring, lower back, and hip flexibility, has been used extensively in field studies.

In a comprehensive review of the literature, Clarke (1975) concluded that general range of motion increases steadily through childhood, with males starting to decline around 10 years of age and females beginning to decrease at about age 12. This conclusion is not decisive, however, as several other reports have indicated both earlier or later initial declines. This is not surprising considering that flexibility is joint specific, rather than a "general" attribute, and these studies have not measured the flexibility of the same body parts. It has also been speculated that activity levels are a better indicator of flexibility than age.

Of the available information, it appears that flexibility begins to decline just after the onset of puberty, with the trend continuing through adulthood (e.g., Boone & Azen, 1979; Clarke, 1975). A rather consistent finding complementing this trend is that, in general, females are more flexible than males from age 5 to adulthood (e.g., Beunen et al., 1988; Branta, Haubenstricker, & Seefeldt, 1984; Clarke, 1975; Jones, Buis, & Harris, 1986; Simons et al., 1990). Explanations for the difference have only been speculative but include relative differences in physical activity patterns, limb length, body size, and specific body composition and hormone levels. Krahenbuhl and Martin (1977) suggest that as body surface area increases, flexibility decreases; thus, during adolescence males would be at a disadvantage. This seems plausible, providing that the males do not participate in exercises that stimulate range of motion.

Physiological maturity and motor performance peak between the approximate ages of 25 and 30 years. In the age span of 30 to 70 years, most physiological functions decline at the rate of 0.75% to 1.0% per year (Smith & Serfass, 1981). The rate of decline in some functions depends to some degree on lifelong health practices, including physical activity. The difference between physiological age and chronological age may be considerable. Figure 4.11 shows hypothetical curves of physiological function of active and sedentary males across the life span. The following information on specific types of change with advanced aging comes primarily from the works of Van Camp and Boyer (1989), McArdle, Katch, and Katch, (1991), Whitbourne (1985), Brooks and Fahey (1985), and Smith and Serfass (1981).

Maximal oxygen uptake follows a clear trend of decline with a drop of approximately 30% between 30 and 70 years of age. The closely related function of cardiac output, consisting of stroke volume and maximal heart rate, also declines

CHANGES WITH AGING

■ *Objective 4.9*

Change in Cardiorespiratory Function

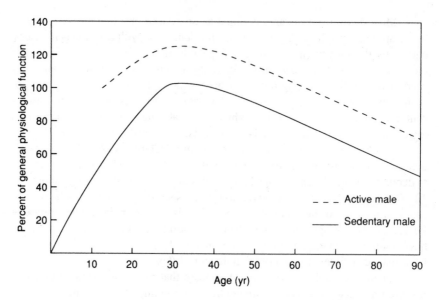

Figure 4.11
A hypothetical curve representing physiological function of active and sedentary males over the life span.

about 30% over this time span. It should be noted that "resting" cardiac output (stroke volume and resting heart rate) is not markedly affected by age. Parallel to the loss in aerobic capacity is a slightly greater decline in anaerobic power—about 40% by age 70. Although there is little change in diastolic blood pressure, systolic values may increase significantly. Aging also produces an increased resistance to blood flow due, in part, to a stiffening of the blood vessels and/or fatty deposits on the vessel walls.

Respiratory function, in general, also declines with age. The main outcome of the aging process in terms of lung function is the reduction of the amount of oxygen delivered from the outside air to the arterial blood. Along with changes in the cardiovascular and muscular systems, the effects of aging in the respiratory structures compound the limitations on the ability to perform physical work. Vital capacity may diminish by up to 40% while residual lung volume (the amount not expired) increases 20% to 50%. The aging process has also been associated with a decrease in the surface area available for the exchange of gas between the alveoli and capillaries due primarily to loss of functioning alveoli and the associated capillary network. Perhaps the most significant functional change with aging is a decline in the static-elastic-recoil force of the lung and the decreased resistance of the chest wall to deformation. Under normal conditions of respiratory function, the lung recoils inward and the chest wall expands outward, in balance, at the end of expiration. As the lung expands inspiration, the elastic-recoil force (pressure) increases, thereby assisting in expiration. The loss of recoil force means that less pressure by the chest wall is needed to produce a change in lung volume. In practical terms, this change means that an older individual performs about 20% more work to overcome the elastic resistance compared to a young adult. Loss of pulmonary elasticity is also related to impaired ventilation due to the premature closing of the airways.

Table 4.3 Summary of Selected Physiological Changes with Aging

Physiological Characteristic	
Cardiorespiratory	
Maximal oxygen uptake	Decreases
Stroke volume	Decreases
Max heart rate	Decreases
Vital capacity	Decreases
Ventilation	Decreases
Functioning alveoli	Decreases
Blood pressure (systolic)	Increases
Blood flow resistance	Increases
Basal metabolic rate	Decreases
Muscular strength	Decreases
Flexibility	Decreases

Change in Basal Metabolic Rate

Closely associated with the decline in work capacity and muscle mass is the decreased ability of the individual to function within the environment. This decline affects basal metabolic rate with a decline of about 10% to 15%.

Change in Muscular Strength

With aging comes a deterioration in muscle mass of approximately 25% to 30%. Accompanying this loss are decreases in the size and number of fibers and the functioning of individual motor units. As this takes place, less contractile force is available when a motor unit is recruited for muscle contraction, and the action potential of the muscle is decreased. In addition, type II muscle fibers, which are associated with strength and power, are lost. These fast-twitch fibers are thought to atrophy because the motoneurons that activate the muscles themselves atrophy, and the muscle fibers die because of the loss of innervation. After the period of peak strength between the ages of 25 and 29, it is not surprising to see muscular force decrease by about 25% to 30% with aging.

Change in Flexibility

Flexibility generally declines across adulthood, but after about age 50, the trend appears to be more significant (Alexander, Ready, & Fougere-Mailey 1985; Einkauf et al., 1987; Murray, 1985). Closely linked to decreased flexibility and range of motion with aging are changes in the joints and muscles and the presence of degenerative joint diseases. The smooth functioning of the body's joints is made possible by the strength and elasticity of the tendons and ligaments and the condition of the synovial fluid. As individuals age, their joints become less stable and their mobility diminishes as collagen fibers and synovial membranes degrade and the viscosity of synovial fluid decreases. As a result, the joints become stiff, possibly decreasing their range of motion. However, there is no conclusive evidence that biological aging inherently causes decreases in flexibility. A considerable amount of research also links loss of flexibility with degenerative joint diseases such as osteoarthritis and osteoarthrosis. Kellgren and Lawrence (1957) report that over 80% of individuals 55 to 64 years of age show signs of osteoarthrosis in some joint.

Table 4.4 Summary of Changes with Aging after Peak Maturity (30 Years of Age)*

	Decrease (↓) Increase (↑)	
Cardiorespiratory system		
Maximal oxygen uptake (VO₂max)	↓	30%
Cardiac output	↓	
Stroke volume	↓	
Heart rate (max)	↓	
Anaerobic power	↓	40%
Blood pressure (systolic)	↑	
Elasticity of blood vessels	↓	
Blood flow (resistance)	↑	
Respiratory		
General pulmonary function	↓	
Vital capacity	↓	40%
Lung residual volume	↑	20%–50%
Functioning alveoli	↓	
Ventilation	↓	
Basal metabolic rate	↓	10%–15%
Nervous system		
Brain cells	↓	
Synaptic delay	↑	
Motor unit size	↓	
Nerve conduction velocity	↓	10%–15%
Reaction time	↑	
Muscle		
Mass	↓	25%–30%
Fiber number/size	↓	
Biochemical capacity	↓	
Strength	↓	25%–30%
Height	↓	
Body weight	↑	
Body composition		
Body fat	↑	
Lean body weight	↓	
Skeletal (bone loss)	↓	female 30% male 10%–15%
Flexibility	↓	
Joints		
Stiffness of connective tissue	↑	
Accumulated mechanical stress	↑	

*Overall decline of functional capacity 0.75% to 1.0% per year.

Table 4.3 provides a summary of selected physiological changes that occur with aging. Table 4.4 is a more comprehensive outline of topics covered in the last three chapters on the effects of aging on various growth and development functions.

SUMMARY

Growth and development of the cardiorespiratory system begins with a single, noncirculating heart tube and the appearance of respiratory buds during the early stages of fetal development. From these beginnings, the system experiences tremendous growth and development prior to birth that continues across the life span.

Maximal aerobic power increases through childhood, with males showing substantially higher levels after puberty. Much of the difference between males and females can be explained by variations in body weight. Peak capacity occurs around the age 25 and declines steadily thereafter. Anaerobic power also rises steadily to early adulthood. Females, in general, do not have the power of males, especially after puberty.

Heart rate changes with age are described in terms of resting, submaximal, and maximal values. Resting and submaximal heart rate levels normally decrease with age, with females having slightly higher levels than males. Maximal heart rate also decreases with advancing age, but there are no differences between the sexes.

Cardiac output is the major indicator of the functional capacity of the system and a primary determinant in maximal oxygen uptake (aerobic power). In general, as heart volume increases with age, maximal cardiac output also increases. The trend is similar to that associated with maximal oxygen uptake; it increases until early adulthood and declines shortly thereafter. Vital capacity tends to increase as body size increases. As with a number of other physiological functions, the measured values of both sexes are similar until after puberty after which the larger body size of males produces the greater absolute values found.

The production of red blood cells begins by the second week of gestation and increases, with only a slight interruption, throughout adolescence. As body weight increases, comparable increases in total blood volume take place. With this increase come higher numbers of red blood cells and higher levels of hemoglobin, which function to collect oxygen from the lungs and transport it to the tissues. Because of the greater body mass in males, especially after puberty, males have a greater blood volume than females. Systolic blood pressure values rise steadily during childhood with a much faster rate of increase during adolescence. Quantitative differences related to age and size are observable in systolic and diastolic blood pressure. Physical activity can have a marked influence on blood pressure values.

Basal metabolic rate declines continuously from birth to old age; males have slightly higher values at all ages. The sex difference is due primarily to greater levels of fat tissue in females, which is metabolically less active than lean body tissue.

Muscular strength normally increases with age in both sexes up to a peak that takes place between 25 and 29 years of age. Due largely to higher levels of androgen and larger lean body mass after puberty, males are typically 30% to

50% stronger than females in most muscle groups. Decline in strength after the peak years is closely related to the loss of muscle mass.

The literature on flexibility is perhaps the most inconclusive of all the developmental characteristics discussed. In general, range of motion increases steadily through childhood, then during early adolescence begins the continued life-span trend of decline. Most reports suggest that females are more flexible than males. However, no clear explanation for the difference has been established.

After peak physiological maturity between the approximate ages of 25 and 30 years, functions begin to show a decline at the rate of 0.75% to 1.0% per year between the ages of 30 and 70 years. Although there are wide differences among individuals, due to factors such as physical activity, the effects of age are evident. By the age of 70, substantial losses are observable in cardiorespiratory function, basal metabolic rate, muscle mass and muscular strength, and joint mobility (flexibility).

SUGGESTED READINGS

Bale, P. (1992). The functional performance of children in relation to growth, maturation and exercise. *Sports Medicine, 13* (3), 151–159.

Bar-Or, O. (1983). *Pediatric sports medicine for the practitioner.* New York: Springer-Verlag.

Boileau, R. A. (Ed.). (1984). *Advances in pediatric sport sciences.* Champaign, IL: Human Kinetics.

de Vries, H. A., & Housh, T. J. (1994). *Physiology of exercise.* Madison, WI: Brown & Benchmark.

McArdle, W. D., Katch, F. I., & Katch, V. L. (1991). *Exercise physiology* (3rd ed.). Philadelphia: Lea & Febiger.

Meisami, E., & Timiras, P. S. (Eds.). (1990). *Handbook of human growth and developmental biology* (Vol. 3). Boca Raton, FL: CRS Press.

Rowland, T. W. (1990). *Exercise and children's health.* Champaign, IL: Human Kinetics.

Whitbourne, S. K. (1985). *The aging body.* New York: Springer-Verlag.

Zauner, C. W., Maksud, M. G., & Melichna, J. (1989). Physiological considerations in training young athletes. *Sports Medicine, 8* (1), 15–31.

Factors Affecting Growth and Development

teratogen
fetal alcohol syndrome

growth hormone (GH)

KEY TERMS

A variety of factors and conditions may affect the normal course of human growth and development, including motor behavior. Common to the research and discussion in this field of study are the risk factors that are either suspected or have been proven to affect prenatal development in a negative way. This information is vital to a comprehensive understanding of the genetic and environmental influences on early development. Other factors come into play to affect the growth and development processes after the birth process; the most important of these are nutrition, exercise and physical activity, and hormones.

PRENATAL DEVELOPMENT

■ *Objective 5.1*

Several factors and conditions can affect the normal development of an unborn infant. Some of these are classified as *internal (maternal) factors* because they are linked with genetic influences or the maternal conditions of the prenatal environment.

Although the fetus is well protected by the placenta, it is not impervious to the outside environment. Through this membrane, the fetus is affected by *external (environmental) factors*—by what the expectant mother eats, breathes, drinks, or smokes. Certain substances are known to penetrate the placenta and cause negative reactions in the form of physical deformities and behavioral dysfunctions. An environmental agent that can cause a birth defect or kill the fetus is called a **teratogen.** Fortunately, about 97% of all babies are born without serious defects or deficiencies; human beings have a strong tendency to develop normally under all but the most damaging conditions.

Principles of Teratogenic Effects   Although the effects on the developing organism vary with the specific teratogen, several general principles appear to apply to most of them (Bukatko & Daehler, 1995; Cole & Cole, 1993):

1. Teratogens affect prenatal development by interfering with basic biochemical processes.
2. The susceptibility of a developing organism to a teratogenic agent varies with the developmental stage the organism is in at the time of exposure. In general, once the different body systems have begun to form, each is most vulnerable at the time of its initial growth spurt; this is associated with a "sensitive" (critical) period in development.
3. Each teratogenic agent acts in a specific way on specific developing tissue, therefore causing a particular pattern of abnormal development.
4. Not all organisms are affected in the same way by exposure to a given amount of a specific teratogen. Reaction depends to some degree on the organism's and mother's genotype.
5. Susceptibility to teratogenic agents depends on the physiological state of the mother (e.g., age, nutrition, hormonal balance).
6. In general, the greater the concentration of the teratogen to which the organism has been exposed, the greater the risk of abnormal development.

7. Accessibility of a given teratogen to a fetus or embryo influences the extent of damage. This refers to the cultural and social contexts associated with the mother.

8. Levels of teratogens that can produce defects in the developing organism may affect the mother only mildly or not at all. For example, some diseases and drugs that have minimal effect on the mother can lead to serious abnormalities in the developing organism.

9. Some teratogens may cause temporary delays in development with no long-term negative consequences, whereas others will cause problems late ("sleeper effects") in development.

The following material presents a brief note on selected maternal influences and environmental (teratogenic) agents. For a more thorough discussion of factors that can endanger development from conception through infancy refer to Kopp (1989) in the Suggested Readings.

Internal (Maternal) Factors

Maternal Age   The age of the expectant mother appears to have some bearing on the ease of the birth process and the general well-being of the child. Females over 35 and under 16 years of age have a higher risk of infant defect, prematurity, and low birth weight (Moore, 1988; Roosa, 1984). Risks to older women include difficulties in conceiving and delivering and an increased probability of having a child with Down syndrome. Another possible consequence is that the child will have problems in developing fine motor skills (Gillberg, Rasmussen, & Wahlstrom, 1982). In young women the reproductive system may not be mature and general prenatal care poor. The years between 22 and 29 are physiologically the best time to have a baby.

Nutrition   The development of the fetus can be affected by the quality of the expectant mother's diet both before and during pregnancy. Because the mother is the sole source of nutrients for the fetus, an adequate diet is vital. If the mother's diet is deficient in the nutrients the fetus needs, the developing infant may exert such parasitic effects on the mother's body as draining calcium from the mother's teeth. Several studies indicate a relationship between maternal diet deficiencies and low birth weight, prematurity, skeletal growth retardation, below normal motor and neurological development, and poor mental functioning (Bhatia, Katiyar, & Agarwal, 1979; Eckert, 1987; Knobloch & Pasamanick, 1974). An especially vulnerable time is during the first trimester; however, the last three months may also be critical because this is a time of rapid brain growth.

Rh Incompatibility   Because human blood is not interchangeable it must be matched to ensure compatibility. Although the blood of the mother does not mix excessively with that of the fetus, a problem of *Rh incompatibility* could arise if both the mother and the father have blood types from the Rh blood groups. The Rh-positive factor is an inherited protein found in the blood of approximately 85% of the general population. Rh positive is a dominant trait;

Rh negative is recessive. People who are Rh negative are therefore double re-cessives. Difficulty arises only when the father carries the Rh-positive factor, the mother is Rh negative, and the child is Rh positive. There is generally no problem with the first baby, but if, during the birth process, its blood (which is incompatible) happens to mix with the mother's, the mother's body may react to the baby's foreign Rh factor by creating antibodies. Should the woman be-come pregnant again, and if that child is also Rh positive, the antibodies could cross the placenta and possibly harm the fetus. Rh incompatibility can be de-termined through a blood test, and if administered in time, drug treatment can minimize any damage.

Genetic-Related Abnormalities   Each year in the United States, about 125,000 infants are born with a genetic abnormality, representing approximately 3% to 5% of births (Santrock, 1992).

Down syndrome, one of the most common genetic abnormalities, is a con-dition caused by the presence of an extra chromosome in the 21st pair of chromo-somes; thus the zygote has 47 instead of the normal 46 chromosomes. Individuals with this condition generally have retarded mental and motor development, heart defects, short hands and feet, and distinctively abnormal facial features. This con-dition is not inherited, and the risk that a woman will bear a child with Down syn-drome increases markedly with the expectant mother's age, especially after the age of 35. Down syndrome appears approximately once in every 700 live births.

Phenylketonuria (PKU) is a metabolic disorder that is also genetically based. It is a hereditary defect in an enzyme of the liver that is needed by the child to metabolize phenylalanine, a common protein food product. Without this enzyme the child cannot digest many needed foods including milk products. If the abnormality is not detected and treated soon after birth, substances may accu-mulate in the blood and cause damage to the brain. Symptoms include irritability, hyperactiveness, seizures, and perceptual problems. The condition can be de-tected with a blood test and treatment generally consists of a scientifically con-trolled diet.

Other frequently observed genetically related abnormalities include *sickle-cell anemia, Turner's syndrome,* and *Klinefelter's syndrome.* Sickle-cell anemia is a painful and sometimes life-threatening disease that occurs when harmful genes are passed on from parents to their children. Among Americans it is found primarily in the black population; approximately 10% of blacks are carriers of the harmful gene. In people with the sickle-cell trait, red blood cells become sickle-shaped and bind together when the level of oxygen is low, such as in times of exercise, illness, and high altitude. These clumps of cells then block the flow of blood through small blood vessels. Depending on the location and extent of blockage, the condition can cause severe pain or death.

Turner's syndrome is a disorder that occurs only in females as a result of having only one (instead of two) X chromosomes. These females lack normal ovarian tissue and do not develop sexually. Symptoms include distinctive physi-cal features that include a short, thick build, webbed neck, and prominent earlobes, as well as spatial perception difficulties. Most children with Turner's syndrome have normal IQs.

Klinefelter's syndrome is the most common of the sex-linked abnormalities associated with males. These individuals have an additional X chromosome (XXY), generally lack sexual development, and tend to be much taller than average. Intelligence levels are generally close to normal. As in Down syndrome, mothers of children with Klinefelter's syndrome tend to be older than average.

Fetal Position   The position of the fetus in the uterus is another internal factor that may affect its development. Pressure on the skeletal structure, namely the tips of the bones, stimulates ossification. If the fetus is positioned in such a way that this pressure is insufficient, immature skeletal development may result. Some depressurizing may take place as the fetus relaxes to accommodate to its somewhat cramped quarters.

Mother's Emotional State   Today researchers believe that an expectant mother's stress can be transmitted to the fetus (Parker & Barrett, 1992; Thompson, 1990). Due to psychological experiences such as intense fear and prolonged anxiety and depression, physiological changes can occur that may affect the developing fetus. For example, fear can produce hormonal changes such as higher levels of adrenaline, which restricts blood flow to the uterine area, possibly depriving the fetus of the normal flow of oxygen. Another possible complication of stress can occur during the birth process; an emotionally distraught mother may experience greater difficulty in labor, thus affecting the newborn's immediate adaptation to its new world (Sameroff & Chandler, 1975).

As noted earlier, although the fetus appears safe in its protective membrane, it is not impervious to the external environment. Although the placenta does a good job filtering many contaminants, the fetus can still be affected by disease-producing bacteria and by what the expectant mother eats, breathes, drinks, or smokes (teratogenic agents).

External (Environmental) Factors

Infection and Disease   German measles (*Rubella*), if contracted in the first month of pregnancy, gives the expectant mother a 50-50 chance of delivering a defective baby. If the rubella virus crosses the placenta during that time, it can result in miscarriage, a stillbirth, premature birth, blindness, deafness, mental retardation, and defects of the heart, liver, and pancreas. The most common birth defect is a loss of hearing. Fortunately, expectant mothers who have already had rubella or have taken the vaccine are unlikely to contract the virus.

■ *Objective 5.2*

Cytomegalovirus disease (CMV) is another potentially harmful infection. The strain of virus that causes this disease is quite prevalent among adults; however, it rarely creates any symptoms in the adult population. The virus is transmitted by way of the respiratory and reproductive systems and can result in stillbirths, miscarriages, low birth weight, and nervous system irregularities such as microcephaly (excessively small head), oversized livers and spleens, and blood abnormalities (e.g., anemia). To date, no methods of prevention or effective treatment for this disease have been found.

Two other potentially dangerous conditions are diabetes mellitus and syphilis. *Diabetes mellitus* is a noninfectious maternal disease that if untreated raises the probability of fetal death and stillbirth to 50%. A baby that survives birth to inherit the disease may suffer such abnormalities as an enlarged pancreas, respiratory problems, excessive weight, and metabolic disorders such as low blood sugar.

Syphilis is another potentially dangerous disease that if undetected in an expectant mother can attack the nervous system of the fetus, producing complications. The disease has been linked to miscarriages, stillbirth, deafness, blindness, and congenital mental dysfunctions. The signs of damage from syphilis may not appear until the child is several years old.

Other viral infections that have been associated to some degree with fetal defects or even death include the following: toxoplasmosis, herpes 2, mumps, polio, smallpox, chicken pox, infectious hepatitis, regular measles, influenza A, tuberculosis, and scarlet fever. Fortunately, the damage from most of these infections can be eliminated or minimized through preventive drugs and medical treatment.

Drugs and Chemical Substances   This section will address the potential effects of tobacco, alcohol, and drugs on the developing fetus. Included in the category of drugs are over-the-counter remedies, prescribed medicines, and illegal substances.

Smoking   The nicotine from tobacco use has been found to have the same types of effects on the fetus as it does on the mother. But because the same amount of the drug is available to a much smaller body area, the effects of the drug on the fetus are much greater. The carbon monoxide produced while smoking is known to interfere with the oxygen-carrying capabilities of the hemoglobin and, therefore to increase the risk of hypoxia, or oxygen deprivation. Associated with the infants of mothers who smoke are premature deliveries, low birth weight, and poorer language and cognitive development during early childhood (Chasnoff, 1991; Fried & Watkinson, 1990). Cigarette smoking has also been associated with miscarriages, malformation of the heart and other organs, and an increased risk of fatal birth defects. Marijuana smoking produces similar detrimental effects on a developing child (Day, 1991; Fried & Watkinson, 1990).

Alcohol   Alcohol can have serious direct effects on fetal and neonatal development because the substance can cross the placenta. Because the fetus's liver is immature, the alcohol remains in its system for a long time. The fetus can be very sensitive to alcohol and even in small amounts has been shown to cause abnormalities in development. During the last trimester of pregnancy, when brain development is quite rapid, alcohol may be extremely harmful to fetal health (West, 1986).

Fetal alcohol syndrome is a condition suffered by some infants who have been exposed to alcohol during the prenatal period. The syndrome was first identified among infants born to pregnant women who drank heavily. Infants with this condition do not go through the normal "catch-up" period of growth after birth and exhibit signs of mental retardation and immature motor development. Other commonly found characteristics are low birth weight; facial, limb, and organ defects; and repetitive body motions associated with autistic behavior.

Unfortunately, the long-term outlook for children with fetal alcohol syndrome is not a positive one. Barr and associates (1990), using a longitudinal design, found that 4-year-olds whose mothers had consumed moderate levels of alcohol during their pregnancies showed deficiencies in fine motor control and balance. The researchers also reported that alcohol scores from the period prior to pregnancy recognition were the highest predictor of poor motor performance. Doctors at one time thought that pregnant women could drink moderately without harming the fetus, but more recent evidence has brought this assumption under serious question with reports suggesting that there may be no "safe" level of intake during pregnancy.

Drugs   Many drugs are able to cross the placenta barrier to have a direct effect on the developing fetus. Substances included in the category of drugs are medicines prescribed by doctors such as antibiotics; over-the-counter substances like aspirin, vitamins, and cold remedies; and narcotics such as heroin, cocaine, amphetamines, and marijuana. Drugs strike the fetus with such force for two primary reasons: First, a small amount of a drug for an adult may be a dangerously huge amount for the tiny fetus; and second, the liver enzymes necessary for breaking down drugs do not develop until after birth. Therefore, just as with alcohol, drugs remain in the body of the fetus.

One of the most tragic and publicized instances of damage to the fetus as a result of drug ingestion by pregnant women took place during the late 1950s and early 1960s. The tranquilizing drug thalidomide was prescribed to quell the nausea of early pregnancy and induce sleep, and as it turned out, an estimated 10,000 malformed births were attributed to the drug. Of this total, only one-half survived to adulthood. A large number of the infants were born with severe limb deformities, or without arms or legs. This disaster brought to the forefront the need to better understand the influence of drugs on the developing fetus.

In more recent years, with the increase of narcotic use in our society, pregnant women addicted to heroin or cocaine are found to be giving birth to infants with symptoms of addiction. The number of these cases is growing at an alarming rate in the United States. Along with the life-threatening ordeal of going through withdrawal, these babies are often born premature, underweight, jittery, exhibit impaired motor control, and are quite fragile (e.g., Zuckerman, 1989). Some effects are still apparent months after birth.

The full impact of the influence of drugs of any type on fetal development is still being investigated. The general consensus in the medical profession is that expectant mothers should be very cautious about taking *any* medication (including aspirin) that is not prescribed.

Radiation   The radiation from X rays and other sources has been linked to harmful mutations of genetic material in the fetus. Exposure to radiation after conception but before the egg implants in the uterine linings is likely to end the pregnancy. Exposure to radiation *after* implantation may affect development in a variety of ways that include deformities of the central nervous system, which cause the death of the infant soon after birth, malignant tumors, leukemia, stunted growth, mental retardation, Down syndrome, defects of the skull, and death. Extreme doses of radiation from atomic explosions and buried atomic waste have also been linked to the same symptoms. Today X rays are seldom taken during pregnancy except in cases of emergency.

As noted earlier, by no means is this discussion meant to be exhaustive. Numerous additional environmental agents may have teratogenic effects. Other considerations include chemicals commonly found in paints, dyes, solvents, oven cleaners, pesticides, food additives, artificial sweetness, and cosmetic products.

NUTRITION AND PHYSICAL ACTIVITY DURING POSTNATAL DEVELOPMENT

■ *Objective 5.3*

The intake of essential nutrients and physical activity play an influential and complementary role in postnatal growth and health-related physical fitness across the life span. Children whose diets are unbalanced or are lacking in proper nourishment generally show below-average physical growth by age 3. Although diets lacking in protein and calories are by far the most common cause of growth failure during infancy and childhood, this problem is relatively rare in the United States and among other economically developed countries. Current information suggests that the American diet consists of too much fat, refined sugar, and salt as evidenced by the high incidence of childhood obesity and by the link that has been established between diet and cardiovascular disease in adults (Gortmaker et al., 1987; NCYFSI, 1985; Parcel et al., 1987). Apparently children and youth today are fatter than they were 20 to 25 years ago, are more susceptible to heart diseases, and many do not participate in sufficient health-related (aerobic) physical activity. The American Academy of Pediatrics (1987) identified this problem by reporting that up to 50% of today's school-age children are not getting enough aerobic exercise to develop healthy hearts and lungs. This general lack of adequate physical activity has been linked with high levels of television viewing and, in more recent years, the large amount of time spent with computers in the home (Dietz & Gortmaker, 1985). Increasing numbers of studies over the years have supported the importance of physical activity and proper diet to optimal physical health. Refer to the special feature in the *Research Quarterly for Exercise and Sport* (1992) for a comprehensive discussion of fitness in children and youth (Suggested Readings).

Nutrition

Regardless of age, individuals must have proper food for growth and for maintenance and repair of body tissues. Nutrition during the child's primary growth years is vital to physical growth and health, and intellectual development. Exciting research conducted over the last 15 years has revealed that the nutritional practices of infants and children may also have a marked influence on their health as adults. Since no standard for "optimal growth" has ever been defined and because individuals have varying nutritional needs based upon differences in energy expenditure and bodily functions, it is difficult to determine what would comprise the ideal diet for optimal growth.

The term malnutrition, in terms of physical growth, generally refers to an inadequate intake of energy in the form of calories or quality protein. A basic symptom of malnutrition is stunted growth. Shephard (1982) has identified two extreme examples of protein deficient malnutrition: kwashiorkor and marasmus. *Kwashiorkor* occurs after an infant is cut off from the mother's breast milk, a primary source of protein, due to the arrival of another infant. Some of the main symptoms are edema, skin disorders, loss of hair and appetite, diarrhea, and muscle hypotonicity. *Marasmus* is related to protein deficiency and an inadequate intake of total calories. Common characteristics of the syndrome include retarded growth and maturation, muscle atrophy, and a loss of virtually all body fat.

Malnutrition and starvation in adults have very little effect on brain mass, but nutritional deficiencies in children during the early periods of growth can result in underdevelopment of the brain. An inadequate intake of nutrients during prenatal development (especially the last trimester) and in the early stages of life after birth can significantly affect brain weight, brain size, number of cells formed, and amount of myelin developed in the brain, possibly resulting in impaired mental and motor behavior. Adequate nutritional intake is critical to sexual maturation and the development of muscle tissue during the pubertal growth spurt. Potential consequences of an inadequate diet are smaller body mass and delayed menarche.

Garn (1982) notes that during the periods of rapid growth, differences in nutrition levels correspond with differences in rate of growth, tempo of maturation, overall size, and amount of body fat. *Overnutrition* (more than average intake) is usually associated with more rapid growth, advanced skeletal and sexual maturation, larger body size, and greater quantities of fat. *Undernutrition* (less than average intake) reflects itself in lower growth rates, delayed physical and sexual maturation, smaller body size, and less stored fat. It can thus be concluded that growth rates during the growing years are strongly influenced by nutritional mediation, as are the attainment of terminal body size and sexual maturation. Some of the support for this conclusion is provided in research documenting that from the first year of life through adolescence, individuals who are fatter (a possible reflection of above-average caloric intake) tend also to be taller, reach sexual maturity faster, and be more advanced in skeletal age than average or lean individuals.

The following information identifies the basic nutrients and the part they play in the growth and maintenance of the structural and functional integrity of the organism. Table 5.1 shows recommended protein intake, ranges for energy needs, and daily dietary allowances (RDA) for selected minerals and vitamins by age group.

Table 5.1 Recommended Daily Dietary Allowances (RDA) and Energy Needs

Age (yrs) or Condition	Weight (kg)	Height (cm)	Protein (g)	Energy Needs (with range), kcal	Calcium (mg)	Iron (mg)	Vitamin A (µg RE)	Vitamin C (mg)	Thiamine (mg)	Riboflavin (mg)
Infants										
0.0–0.5	6	60	13	kg × 108 (95–145)	400	6	375	30	0.3	0.4
0.5–1.0	9	71	14	kg × 98 (80–135)	600	10	375	35	0.4	0.5
Children										
1–3	13	90	16	1,300 (900–1,800)	800	10	400	40	0.7	0.8
4–6	20	112	24	1,800 (1,300–2,300)	800	10	500	45	0.9	1.1
7–10	28	132	28	2,000 (1,650–3,300)	800	10	700	45	1.0	1.2
Males										
11–14	45	157	45	2,500 (2,000–3,700)	1,200	12	1,000	50	1.3	1.5
15–18	66	176	59	3,000 (2,100–3,900)	1,200	12	1,000	60	1.5	1.8
19–24	72	177	58	2,900 (2,500–3,300)	1,200	10	1,000	60	1.5	1.7
25–50	79	176	63	2,900 (2,300–3,100)	800	10	1,000	60	1.5	1.7
51>	77	173	63	2,300 (2,000–2,800)	800	10	1,000	60	1.2	1.4
Females										
11–14	46	157	46	2,200 (1,500–3,000)	1,200	15	800	50	1.1	1.3
15–18	55	163	44	2,200 (1,200–3,000)	1,200	15	800	60	1.1	1.3
19–24	58	164	46	2,200 (1,700–2,500)	1,200	15	800	60	1.1	1.3
25–50	63	163	50	2,200 (1,600–2,400)	800	15	800	60	1.1	1.3
51>	65	160	50	1,900 (1,400–2,200)	800	10	800	60	1.0	1.2
Pregnant			60	+300 (2nd/3rd tri.)	1,200	30	800	+20	1.5	1.6
Lactating										
1st 6 months			65	+500	1,200	15	1,300	95	1.6	1.8
2nd 6 months			62	+500	1,200	15	1,200	90	1.6	1.7

Source: Reprinted with permission from *Recommended Dietary Allowances*, 10th ed. Copyright © 1989 by the National Academy of Sciences. Courtesy of the National Academy Press, Washington, D.C.

Proteins   The word *proteins* comes from the Greek word meaning "of prime importance." Proteins contain amino acids, the essential "building blocks" of growth. Amino acids are critical for the synthesis of cellular components and growth of new tissue. During the periods of rapid growth in infancy and puberty, more than one-third of the protein intake is retained for new tissue growth. As the rate of growth declines, so does the percentage of protein retained for growth-related processes (table 5.1). Protein intake during the first year of life should be about twice the body weight, the highest protein intake amount recommended for any time during the life span. Protein makes up approximately 15% of the adult body weight.

Carbohydrates   *Carbohydrates* supply the greatest percentage of calories to the body of any of the essential nutrients. The main function of carbohydrates is to serve as a fuel source for the body. The energy they provide is essential for muscle contraction, for the metabolism of fat for heat, and for providing fuel allowing the central nervous system to function.

Fats   *Fats* both provide the body with its largest store of potential energy and insulate it from the thermal stress of cold environments. These nutrients also serve as a cushion for the protection of vital organs. In recent years, fats have been studied a great deal for their link to cardiovascular diseases.

Increasing evidence suggests that saturated fat and cholesterol may be a risk factor in heart disease and that the total intake of fat (saturated and unsaturated) should be reduced. Recommended cholesterol intake ranges from 150 milligrams to no more than 300 milligrams per day. Although cholesterol has been cast in a negative image, it is normally required in many complex bodily functions, including the synthesis of vitamin D and several hormones (e.g., estrogen and androgen).

Minerals   *Minerals* are another type of nutrient vital to normal growth and body functions. The developing child needs at least a dozen different minerals in proper quantities to properly form new tissues and body fluids. The mineral content of the body gradually increases with age from approximately 3% at birth to about $4^1/2\%$ in the adult. Minerals interact with the body by becoming incorporated within the structures and interacting with chemicals. Among the minerals essential for growth and motor development are calcium, phosphorus, iron, magnesium and the electrolytes of potassium, chlorine, and sodium.

Calcium is the body's most abundant mineral. Together with phosphorus it combines to represent about 75% of the total mineral content present in the body. Its primary function is to form bones and teeth during the growing years and to act as a defense against bone loss with age. Calcium, used in conjunction with regular exercise, can be very influential in preventing osteoporosis, which tends to affect individuals, especially women, in their later years.

Besides its work in helping calcium to form bones and teeth, *phosphorus* is an essential component of the high-energy ATP (adenosine triphosphate) and CP (creatine phosphate) systems. These energy sources supply fuel for all forms of biologic work.

Iron is found in the adult body in relatively small amounts—only 3 to 5 grams. Approximately 75% to 80% of this mineral is functionally active in compounds that combine with hemoglobin in the red blood cells. The primary functions of iron are to aid in oxygen transport to the tissues and to support the oxidation processes carried on by the cells.

Magnesium is found throughout all cells of the body with about 75% located in those of the skeleton and another large portion in the muscle tissue. Magnesium plays a vital role in providing energy to the body by participating in the metabolic processes that break down glucose, fatty acids, and amino acids into sources of fuel.

Collectively, the potassium, chlorine, and sodium minerals are referred to as *electrolytes,* meaning that they are dissolved in the body as electrically charged particles known as *ions. Sodium* is found mainly in the extracellular fluids of the body. *Chlorine,* acting as a chloride, is a component of all body secretions and excretions. Sodium and chlorine combined constitute the most important electrolyte in the extracellular fluid by maintaining acid-base and water balances. This balancing process allows for a well-regulated exchange of nutrients and waste products between the cell and its external fluid environment. *Potassium* is the chief intracellular ion. Sodium and potassium function to establish the proper electrical gradients across cell membranes, a process required for transmitting nerve impulses, stimulating muscle contraction, and enabling the glands to function properly.

Another health concern that has been given national attention in recent years is salt (sodium) intake. For most people, a normal salt balance is maintained through urine excretion; however, for some individuals excess sodium is not regulated and this can present a health risk as fluid levels rise and blood pressure increases to cause hypertension. Sources indicate that the typical American daily diet contains about 4,500 milligrams of sodium, almost 10 times more than the body actually requires.

Vitamins   There is little, if any, disagreement that *vitamins* are essential to normal growth and development. General knowledge of the essential vitamins is normally quite well known to educated people in today's society. The major function of vitamins is to act as coenzymes—to activate other enzymes that function in the basic metabolic processes. Although the compounds in vitamins alone provide no useful energy source to the body, they do serve as essential links to aid in the regulation of metabolic reactions that facilitate the release of energy sources. A well-balanced diet that provides an adequate amount of calories will usually ensure an adequate intake of vitamin nutrients, even during the rapid periods of growth (Rudolph, 1982). At times, however, vitamin supplements may be needed, such as with serious or prolonged illness, malnutrition, and significantly increased physical activity. Most cases of vitamin deficiency and below-average height or weight in school-age children have been associated with poverty. However, the number of these cases appears to be diminishing with free-lunch programs that have been made available in most states.

Water   *Water* is a nutrient so essential to life that without it death will occur within days. Water comprises 40% to 60% of total body weight; roughly 60% of water weight is located within the cells, and 40% is found in the plasma, lymph, and other fluids outside the cell.

Approximately 72% of muscle consists of water, whereas the weight of fat is only about 20% to 25% water. The primary function of water is to act as the body's transport and reactive medium. Virtually all bodily processes require the use of water as a medium. The best examples are the diffusion of gases and the transport of all other nutrients and waste products from one part of the body to another. Water also acts as a heat stabilizer by absorbing considerable quantities of heat, if needed, with only a small change in body temperature. It also helps give structure and form to the body and acts as a lubricant for the joints.

The proportion of water in the body remains relatively stable throughout life, although the daily volume of fluid that enters and leaves may vary considerably with age. In absolute terms, the fluid intake of an adult is about two to four times that of an infant (Weil, 1982). On a relative basis, however, the infant requires more water than the adult due to its relatively higher metabolic rate. In general, the energy requirements, including fluids, of younger and smaller individuals are greater per unit of body mass than those of adults. These requirements remain relatively constant per unit of surface area. Excess intake is normally excreted by the kidneys.

Current recommendations for dietary practices of individuals 2 years old and older include increasing the consumption of complex carbohydrates; reducing sugar, salt, and total fat intake; and reducing the intake of saturated fat and cholesterol. The approximate breakdown of the balanced diet is as follows: protein, 15% to 25%; carbohydrates, 45% to 55%; and fat, 30% maximum. For active people and those in exercise training, the amount of carbohydrates should increase by 5% to 10%.

Physical Activity

Physical activities are, in essence, the manifestation of the body's motor capabilities. The effects of physical activity on human development and behavior may be described from several perspectives. Two that are of direct bearing on the study of lifelong motor development are (a) the effects of physical activity and exercise on the structural aspects of physical growth and maturation, and (b) the influence of physical activity on the physiological function of the human body.

■ *Objective 5.4*

Although most professional educators and researchers would agree that a minimum of physical activity is probably necessary for optimal biological growth and development, very little scientific information has been published on the subject, especially with respect to the long-term influence of habitual physical activity on human development. Much of the available information has been of a speculative nature, as may be found in numerous dated physical education sources. The difficulty in identifying the specific benefits of physical activity to growth and development from a scientific perspective lies primarily in the definition of terms. What is meant by the term "optimal growth"? And what *level* of activity is most beneficial to optimal growth?

Level of physical activity may be used to describe individuals as "active" or "inactive," to identify those children who engage in play as "normal," or to classify individuals who participate in regular physical training programs ranging in intensity from moderate to strenuous. Difficulty also lies in attempting to isolate physical activity in the growth process. An inherent weakness of almost any study of a topic that spans the growing years is how to explain the complex interaction of both genetic and environmental variables. In recent years, a considerable amount of information has been gathered on the specific effects of various types and intensity levels of physical activity on health-related fitness characteristics across virtually all ages. However, much work remains to be done to develop a clear understanding of how much physical activity is required to stimulate normal or optimal growth and development of other biological components.

Selected literature reviews and reports on the subject agree, in general, that a minimum of physical activity is apparently essential for specific aspects of normal growth and development to take place, although just how much is necessary is unclear (Broekhoff, 1986; Malina, 1990). The phenomenon of accelerated maturation and the increased body size associated with secular trend over the last 100 years or so has given little mention to the possible contributions of exercise or physical activity. Instead, explanations speak more to better nutrition, improved health care and health status, and contributing genetic factors (Broekhoff, 1986). It is probably best to view secular trend as an interactive result of several genetic and environmental factors, including the possible influence of physical activity.

The following discussion concerning the effects of physical activity on biological growth and development was derived primarily from comprehensive reviews reported by Zauner, Maksud, and Melichna (1989), Broekhoff (1986), and Malina and Bouchard (1991). In the context of these reviews, the term *physical activity* refers to practices that are habitual or additional to normal movement or irregular patterns of play and recreation. Included in this definition are regular physical training programs such as physical education in schools, swimming, running, youth sports, calisthenics or aerobics, and weight training.

Stature   The vast majority of evidence based on studies using equivalent control groups of preadolescent and adolescent children indicates that regular physical activity has no apparent effect on stature. Although some earlier studies suggested that stature increases with physical activity, reviewers of that research have pointed out that the differences were usually quite small and the experimental procedures questionable. In some studies reporting increased growth in stature due to athletic training, the apparent increase was probably not as much the product of intensity of training as it was early maturity. In examining one study of female swimmers, they were found to have been taller than the average in the years prior to the research; apparently they had entered puberty earlier than the reference group with which they were compared. The same problem may have been prevalent among studies comparing young athletes with "normal" reference data. Note that while the evidence does not indicate that training actually *promotes* significant growth in stature, it is also clear that the training has no *negative* effect on this aspect of physical growth.

Body Mass   The existing data are limited for younger and older populations but suggest that although body composition changes considerably during normal growth, particularly during adolescence, it can be further modified by habitual physical activity. Regular physical activity is, in fact, an important factor in the regulation and maintenance of desirable body weight. Optimal change, however, occurs when diet is also controlled. Even during the growing years in which an increase in body weight is characteristic, and across the life span as well, individuals who engage in regular training programs tend to have more lean body mass and less body fat than their nontraining peers. Ample evidence exists to verify that carefully designed physical activity programs can be effective in reducing the percentage of body fat in children and adults. The magnitude of change, however, may vary considerably depending on the intensity and duration of the program. Changes in body composition are highly dependent on continued activity.

Evidence also suggests that individuals who maintain a program of habitual physical activity, especially endurance training, throughout adulthood have less tendency for increasing body fat with advancing age. The available information on individuals who began training in older adulthood provides mixed results. In some cases minimal changes occurred in body fat, body weight, or lean body mass, whereas other studies reported significant losses of body fat. Further research is needed to arrive at more definitive conclusions.

Related to the "fat cell theory" mentioned in an earlier chapter, the results of a series of experimental studies using rats (Oscai et al., 1972, 1974) indicated that there is a critical time in early infancy in which fat cell proliferation may be enhanced by overfeeding to result in an abnormal number of cells that remain with the body for life. Evidence suggested that increased physical activity and control of diet initiated in the early stages of development can effectively limit fat cell proliferation and reduce body fatness in later life. These results, although not decisive, have some very practical implications for infant care and weight control.

Body Proportion and Physique   In general, there is little evidence to suggest that physical activity at any level of intensity affects body proportions or overall body shape. Interestingly, although there have been significant increases in the secular trend of height and other dimensions over the last 100 years or so, changes have been approximately proportional. Thus, if any alternation in body proportionality occurs it is minimal. Studies that have focused on this aspect of physical activity and growth also indicate minimal influence due to training.

In one of the more comprehensive longitudinal studies of physical activity and growth, Parizkova and Carter (1976) found that boys engaged in various levels of physical activity, from intense to untrained, over a seven-year span from 11 to 18 years, showed no significant difference in the distribution of somatotype ratings among the activity groups. It was also noted that individual ratings changed considerably over the period and occurred in a random manner among the groups, suggesting that such changes cannot be attributed to physical activity. All participants changed somatotype rating at least once and the majority changed in component dominance.

The short-term benefits of training for muscular development through weight training to increase the mesomorphic component of the somatotype have been verified with numerous studies of teenagers and adults.

Skeletal Growth and Development   Evidence currently suggests that regular physical activity has no apparent effect on skeletal maturity; however, it does enhance skeletal mineralization and density. Data from the Parizkova longitudinal study of change in physique also provide information about the effects of physical activity on skeletal maturation. Findings indicated that there were no significant differences among the various groups before, during, or after the study as assessed using ossification ratings of the hand and wrist. Results also revealed that variation was greater within specific activity groups, rather than between them. Other longitudinal studies have reported similar findings, suggesting that skeletal maturity is not affected by regular physical training.

Numerous studies have suggested that greater skeletal mineralization and bone density and wider, more robust bones are associated with prolonged physical training. Although several of the studies have been conducted with animals, observations on adult humans have indicated similar results. In general, the effects of prolonged activity provide localized results and are activity related. For example, studies involving athletes have indicated that the more actively preferred limb of, for example, a tennis player, pitcher, or soccer player will exhibit greater mineralization and density than a nonpreferred limb. Studies that have observed bone density changes in runners relative to nonrunners also suggest greater values as a result of training. As noted earlier, it has been suggested that these effects are related to regular physical training over several years' time. The data related to children are quite limited, and perhaps problematic due to the confounding growth factor and lack of a long-term perspective.

The information on training and its effects on bone length is limited. Evidence has been reported indicating increased linear growth in dominant compared to nondominant limbs; however, other studies using animals have reported reduced bone lengths as a result of training. The apparent conflicting results center on the effects of pressure on bone growth plates. Theoretically, the pressure effects of physical activity can either stimulate growth or, in excessive amounts, can retard linear growth. The difficulty lies in identifying and defining what is perceived as required, optimal, and excessive pressure. A theoretical notion known as *Wolff's Law* may have implications for bone growth in general (Rarick, 1973). The law simply states that bone structures will adapt to suit the stresses and strains placed on them. Obviously more research and information are needed relative to the effects of varying levels of physical activity on growth plate genesis.

This body of information adds significantly to our general understanding of the influences that physical activity may have on skeletal growth and development. Lack of clear definition for functional effects and for minimal and optimal activity levels in individuals makes it difficult to answer key questions. One of the more practical questions that this data cannot answer clearly is whether "active" children have any skeletal advantages over "relatively active" children in later years. At this time, it appears that a lot more is known about the effect of physical inactivity on bone structure changes than about the effects of varying levels of physical activity.

Sexual Maturation   Earlier it was pointed out that individuals today mature earlier than people did 100 years ago. The secular trend for menarche has decreased approximately four months per decade since 1850. There is no evidence that this trend is linked to changes in physical activity patterns nor has there been any documentation in more recent years supporting the notion that habitual physical activity has a positive influence on sexual maturation of the reproductive system. There have been indications that high-intensity physical training and athletic competition may be a factor in delayed menarche, however (Malina, 1983, 1984).

A rather consistent body of evidence indicates that selected athletes (e.g., gymnasts, long-distance runners, ballet dancers, figure skaters), on the average, experience menarche later than the general population. Observations also indicate that the number of years of training as well as the level of competition, prior to maturation, are linked to delayed menarche. It should be noted that most of the data on this subject are quite limited, associative, speculative, and do not control for other factors that may have influenced the onset of menarche. It is generally agreed that it is very difficult, at best, to design a study in which individuals are randomly assigned to athletic activities and then to isolate the training factor as the primary cause for delayed menarche.

Of the proposed theories related to the possible influence of intense physical training on the timing of menarche, one in particular called the *critical-weight theory* (Frisch, 1991, 1976) has received considerable attention. The researcher proposed that a critical percentage of body fat of approximately 17% must be reached before menarche can occur. It is also suggested that a body fat level of 22% is optimal for maintaining regular menstrual cycles. Thus, the association between female athletes who stay lean (i.e., below 17% body fat) throughout their training and the incidence of amenorrhea (irregularity) and delayed appearance of menarche provides the evidence to support this theory. Survey research (Frisch et al., 1981; Marker, 1981) also suggests that a relationship exists between the number of years of training prior to menarche and the time at which it appears. Marker noted that delayed menarche occurs more frequently in gymnasts (who have the highest mean age), figure skaters, and divers than among other athletes. An inferred relationship can be drawn, since females in these sports usually begin training early in life. Both studies reported mean ages for menarche of slightly above 15 years for those females who had extensive training before menarche; the average for the general population is around 12.5 years.

The critical-weight theory has not found universal support and the mechanisms by which intense strenuous physical training may affect the timing of menarche have not yet been clearly discerned. The major problem with interpreting the available data is the number of other possible factors that may be at work around the time of menarche.

Other proposals also warrant consideration. For example, Warren (1980) concluded that along with a low percentage of body fat, an energy drain as a result of the intense training may significantly alter the hormonal processes that affect the set point of menarche. This alternation may result in a prolonged prepubertal phase and induced amenorrhea. A sociological explanation has also been suggested (Hata & Aoki, 1990); females who mature and reach early menarche may be socialized away from sports participation, whereas females who mature late tend to be drawn to sports participation.

Muscular Development and Strength   The evidence is overwhelming that specific strength training increases muscular force capabilities in adults and post-pubertal children. Regular physical activity also results in some degree of skeletal muscle hypertrophy; the degree of hypertrophy varies with the intensity of training. The characteristics that generally accompany muscle hypertrophy are increases in enzymatic activity, myofibrils, contractile substances, and muscular strength. Muscles are strengthened by increasing their size and by enhancing the recruitment and firing rates of their motor units. Both of these processes are involved in the body's response to resistive physical activity, or exercise.

It appears that older males, prepubescent children, and females increase muscular strength primarily by neural adaptation (with some hypertrophy), whereas young males rely more on increases in muscle size. An ongoing debate in exercise physiology is whether resistance training induces the muscle to both increase the number of muscle cells (hyperplasia) and increase the cells in size (hypertrophy). For a long time the consensus was that muscles increased in size only after the hyperplasia period that ends shortly after birth. However, some recent evidence, while controversial, suggests that as a result of resistance training, muscle fibers may split as a means of increasing cell number. Even with this new hypothesis, it is generally agreed upon that the primary response to overload stress is muscle hypertrophy (McArdle, Katch, & Katch, 1991). Since the effects of intense (resistance) physical training on muscular development and increases in strength in postpubertal teenagers and young adults are well known, the remainder of this discussion will focus on the young and old.

We learned in chapter 3 that adult men have approximately 10 times the androgens of children and adult women and that the bulk of this amount is produced during the adolescent growth spurt. At this time the greatest changes in male strength seem to appear and evidence strongly suggests that improvements in muscle force are related to greater testosterone levels. Often asked is whether prepubescent children can make significant gains in strength with training. Until recently it was thought that children of this age could obtain only minimal benefits from resistance training due to an insufficient amount of circulating androgens (American Academy of Pediatrics, 1983). Reviews of recent research reported by Weltman (1989) and Duda (1986) confirmed that prepubescent children are capable of making significant gains in strength, even though they possess relatively low hormone levels.

Weltman and colleagues (1986) support this conclusion, noting that circulating androgens may not be the only primary influencing mechanism for strength development in this case. It is well documented that women, who also have low androgen levels, can make significant gains in strength without appreciable muscle hypertrophy. Other influencing factors include the number of motor units recruited and their level of synchronization. The conclusions of these studies indicate that increases in strength in prepubescent children are probably the result of neural adaptions. In the Weltman study, strength gains took place in the absence of muscular hypertrophy. It also suggested that the risk of short-term musculoskeletal damage is minimal if training is supervised; the use of hydraulic machines reduces the risk even more. However, more research is needed in regard to the effects of resistance training on articular cartilage.

The physiological responses of weight-resistance training to strength development in children appear to be beneficial. A similar question can be asked with regard to the effects of regular participation in sports. Bailey, Malina, and Rasmussen (1978) found that children and adolescents who were involved in youth sports on a regular basis were stronger than those individuals who did not participate. Although this result may be an indication of the strength-promoting benefit of sports, it could also be due to a weak comparison between groups because young athletes are often early maturers and thus are bigger for their age. It would seem reasonable to assume that any benefits to strength development would depend primarily on the specific physical activities associated with a particular sport.

In regard to older individuals, it appears that they are not able to improve their strength capacity to the same extent as young people; however, significant improvement from regular vigorous training can be expected regardless of age (McArdle, Katch, & Katch, 1991). The reasons for the decrease in "trainability" with age are not well understood, but it has been speculated that they are due, in part, to the general decline in neuromuscular functions. As with prepubescent children, studies of older individuals indicate that their strength increases not primarily by muscle hypertrophy but by increased neural stimulation, or motor unit recruitment (Brooks & Fahey, 1985; Whitbourne, 1985). As discussed in the last chapter, a general decrease of strength with age is expected because individuals generally experience a loss of muscle mass followed by a loss of motor unit fibers. A review of several studies suggests that significant gains in muscular strength can be achieved with vigorous training, at least into the seventh decade of life (McArdle et al. 1991). Very little information is available on older females from which to comment.

Cardiorespiratory Development McArdle et al. (1991) state that "regular vigorous physical activity produces physiological improvements regardless of age" (p. 708). However, just as in the case of muscular development, the magnitude of the potential improvement depends on several factors, including initial fitness level, type of training, and age. In regard to age, the situation is again quite similar to that describing the effects of training (physical activity) on strength development. Older teenagers and adults who engage in vigorous aerobic activity for 20 minutes or longer at least three times per week are likely to experience physiological changes that include an increase in heart volume and cardiac output, a lower resting heart rate, and a faster return to resting heart rate.

In chapter 4 it was noted that the aging process can decrease general cardiovascular function by approximately 1.0% per year between the ages of 30 to 70. Along with the general decline in cardiorespiratory function as an individual ages, there also appears to be decreased "trainability"; that is, older individuals are not able to improve to the same extent as young people.

In the final report of a longitudinal study of active men aged 45 to 65 years, evidence was presented suggesting that physical training may delay the decline in maximal oxygen uptake until at least age 65 (Kasch et al., 1988). Instead of the typical decline of 1% to 2% per year over the 20-year span, aerobically active men declined an average of only 12% over that time and revealed essentially no differences in resting heart rate. The authors did associate their findings with aerobic training practices of sufficient intensity, frequency, duration, and energy expenditure and the fact that many of their subjects had experienced a lifetime of physical activity. Complementing these data is a finding by Bruce (1984) that sedentary individuals have nearly a twofold faster rate of decline in VO_2max as they age.

Skill-Related Fitness The contention that regular instruction and practice of motor skills results in improved balance, coordination, speed, power, agility, and other skill-related fitness is well supported in the literature (Malina & Bouchard, 1991; Vogel, 1986). Such findings explain why regular participation in physical education programs is emphasized.

GLANDS AND HORMONAL ACTIVITY

■ *Objective 5.5*

Hormonal activity has been discussed in terms of its relevance to the adolescent growth spurt. Though it is beyond the scope of this text to detail the biochemical mechanisms of the endocrine glands, a brief review of hormonal activity as it relates to more general growth and development across the life span is appropriate.

Remember from chapter 3 that the endocrine system consists of several different glands that secrete hormones directly into the bloodstream. The endocrine glands regulate such vital functions as growth, sexual reproduction, metabolism, immunity to stress and disease, and the aging of cellular tissues. As with most physiological mechanisms in the human body, aging also affects the endocrine system. Characteristic of these changes are a decrease in the secretion of hormones, diminishing responsiveness of specific cells to hormones, a change in the chemical transmitters that carry hormonal messages into cells, and a change in the levels of enzymes that respond to hormones (Marx, 1979). Hormones, in general, are highly interactive and affected by numerous factors, many of which have not been identified; consequently, the study of the effects of hormonal activity on growth and development is a complex undertaking.

It is not known for certain which hormones can pass from the mother to the fetus through the placenta, but there is evidence that estrogen, progesterone, and gonadotropin make this prenatal journey (Lowrey, 1986; Shephard, 1982). The fetus must produce the other hormones on its own. These and the glands that produce them are described in the following material.

The pituitary gland is known as the master gland of the endocrine system. It produces one of the most important hormones involved in the process of human growth—the **growth hormone (GH).** The only major portion of the body not affected by this hormone, in terms of growth stimulation, is the central nervous system. The fetus has the capacity to synthesize GH by the end of the first trimester. Blood serum levels of the hormone in the neonate are significantly elevated above levels found during childhood. The importance of GH to the adolescent growth spurt is well documented; its relevance to fetal and infant growth is relatively unknown, although it has been determined that normal fetal development does occur in the absence of GH. Interestingly, the maximum production and secretion of GH occurs in association with periods of deep sleep.

As noted in chapter 3, GH promotes the incorporation of amino acids into tissue protein and stimulates the development of muscle and bone rather than fat tissue. Growth hormone is also known to stimulate DNA synthesis and cell multiplication. Its influence on increasing the number of cells of the body lasts from late infancy until adulthood, and its metabolic functions continue until death. The thyroid hormones and androgens interact with GH to perform functions vital to cell multiplication.

Although the size and weight of the pituitary gland does not change with age, its blood supply gradually decreases after puberty. After about age 30, secretion of GH tends to decline, which in older adults is responsible in part for the decrease in lean body mass, increase in adipose tissue mass, and thinning of the skin (Rudman et al., 1990).

Pituitary Gland

The thyroid gland is second only to the pituitary in terms of influence on human growth and development. Thyroid hormones are important to cell multiplication and growth after birth. Their presence is so important, in fact, that regardless of other hormones, their absence would result in retarded and abnormal growth. Thyroid hormones are vital to the production of all forms of RNA, and their presence stimulates ribosome production and protein synthesis. Other functions include metabolizing fats and carbohydrates, stimulating the cell's use of oxygen, and, along with sex hormones, facilitating skeletal development through calcium and phosphate absorption. The pituitary GH is relatively ineffective without thyroid hormones. In contrast to GH, which has little if any effect on fetal and early infant growth, thyroid hormones are very influential.

Thyroid Gland

As people age, the cells within the thyroid gland undergo several structural changes and collagen fibers appear. With increasing age, the blood levels of thyroid hormones do not drop even though the thyroid itself produces fewer hormones. The thyroid gland continues to function normally in healthy adults throughout the life span.

Pancreas

The pancreas, which is well developed by the end of prenatal development, secretes a powerful anabolic hormone called insulin. Its primary functions are to metabolize sugars in the diet and increase the transport of glucose (blood sugar) and amino acids across the plasma membrane. Insulin also enhances protein synthesis, although tissue hypertrophy can occur in its absence, and appears to increase the synthesis of DNA and RNA in muscle and bone tissues. The commonly known disorder of diabetes is a result of an abnormality in the body's ability to metabolize blood sugar. With increasing age, many people experience longer and higher levels of blood sugar as a result of the aging cell's tendency to absorb less blood sugar and decrease its active insulin secretion.

Adrenal Glands

Of the several groups of hormones that the adrenal glands produce, the androgens are the most important to a discussion of factors influencing growth and development. Although androgens are secreted by the adrenal cortex in both sexes, they are also produced by the testes in the male; their release is triggered by actions originating from the pituitary gonadotropins.

Chapter 3 discussed the importance of adrenal hormones to the development of muscle tissue and skeletal growth. Other functions include causing the release of epinephrine and norepinephrine to mediate the response to stress, mediating the kidney's absorption of sodium and chloride, and metabolizing fat and carbohydrates. Although a major portion of the adrenal cortex is developed and differentiated by the third fetal month, it does not reach its adult character until about the third postnatal year (Lowrey, 1986).

As the system ages, the production of some adrenal hormones, including androgens, declines. The hormone secretions that regulate fat and carbohydrate metabolism do not decline with aging, although it will take individuals longer to complete this metabolic process as they age. Andres and Tobin (1977) estimate a 40% difference in metabolism time between a 75-year-old and a 35-year-old male. Hormones affecting fluid balance in kidney function remain relatively stable well into old age.

Other Glandular Functions

Closely associated with sex differentiation, physical growth, and sex hormones are the functions of the *gonads* (the testes and ovaries). Adrenal and gonadal tissues originate from similar cells; both are known as *steroid-producing glands*. Exposure of the brain to sex hormones during early fetal development is associated with future reproductive physiology and sex-oriented behavioral characteristics. For example, during the early stages of fetal development, androgens interact with the immature central nervous system to imprint future male characteristics. Ovarian estrogens in females, like testosterone in males, normally are not secreted in physiologically effective quantities until shortly before puberty.

Hormones secreted by the *parathyroid glands* also play a major role in normal growth and development. At birth the parathyroid glands are morphologically similar to those of the adult. These hormones are responsible for normal skeletal development by regulating calcium and phosphorus metabolism. In addition to its function in skeletal development, these hormones also enhance calcium absorption in the presence of vitamin D. The parathyroid glands, located at the base of the throat, grow heavier until males turn 30 and females reach 50 years of age. After these peaks, their size and rate of hormone production remains stable.

Most experts agree that physical activity probably influences many of the relationships among hormonal activity, growth, and behavioral characteristics. Although varying levels and types of exercise have been shown to affect specific hormone concentrations, the consensus is that it does not significantly influence physical growth beyond that which would occur naturally (Shephard, 1982).

SUMMARY

Several factors may affect the normal course of growth and development throughout the life span. A number of maternal and external factors have the capability of producing a profound effect on development of the unborn infant. Among the potentially harmful maternal factors are age of the mother, nutrition, Rh incompatibility, genetic-related abnormalities, fetal position, and the mother's emotional state.

Although the fetus appears safe in its protective membrane, the fetus can still be affected by external environmental factors through the placenta. Broadly, external factors can include anything the expectant mother breathes, drinks, eats, or smokes or the disease-producing bacteria with which she comes in contact. Exposure to harmful agents (teratogens) can cause negative reactions in the form of birth defects and behavioral dysfunctions. Among them are infections such as rubella, cytomegalovirus disease, diabetes mellitus and syphilis, tobacco, alcohol, drugs, radiation, and various chemicals. In spite of the possible dangers, human beings have a strong tendency to develop in what appears to be a normal manner under all but the most damaging conditions.

The intake of essential nutrients such as proteins, carbohydrates, fats, minerals, vitamins, and water is vital to growth and development throughout the life span. In general, infants with malnourished diets (especially if they are low in protein and total calories) show below-average physical growth by age 3. However, the primary concern for children in the United States appears to be the overconsumption of fat, refined sugar, and salt rather than a deficiency of protein and calories.

Although the literature is unclear regarding what level of physical activity is beneficial or harmful to physical growth and development, its effects on specific characteristics have been documented. In general, habitual physical activity has been shown to have a positive influence on body mass (body weight, body fat, lean body mass), skeletal mineralization and density, muscular development and strength, and cardiorespiratory development. Most of the literature, however, does not claim that habitual physical activity has any significant influence on stature, skeletal maturity, or body proportion and physique. Some research has suggested that intense exercise may be a factor in delayed menarche, but there is no evidence that physical activity has a positive influence on sexual maturation.

One of the most influential factors in physical growth and development throughout the life span is hormonal activity. Primary producers of hormones that have a profound effect on physical growth and development are the pituitary gland (growth hormone), thyroid gland (thyroid hormones), pancreas (insulin), adrenal glands (androgens), gonads (androgens and estrogens), and parathyroid glands. Hormones, in general, are highly interactive and affected by numerous factors, many of which are unknown.

SUGGESTED READINGS

Kopp, C. B., & Kaler, S. K. (1989). Risk in infancy: Origins and implications. *American Psychologist, 44* (2), 224–230.

McArdle, W. D., Katch, F. I., & Katch, V. L. (1991). *Exercise physiology* (3rd ed.). Philadelphia: Lea & Febiger.

Malina, R. M., & Bouchard, C. (1991). *Growth, maturation, and physical activity.* Champaign, IL: Human Kinetics.

Special feature. Are American children and youth fit? *Research Quarterly for Exercise and Sport, 63* (2), 95–136.

PART ○ *three* -----------------------------

■ *Objective 6.1*

Part Three takes a developmental perspective to focus on the way people perceive and act upon sensory information that derives from the external and internal environments. This general process is called **information processing.** From a motor behavior perspective, information processing involves both *perception,* monitoring and interpreting sensory information, and *motor response,* deciding upon and organizing response activity. Figure 6.1 illustrates the

Perception and Information Processing

information-processing, or *perceptual-motor process,* model. The model contains four primary components: sensory input, reception of information, interpretation of information and decision making, and overt motor response.

The processing stage will be studied in two separate chapters. Chapter 6 looks at the developmental aspects of the perceptual systems in regard to receiving, monitoring, and interpreting sensory data. Chapter 7 covers the developmental processes involved in processing the information provided by the perceptual systems, giving special emphasis to attention, memory, processing speed, and cognitive strategies.

Perceptual Development

KEY TERMS

information processing
perception
sensation
visual acuity
perceptual constancy
spatial orientation
figure-ground perception
whole-part perception
depth perception
field of vision
coincident timing
visual-motor coordination

kinesthesis
kinesthetic acuity
kinesthetic memory
body awareness
spatial awareness
directional awareness
vestibular awareness
rhythmic awareness
tactile perception
perceptual integration
constructivist approach
ecological approach

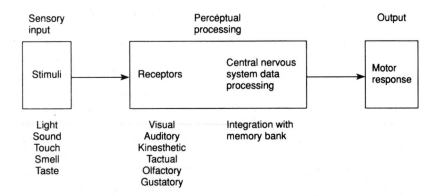

Figure 6.1
The general
information-processing
(perceptual-motor)
model.

The essence of human behavior and motor performance is based on the ability to receive and interpret sensory information. Human beings live in a vast sea of sensory information, yet they thrive rather than drown because of an elaborate network of perceptual systems. The human perceptual system, in fact, has a constant need to receive sensory input from the external world and from its own internal environment. When sensory input is reduced or eliminated, the system reacts negatively. This same negative reaction can also occur when the system is overloaded by too much sensory input, as happens when you try to carry on a telephone conversation while reading a paper or when a beginner in dance class tries to stay in step with the teacher's instructions during a fast-paced song.

Perception refers to the processes used to gather and interpret sensory information from the external and internal environment. Although external information is perceived by the fetus, the amount of sensory data received is magnified tremendously from the moment of birth. The two related processes of sensation and perception enable the body to receive stimulation and organize it for further processing.

■ *Objective 6.2*

Sensation refers to the stimulation of sensory receptors (fig. 6.1) by physical energy received from the internal and external environment. The retina of the eye, for example, reacts to light rays (sensory input) and translates the message into nerve impulses (reception). The nerve impulses are then transmitted via the perceptual modality to the brain, which reacts to the stimulation in various ways, the point at which perception occurs. Basically perception involves the monitoring (reception) and interpretation of sensory information. What interpretation is given to the information is based on past experience (memory) and cognitive analysis (judgments) of the information. Intertwined with interpretation is the ability to discriminate among various types of sensory stimuli with varying degrees of precision—the similarities or differences, arrangements, and organization.

Another complex function of the brain in this process is to integrate information coming from different perceptual modalities. This information provides the basis for formulating a motor response, which completes the information-processing cycle. This sequence may be thought of as the information-gathering segment of the general information-processing model.

Of the six perceptual modalities, perhaps the two that provide the most relevance to the study of motor behavior are vision and kinesthesis. This is not to minimize the importance of audition and tactile information; these modalities will be discussed briefly along with system integration. However, human movement is universally described in terms of its spatial-temporal characteristics—the spatial coordinates and timing of the body with moving stimuli, as well as internal rhythmic awareness. Although individuals can move in the environment without visual information (as the blind do), this modality is considered to be crucial in determining reference points and judging the movement of objects. The visual modality provides spatial-temporal information about the external environment, and the kinesthetic system may be described as the body-knowledge network that provides information about the body's movements without reference to auditory or visual cues.

VISUAL PERCEPTION

■ *Objective 6.3*

In many ways, vision may be considered the dominant perceptual modality. The visual world is the richest source of information about the external environment for human beings. Experts estimate that approximately 80% of all sensory information derived outside the body is channeled through the visual system. In terms of movement, visual input provides one of the primary sources of information from which motor behavior is organized and carried out. Visual information is used to formulate a motor program, monitor movement activity, and provide the feedback that allows for immediate correction to be made. Visual input is such a dominant source of information that people tend to rely on it even though, in specific instances, other sensory information may be more useful.

Most of the available data on this subject focus on infants and children during the growing years. Up until about 25 years ago, it was believed that the newborn infant was incapable of processing information meaningfully. Today most experts agree that the visual system of the newborn is relatively well developed at birth and that the neonate demonstrates considerable visual ability. On the other end of the lifelong developmental continuum several deficits attributed to aging have been documented, though the information on perceptual changes during adulthood and old age is sparse.

The Visual Process

Three basic functions occur in the course of visual perception. First, the eyes receive light and generate messages through nerve impulses about that light; second, visual pathways transmit those messages through the optic tract from the eye to the visual centers of the brain; and finally, the visual centers make interpretations of those messages.

Light enters the eye through the pupil, passes through the lens and chambers, and terminates in the retina. The retina is a complex structure consisting of numerous photosensitive receptors and other cells that assist in optic transmission of visual nerve impulses. After nerve impulses are transduced from light energy in the retina, they are transmitted over the optic nerve. The optic nerve from each eye resembles a cable that contains approximately 1 million individual wires and acts as a huge communication line. All of the visual information that comes into the brain for neural processing must be input by the optic nerves. Optic nerves from the two eyes pass backward to unite at the optic chiasma and from there project posteriorly even further as the optic tracts. Fibers from these tracts make synaptic connections with several subcortical structures in the brain. Signals then enter the visual projection areas located in the occipital lobes at which point impulses are dispersed in various directions to the association areas where interpretation, or perception, takes place.

Every structure in the visual system undergoes some early postnatal change (Banks & Salapatek, 1983). Although all of the visual structures are intact at birth, several are immature in terms of myelinization and synaptic potential. There is also great variability in the growth rate of different parts of the eye. The size of the eye approximately doubles between the time of birth and maturity.

Developmental Change in Visual Structures

The newborn retina, the part of the eye that turns light into nerve signals to the brain, is similar in structure to that of the adult but is thicker and has no distinct fovea. The fovea is the area in which central visual images are formed. Without this structure, the processing of spatial resolution is poor in the central retina; thus the ability to see images clearly is relatively weak in comparison to the mature structure. Development does proceed rapidly and by the end of the first year the major structures of the retina are like those of the adult.

The part of the nervous system that relays visual impressions between the retina and the cortex of the brain also develops quickly and reaches a large part of its ultimate efficiency during the first 11 to 12 months. In the visual cortex, dendritic development at the ends of the nerve cells forms more branches, and by 4 to 5 months of age, the myelinization process is almost complete. At about 2 months, the infant's visual cortex begins to process visual stimuli.

Compared to the adult, the newborn's eyeball is short and the distance between the retina and the lens reduced, resulting in a temporary form of farsightedness. Many infants will also have astigmatisms (difficulty in focusing) during the first year because their cornea is not yet symmetrical. Another developmental difference between the neonate and the adult is in the strength and control of eye muscles. Because of weak ciliary muscles attached to the lens, few newborns can change the shape of the lens to accommodate to the shifting plane of focus for visual targets as occurs in the eye of the adult. Until the end of the first month, infants can adjust their focus only for targets that are between 5 to 10 inches away. By 4 months of age, infants can accommodate to shifting planes of focus equally as well as adults. Illustrative comparisons of the adult and newborn eye are presented in figure 6.2.

Visual Acuity **Visual acuity** refers to clearness of vision and the capacity to detect both small stimuli and small details of large visual patterns. From a motor behavior perspective, visual acuity is often classified as static or dynamic. *Static*

Developmental Change in Visual Functions

Figure 6.2
A comparison of an adult
and newborn eye
indicating limitations of
the infant's visual
system.

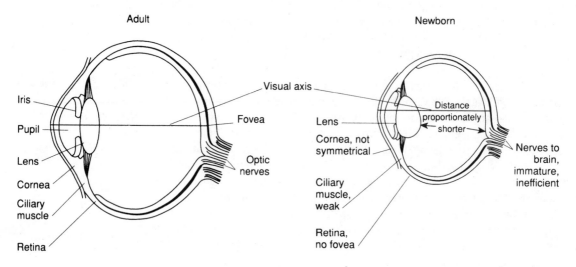

■ *Objective 6.4*

visual acuity, the most common form of assessment, is concerned with the ability
to detect detail in a stationary object, whereas *dynamic visual acuity* refers to the
ability to perceive detail in a moving object.

At birth the neonate can see; however, it is not known with what degree of
clarity. Although at one time it was assumed that the newborn could not see more
than a blur until 3 or 4 weeks of age, researchers now know that soon after birth
a baby can focus on a near object and respond visually to a moving light (Macfarlane, 1977). Visual acuity of newborns and young infants is frequently assessed by measuring contrast sensitivity in discriminating black-white stripe
widths. Allen (1978) reports that at 2 weeks, infants can make out stripes that are
one-eighth inch wide, and by 6 months acuity improves to discriminate stripes
one-fortieth of an inch in width.

Using Snellen values as a reference, the acuity of a newborn has been estimated to be between 20/200 and 20/600, meaning that an individual sees an object
that is 20 feet away as if it were 600 feet away (Haith, 1991). The 20/600 value is
about 30 times lower than adult visual acuity (20/20). By the end of the first year
most infants will acquire acuity levels that are close to the adult range of 20/50 to
20/20 (Dworetsky & Davis, 1989); however, values of 20/100 to 20/200 are also
common. By the age of 3 to 4 years, a more precise assessment of acuity can be
determined through the use of the Snellen method or standardized pictorial techniques. A review of this information along with the previously reported data

suggests that visual acuity gradually increases until the age of 5 years when adult levels (20/20) are attained (Apt & Gaffney, 1982; Lowrey, 1986).

Closely related to visual acuity is *visual accommodation,* or the ability to maintain visual acuity over a range of viewing distances. It was once believed that infant acuity was optimal when objects were about 5 to 10 inches away and decreased when objects were nearer or farther away. Indeed, focusing errors are much greater for the newborn than those of a 3-month-old infant. Interestingly, however, though it would seem logical to expect a 1-month-old infant to see objects more clearly at a distance of 5 to 10 inches, this is not the case. Apparently visual acuity in 1-month-old infants is not related to viewing distance (Salapatek, Benchtold, & Bushnell, 1976). It seems that the infant's *depth of focus,* the distance that an object can be moved without a perceptible change in clarity, is quite large because of other adequacies and keeps even substantial focusing errors from causing an appreciable increase in blurring. As noted earlier, visual accommodation approaches an adult level by 4 months of age.

Although these comparative figures might lead one to speculate that a young infant cannot see very well, it should be noted that in practical terms, infants are far from blind. Banks and Salapatek (1981) suggest that though the young infant's visual "window" is quite limited, it can detect the contours of many common objects. For example, young infants can see the contrast between hair and skin on a parent's face and can detect large objects at close range.

The perceptual component of dynamic visual acuity reflects the ability of the central nervous system to estimate the direction and velocity of an object, and the ability of the ocular-motor system to catch and hold its image on the fovea long enough to permit resolution of its detail. Although different estimates of age-related acuity levels have been reported, it is quite evident that static visual acuity matures before dynamic acuity and that the ability to focus on a moving object is a function of the speed at which the stimulus is moving (i.e., efficiency decreases as speed increases). Dynamic acuity increases gradually during childhood with notable periods of improvement at 5 to 7 years, 9 to 10 years, and 11 to 12 years (Williams, 1983). Efficiency remains relatively stable from 12 years of age to the beginning of middle-age adulthood. Males demonstrate superior abilities in static and dynamic acuity over females at all ages.

As noted earlier, among the primary factors affecting visual acuity are development of the fovea, degree of myelinization, number of neural connections in the visual cortex, shape and structure of the eye, and strength of the ciliary muscles. There is also some indication that the quality of visual acuity coincides with the development of binocular vision (the ability to use both eyes together and fixate on an object), which is evident by 2 months of age.

Perceptual Constancy   Of the various aspects of **perceptual constancy,** the one most closely related to the study of motor behavior is the perception of size constancy. Other constancies include shape and color. When an object becomes known to an infant, it becomes recognizable as that specific object regardless of the angle or distance from which it is observed. The object has "constancy" in that it remains the same form even though the retinal image changes. Hence, the ability to perceive constant size is referred to as perceptual constancy.

Another way of describing constancy is the notion that the perceived size is "scaled." That is, constancy is a scaling process because one unit of measurement is converted, or scaled, into another (perceived size). Road maps typically include scales that transform inches into miles; theoretically, the visual system contains a mechanism that scales an image into object size.

Sekuler and Blake (1994) provide an excellent example of how constancy seems to occur automatically. Hold up both thumbs side by side with nails facing you, approximately 8 inches (20 cm) from your eyes. At this point both thumbs look identical in size, as well they should. Next, extend one arm until the thumb is about 16 inches (40 cm) away. At first glance the two thumbs should continue to appear identical in size; even though the image size of the closer thumb is now twice that of the farther one; this is perceptual constancy.

Another example is perceiving that the airplane on the ground before take-off is the same object that, minutes later, is several thousand feet in the air. The importance of constancy is that it represents perceptual stability and underlies the ability to interpret the environment consistently and accurately.

Bower (1966) reported object size constancy in babies at 6 weeks of age. At later ages, judgment of object constancy is influenced by background cues such as color, texture, and gradients. Evidence indicates that 7-year-olds are somewhat influenced and adults are markedly persuaded by background cues. In general, perceptual constancy gradually improves with age and approaches maturity by 11 years (Collins, 1976).

Spatial Orientation   Perception of **spatial orientation** refers to the ability to recognize an object's orientation or position in three-dimensional space. Closely associated with this visual ability is a component of kinesthetic perception known as *spatial awareness.* Spatial awareness is the internal awareness of the position (orientation) of objects and of the body in three-dimensional space. This discussion will concentrate primarily on the visual perception of objects rather than on aspects of internal awareness of spatial orientation and movement in the external environment.

Depending on the situation, it is important at times to recognize the differences in spatial orientation of objects. Most older children and adults possess the ability to immediately notice these differences, whereas young children do not have this facility. Young children must first learn to recognize the spatial characteristics of objects before they can ignore irrelevant information (Pick, 1979). In some everyday tasks and motor skill activities it is important to recognize that an object is identical even if it is upside-down, rotated, or angled to one side. In the case of recognizing letters such as *d* and *b,* and *p* and *q,* proper identification of spatial orientation is critical. The successful performance of certain target projecting tasks, gymnastic events, and game activities by recognizing boundary lines and designated spaces also necessitates the accurate perception of spatial orientation.

By 3 to 4 years of age, children are aware of most basic spatial "dualisms" such as over/under, high/low, front/back, in/out, top/bottom, and vertical/horizontal. However, until these characteristics become commonplace and perceptually refined in their spatial worlds, children are likely to get things reversed, inverted, and confused. It has also been observed that children appear to master spatial dimensions in a fairly orderly sequence—vertical to horizontal, then to diagonal or oblique (Williams, 1983). Although 3- and 4-year-olds have some ability to distinguish vertical from horizontal, they still lack the ability to effectively perceive mirror images (left-right reversals) and oblique lines and diagonals. By 8 years of age, most children can perceive the spatial orientation of objects, although some individuals continue to have problems with right-left or mirror image reversals.

Figure-Ground Perception   **Figure-ground perception** refers to the ability to distinguish an object from its surrounding background. This ability, as with most meaningful motor behavior tasks, necessitates that the individual be able to concentrate on and give selective attention to a visual stimulus. Examples of figure-ground perception include the ability to track and/or intercept objects as takes place in softball, golf, soccer, and ice hockey. To catch a softball that has been hit into the sky, the outfielder must first focus on and track the ball (figure) against the backdrop (ground) of white sky or lights (if at night).

Figure-ground perception appears to improve rapidly between the ages of 4 and 8 (Temple, Williams, & Bateman, 1979; Williams, 1983). Evidence also suggests that the refinement of figure-ground perception continues to improve between the ages of 13 and 18. Individuals at all ages are more proficient at identifying familiar figures embedded in distracting backgrounds than they are at perceiving unfamiliar stimuli from similar backgrounds.

Whole-Part Perception   The ability to distinguish the parts of an object or a picture from its whole is referred to as **whole-part perception.** Young children generally focus on the entire object or picture, whereas in later development concentration shifts to recognition of the parts that make up the whole. Figure 6.3 shows examples of items used to assess whole-part perception. Using the happy face as an example, a typical 5-year-old child would indicate seeing just the happy face or just the various household items. At 7 or 8 years, children are likely to report seeing both the parts and the face, but not simultaneously. In comparison, the vast majority of 9-year-olds and adults indicate seeing "a happy face made of household items" (Elkind, 1975).

Depth Perception   **Depth perception** is the ability to judge the distance of an object from the self. This perception is an automatic occurrence that takes place every time an individual judges the distance to other vehicles when driving or attempts to shoot a basketball. Reacting quickly to such tasks requires that the individual know where the object is located in three-dimensional space.

Depth perception can be referred to in absolute distance and relative distance. Absolute distance refers to a rather precise judgment of the space from the

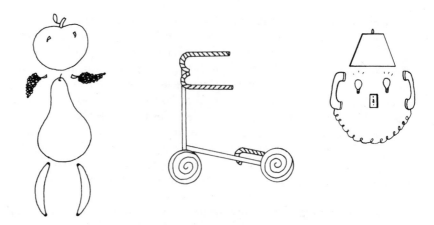

Figure 6.3
Examples of items
used to assess
whole-part
perception.

person to an object. For example, a shortstop attempting to throw out a runner at first base must judge the distance between himself and the base. An infant's attempt to place a cap onto a bottle and an adult reaching for a cup of coffee are other examples of judging absolute distance.

On the other hand, relative distance refers to the estimation of distance between one object and another or between different parts of a single object. Examples include discriminating the distance between two rattles or determining whether an opponent or partner is closer to the goalpost. The ability to judge relative distance is much more precise than the detection of absolute distance.

The perceptual system uses two general types of cues to detect depth: oculomotor cues and visual cues. The development of each of these abilities is tied to independent perceptual systems (i.e., visual and kinesthetic). Oculomotor cues are kinesthetic in nature, meaning that cues are derived from the various kinesthetic receptors located in the muscles and joints. An example of this is evident in the absolute distance task of using the arm and hand to reach for a glass of water. Visual cues, in contrast, provide good relative depth information but might also offer additional information about absolute distance. Information is gathered through visual perception either with binocular cues (using both eyes) or when just one eye is used (monocular cues). The independent development of kinesthesis (kinesthetic sense), monocular vision, and binocular vision, all have developmental implications for the individual's ability to judge distances.

It is generally agreed that depth perception is absent at birth and is relatively rare in 6-week-old infants (Pettersen, Yonas, & Fisch, 1980). Most sources indicate that by the age of 6 months, most children are capable of judging depth with fair accuracy. However, a number of studies of infants as young as 2 months of age suggest that they are aware of and perceive to some degree the spatial dimensions of depth. By dropping objects toward an infant's face (protected by a sheet of glass) to evoke defensive blinking, studies have suggested that some aspects of depth perception are developed by the age of $2^{1}/_{2}$ to 3 months (Pettersen et al., 1980; White, 1971). Supporting evidence for the presence of depth perception in this age range is based on the fact that true binocular vision, a primary component of accurate depth perception, appears at about 2 months, suggesting the basis for refinement of the ability. Binocularity continues to improve during the early childhood years (age 2 to 6) and is closely tied to improvements in depth perception. By the age of 6 or 7 years, children can judge depth as accurately using monocular cues as with binocular (using both eyes).

A simple observation of 4-month-old infants in grasping objects reveals the presence of some degree of depth perception. When an object is placed within grasping range, most infants will attempt to grasp the object, whereas when the item is placed out of reach infants rarely grab for it.

The classic study of depth perception in children involved the use of a "visual cliff" (Gibson & Walk, 1960). The apparatus consisted of a table several feet high covered with a heavy sheet of glass (fig. 6.4). On one side of the table (runway), a checkerboard pattern was fixed just beneath the surface of the glass. On the other side, the same checkerboard pattern was set 3 1/2 feet below the surface of the glass, thus creating a visual cliff. Infants were placed on the shallow side and beckoned by their mothers to cross the glass over the "cliff." The investigators found that infants as young as 6 months of age clearly paid attention to depth cues and would not crawl to their mothers over the deep end even when coaxed.

Although the data suggest that depth perception capabilities are present to some extent in infants as young as 2 months and that improvement proceeds quite rapidly up to 6 months, refinement (especially of absolute distance) continues until approximately age 12 (Lowrey, 1986; Williams, 1983).

Field of Vision   Also known as *peripheral vision,* **field of vision** refers to the entire extent of the environment that can be seen without a change in the fixation of the eye. References to peripheral vision are frequently described in terms of lateral and vertical capabilities. Most animals can visually take in the majority of their surroundings all at once. In contrast, humans have a rather limited field of vision. Humans can compensate for this relative limitation by being able to adjust where the eyes are directed (e.g., looking both ways before driving or cycling through an intersection). According to Sekuler and Blake (1994), normal adult lateral peripheral vision is approximately 90 degrees from straight ahead, hence creating a visual field of about 180 degrees. In contrast, vertical peripheral vision in adults is approximately 47 degrees above and 65 degrees below the visual midline (Sage, 1984).

Information on the development of peripheral vision, especially the vertical aspects, is quite limited. At 7 weeks, lateral peripheral vision is about 35 degrees, roughly 40% of adult capability (Macfarlane, Harris, & Barnes, 1976). According to a review (Cummings et al., 1988), most sources indicate that by the end of the first postnatal year, visual field extension is still immature and that only after age 5 are visual fields equivalent to those of adults. However, more recent research using a light-emitting diode perimeter technique with a relatively small target found no significant differences in visual field between children (2 to 5 years old) and adults (Cummings et al., 1988). A person's lateral field of vision is at its peak at about age 35 (Burg, 1968). Although the role of peripheral vision

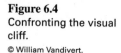

in the motor performance of children is not clear, its importance in sports at the adult level is evident. In activities that involve quick responses and numerous visual cues (e.g., basketball, tennis, hockey, soccer), central and lateral peripheral vision are essential for spatial orientation.

Perception of Movement   One of the most important and complex perceptual abilities related to motor behavior is the detection, tracking, and interception of moving objects. Many more demands are placed on the perceptual system when objects begin to move in space and individuals are confronted with dynamic rather than static conditions. Just a few of the difficulties are evident when one considers the differences between static (stationary) objects and dynamic visual acuity, and the perception of depth in reference to moving (dynamic) objects, in comparison to those that are stationary.

In addition to the functional aspects of these perceptual abilities, the developmental dynamics of eye movement changes are also factors that influence the ability to perceive moving objects. *Saccadic eye movements* are rapid movements between one point of visual fixation (i.e., the process of focusing one's gaze on something) and another. These movements are slower and more numerous in young infants than in older children and adults. At maturity (approximately 12 years), a saccade moves the eye about 90% of the distance to the visual target, but in infants several saccades are required to reach their visual target. An infant's eyes also move less smoothly in tracking a moving object and will tend to refixate frequently, or focus first on one spot and then another. It is difficult for the newborn to focus both eyes on the same point (convergence) and so to see only one stimulus.

By 48 hours after birth, most infants can track a very slow moving object, although this response is uneven and focus is easily lost (Haith, 1966). Smoother, more accurate tracking is evident by 2 months, and by 4 months the more complex task of tracking and predicting the path of a slow-moving object is within the capabilities of most infants (Von Hofsten, 1979, 1980). By 4 to 5 months the infant has reached an unrefined visual-motor stage, exhibiting approaching movements with the upper body as if attempting to receive the object.

A comprehensive review of the subject by Williams (1983), which includes much of the researcher's own work, suggests that the ability to accurately perceive moving objects continues to develop through the childhood years until relative maturity at approximately 12 years of age. However, between the ages of 6 and 11 years, a number of developmental trends have been observed. When asked to judge the speed and direction and interception point of a moving object, children 9 to 11 years of age judged the flight of the moving object with significantly more accuracy than the younger children. Although the younger children generally responded quickly to the object, they were not able to effectively integrate the available sensory information about its flight path with their motor behavior. It does appear that the upper elementary school years is a period of transition during which the ability to integrate the visual and motor mechanisms increases. By the sixth grade, children make judgments much faster and more accurately about moving objects than do younger children. This behavior suggests that the visual-motor processes are nearing maturity.

Coincident timing, another aspect of tracking and intercepting objects, also appears to improve with age. Coincident timing tasks are commonly used in sport skill activities that involve the ability to coordinate visual and motor behavior to a single coincident point (interception) such as in catching a ball, kicking a moving ball, and passing a ball to a running player. An important determinant in coincident timing performance is the relative direction of the moving object. In general, the ability of the individual to time movements to an interception point with a moving object improves with age up to young adulthood (e.g., Fischman, Moore, & Steele, 1992; Keogh & Sugden, 1985). And, as might be expected, accuracy tends to decrease as velocity increases. Proficiency is generally achieved during the early teenage years. Other than the fact that coincident timing abilities seem to improve with age, the specific developmental mechanisms involved are not well understood. This ability is an excellent example of "complete" information processing; much more is involved than the simple perception of moving stimuli. Coincident timing involves a very complex interplay between visual-perception proficiency and motor control. A more complete treatment of visual-motor abilities is provided in the discussion of motor performance.

Visual-Motor Coordination   The ability to coordinate visual abilities with movements of the body is termed **visual-motor coordination.** This aspect of movement combines both visual and kinesthetic perceptions with the ability to make controlled and coordinated bodily movements usually involving eye-hand or eye-foot integration. Development of these actions generally follows the proximodistal (midline to periphery), cephalocaudal (head to toes), and gross-to-fine motor order. In fine motor tasks such as lacing shoes or writing, the small muscles of the hand or foot are synchronized and, in turn, are coordinated with

Table 6.1   Selected Fine Visual-Motor Characteristics of Children 3 to 8 Years Old

Age (Yr)					
3	4	5	6	7	8
Uses hand constructively to direct visual responses	Visual manipulation; does not need support of hands	Understands horizontal and vertical concepts	Copying tasks come easily	Prefers pencil over crayons	Can copy a diamond shape
Skillful manipulations	Discoveries made in depth perception	Copies squares and triangles	Fair ability to print	Ability to make uniform size letters, numbers, etc.	Attempts cursive writing
Draws lines with more control	Laces shoes	Increase in fine finger control		Drawing of person more accurate	Uniformity in alignment of letters
Can copy circles and crosses	Buttons large buttons	Colors within the lines			
Stacks 1-inch cubes	Orients movements from center of periphery	Fairly accurate cutting abilities			
Writing utensils handled like adults	Draws a recognizable picture with some detail	Draws a recognizable person			
Strings beads					
Cuts with scissors					

vision. Table 6.1 presents a selection of fine visual-motor behaviors characteristically exhibited by children during the developmental period of 3 to 8 years of age. The development of gross-motor abilities requiring visual-motor coordination will be discussed in subsequent sections.

KINESTHETIC PERCEPTION

■ *Objective 6.5*

Often referred to as the "sixth sense," kinesthetic perception is a comprehensive term encompassing the awareness of movement and body position. The word **kinesthesis** derives from two Greek words meaning "to move" and "sensation." It involves the ability to discriminate positions and movements of body parts based on information that derives from the individual's internal environment. Unlike the visual, auditory, and tactile perceptual modalities that receive sensations from outside the body, the kinesthetic system is supplied with sensory input from receptors located in muscles, tendons, joints, and the vestibular (balance) system. Although kinesthetic information is not the exclusive determinant of the way movement patterns are organized, it does provide body and movement knowledge without reference to visual and verbal cues. This modality is basic to all movement and, along with the visual modality, dominates the learning and acquisition of motor skills.

In spite of several years of inquiry on the subject, kinesthesis is not fully understood either physiologically or psychologically. Its complexity is due, in part, to the fact that kinesthetic perception is not a single sensory system. Rather, the kinesthetic system consists of several different sensory receptors.

Structure and Function of the Kinesthetic System

The group of kinesthetic receptors are associated with the *somatosensory system* handling sensations from the body. Somatic receptors are located in the skin, muscles, joints, and the vestibular apparatus of the inner ear. Somatic receptors may be further subgrouped into cutaneous receptors and *proprioceptors*. Cutaneous receptors are located near the skin and perceive touch, pressure, temperature, and pain. Sensations picked up by these receptors derive primarily from outside the body and do not appear to play a primary role in motor behavior. However, because there is such a strong relationship between the two perceptions, they are often referred to as tactile-kinesthetic perception. Collectively the proprioceptors of the muscles, joints, tendons, and inner ear form the perceptions known as kinesthesis (kinesthetic perception). All are sensitive to stimuli arising from movement. Insufficient information is available to clearly document the developmental aspects of the neuromuscular mechanisms involved, but the location and function of these receptors is described in the following paragraphs.

■ *Objective 6.6*

Vestibular Apparatus   Located in the membranous labyrinth of the inner ear, the vestibular apparatus is made up of two kinds of receptors: the *semicircular canals* and *otolith organs*. The function of the vestibular apparatus is to provide information to the brain about the movements of the head. The saccule and utricle are housed in the otolith organ of each ear. They signal information about the orientation of the head in regard to linear acceleration, or line of gravity, such as that sent when your car accelerates or decelerates or when you go up or down an elevator. If it feels as though your head is spinning, information about the rate and direction of spin is transmitted from the macula (the neurosensory portion of the otolith organ) via stimulation of hair cells to the brain.

The semicircular canals, oriented in each of the major planes (frontal, saggital, horizontal), are most sensitive to angular acceleration, or change in directions of movement and rotation. The crista (the neurosensory portion of the canals) transmits information to the brain as a result of changes in the flow of canal fluid. Although primarily related to the perception of balance, these structures also provide critical information about accelerations and forces applied to the head (e.g., while doing a cartwheel or forward roll). The vestibular system is also important in maintaining visual fixation and contributing to some aspects of reflexive behavior.

Muscle Spindle Receptors   The muscle spindle receptors are cigar-shaped structures located between the fibers of the main muscles of the body. Because they are parallel in structure with the muscles, they stretch when the main muscle is stretched. The muscle spindle acts as both an auxiliary motor unit and sensory organ to gauge the degree of tension in a muscle.

Joint Receptors   Embedded within the joint capsules of the various limbs are different types of receptor cells known as the joint receptors. These receptors provide a variety of sensory information about the joint to higher neural centers. Information is sent with regard to static position; onset, duration, and range of movement; velocity and acceleration; pressure and tension, and joint torque; and pain.

Golgi Tendon Organs   These receptors are located in the muscles adjacent to the musculo-tendinous junction, where the muscle merges with the tendon. Like the muscle spindles, these receptors provide information about muscle tension but with specific reference to the force-producing contractile elements surrounding muscle-tendon junctions.

As noted in chapter 2, spinal tracts are the major pathways for transmitting signals from these receptors via the periphery to the specific brain interpretation centers. Spinal tracts are located alongside the vertebrae that make up the spinal column. Input to the CNS, except for kinesthetic input from the structures in the head (vestibular apparatus), is through roots (nerve bundles) that collect and guide the input to the spinal cord at each segment; each segment serves a specific region of the body. One of the most important concepts concerning kinesthesis is the understanding that any one of the kinesthetic receptors in isolation from the others is generally ineffective in signaling information about the movements of the body (Schmidt, 1988). The various receptors are often sensitive to a variety of aspects of body motion at the same time.

Development of Kinesthetic Perception

■ *Objective 6.7*

The developmental aspects of kinesthetic perception may be measured and described in terms of their basic physiological qualities and in their applied context involving knowledge of the body and movement awareness.

Keogh and Sugden (1985) discuss what may be described as two basic physiological aspects of kinesthetic perception: *kinesthetic acuity* and *kinesthetic memory*. **Kinesthetic (discrimination) acuity** refers to the ability to proprioceptively detect differences or match qualities such as location, distance, weight, force, speed, and acceleration. Examples include the ability of an individual to discriminate the difference between two objects of slightly different weights when placed in the hands while blindfolded (weight discrimination) and the ability to detect when blindfolded which arm has been raised slightly higher than the other (location).

If the stimulus is removed after presentation, then **kinesthetic memory** is involved. Tests of kinesthetic memory usually involve a reproduction of movements. For example, while blindfolded an individual is asked to move one hand to a distance of choice or at a measured speed and, after a pause, to reproduce the movement. In another more passive variation, the experimenter would move the subject's limb and then ask the subject to reproduce the movement. Less motor information is available to the subject in the passive situation.

A review of the studies on this topic suggests that kinesthetic acuity improves faster and earlier than kinesthetic memory. The ability to discriminate kinesthetic information approaches adult levels by age 8, whereas the mature stage of kinesthetic memory is generally not reached until after 12 years of age. Although very little information is available on young children, research findings note substantial yearly decreases in performance errors after age 5 up to the age of 8.

Several different descriptors may be found in the literature concerning the more applied developmental aspects of kinesthetic perception. Some sources use single descriptors such as body knowledge, body awareness, or spatial awareness, whereas others have identified several subcategories. All of these terms have been used to identify movement awareness (both cognitive and motor) abilities that have theoretically been associated with kinesthesis and one's internal awareness of various movement parameters. The following are basic movement awareness components related to kinesthetic perception: body awareness, spatial awareness, directional awareness, vestibular awareness, and temporal awareness. These awarenesses are closely interrelated and in most movement conditions are inseparable. The components have been identified to aid in categorizing the various kinesthetic abilities based on the notion that presenting the "parts" to such a complex process will produce a better understanding of the "whole."

■ *Objective 6.8*

Body Awareness   Sometimes referred to as body concept, body knowledge, or body schema, **body awareness** in this context involves an awareness of body parts by name and location, their relationship to each other, and their capabilities and limitations. As children increase in age, they become increasingly more cognizant of their body parts in reference to the aspects described. With this basic movement knowledge they gradually improve their ability to perform desired movements efficiently.

The knowledge of body parts and their functions is one of the most basic aspects of kinesthesis and one of the more researched topics. At birth the newborn is not capable of true conscious body awareness; however, during the first month a relatively crude awareness appears to emerge that the body is distinct from the surroundings. With growth, the infant then becomes increasingly cognizant of the capabilities of various body parts in moving the eyes, head, trunk, arms, and legs. By the end of the first year or so (12 to 15 months), correct reactions to verbalizations of body parts become more evident. Even though infants at this age cannot identify specific parts verbally, they are capable of indicating a few body parts (tummy, nose, ears, eyes, feet) correctly when given verbal cues by another person. They can do this first on their own bodies and then later on other people. As their body-part vocabulary increases, so does their accuracy at physically touching and verbally identifying specific body parts. At age 5 or 6, the majority of children can accurately identify the major parts, and by 7 years the ability to distinguish the minor parts (ring finger, wrists, heels, ankle, elbows) has also been mastered (Williams, 1983).

Though it may not be surprising, the ability to identify various body parts depends heavily on both conceptual and language abilities, as well as other sensory perceptions. For example, a response to a verbal command such as "Where is your left ankle?" involves both auditory memory and the ability to plan a motor response. If a series of responses to such commands such as "Touch your right elbow to your left knee" is required, the demands of the situation are significantly increased. It would seem reasonable to assume that children who have been tutored in body-part identification would show superior abilities in comparison to the norm described. In an unpublished study by Spoinnek and reported in Cratty (1986), it was shown that by 5 years of age children can be taught to correctly identify their left and right body parts; this is about two years before such accomplishments are normally expected.

Spatial Awareness As noted earlier, the spatial awareness aspect of kinesthetic perception interplays with the visual perception of spatial orientation. Both perceptions involve the ability to recognize objects in three-dimensional space. The two differ in that **spatial awareness** involves the ability to draw inferences in relationship to self space or position *as well as* object recognition. In essence, it is the sense of the location of one's own body in space in relationship to the environment. This aspect of kinesthesis has also been viewed in the developmental psychology literature as the general sense of direction required for spatial mobility.

Spatial awareness may be described in terms of *egocentric localization* and *objective localization*. Egocentric localization refers to the ability to locate objects in space in reference to the self, which is characteristic of younger children. The objective form of this awareness generally follows egocentric perception and is the ability to locate objects using something other than the self as a reference. Such behavior is exhibited by older individuals who can locate an object relative to its nearness to other objects. Knowledge of the spatial environment and of the body is the information necessary for an individual to formulate a motor program (plan) and project the body effectively.

Before individuals can effectively project their bodies into space they must have a sense of where objects are located. The ability to make such perceptions is based somewhat on the frame of reference that is used. This general "sense of direction," as it is sometimes referred to, may include several perceptions in addition to kinesthesis—visual, auditory, tactile, and even smell (olfactory). For example, an individual might perceive or learn the location of an object relative to him- or herself (e.g., second base is directly in front of me when I am batting) to demonstrate the use of an egocentric frame of reference, or egocentric localization. If an objective frame of reference or objective localization were in use, the individual might learn the location of second base relative to various landmarks in the area (e.g., second base is between first and third; or second base is next to shortstop). The strategy used in objective localization is quite similar to that required in map reading and orienteering.

Current research findings suggest that adults have the ability to use and exercise both egocentric and objective frames of reference, depending on the situation (Mandler, 1983). For example, an adult may learn a new pathway through a mountain bike course in an egocentric way ("Go to the first sign and turn right, then go straight ahead until you get to the fork and turn left, etc."). In other situations, however, the cyclist may use objective localization in checking out the possibility of a shortcut on the second run or in looking at a map and then using his or her sense of direction to proceed. Apparently, young infants under 12 months of age do not have the ability to use objective frames of reference and are restricted to egocentric orientation (Acredolo, 1978; Piaget, 1971). Objective localization appears sometime between the ages of 12 and 16 months and seemingly is used with increasing frequency through childhood. In a study of children ages 6 to 12 years, Long and Looft (1972) note that 7-year-olds can describe an object as "between" or "in the middle" of two other objects.

One interesting study (Thomas et al., 1983) investigated the strategies that children 4 and 9 years of age used to recall spatial locations and distances in their environment while jogging. As one might expect, the older children recalled both spatial location and distance better than the younger children. However, both groups had difficulty in recalling distance even after cues were presented. Approximately 30% of the 9-year-olds used a strategy like counting steps to judge distance, whereas the younger group in general did not exhibit such maturity. These researchers also found that children 5 to 12 years of age can be taught to effectively use a strategy to recall jogging distances. Interestingly, tutored 5-year-olds were as good as 9-year-olds and only slightly less proficient than 12-year-olds who were not provided with instruction.

Directional Awareness   In theoretical terms, **directional awareness** is that aspect of kinesthetic perception assumed to be an extension of body and spatial awareness. Directional awareness refers to the conscious internal awareness of two sides of the body (*laterality*) and the ability to identify various dimensions of external space and project the body within those dimensions (*directionality*). The term *lateral preference,* which involves hemispheric control and preferential use of one eye, hand, or foot over the other, is also associated with directional awareness and will be discussed in chapters 8 and 9 in reference to asymmetries.

Laterality is the basis for directional awareness and the ability to identify the dimensions of space and project the body purposefully through those dimensions. The internal awareness that the body has two sides and that those sides can be differentiated as "right" or "left" is thought by many to be the essential foundation on which the delineation of other spatial dimensions are derived. The first observations of laterality may be noted quite early in life. During the first six months of postnatal life, babies exhibit this perceptual distinction by using their hands independently and in a coordinated manner. For example, they may use one hand to hold a cube and the other to explore it, then interchange the function, or even use both hands. Seemingly, babies recognize that they have two limbs that can perform independent functions or can work together.

Refinement of this awareness continues with other body parts as the child realizes that, though a number of these parts come in pairs, each part of the pair can be positioned and moved independently. This awareness continues to improve steadily until approximately 4 to 5 years of age when the understanding that the body has two distinct sides is well developed (Cacoursiere-Paige, 1974; Hecean & deAjuriaguerra, 1964). Around this same time period, a more advanced form of laterality begins to emerge—the ability to label specific body parts as right or left (i.e., left-right discrimination). As this aspect of laterality develops the child clearly moves from the perceptual-motor phase of development to the conceptual-cognitive phase. Children at this age may realize that right and left are opposite sides of the body, but they may not perceive which is which or will oftentimes reverse the two. Although there is a wide variance among individuals, children younger than age 5 are likely to give correct responses to right-left discrimination questions only by chance. By the age of 8 years most children can respond accurately to questions regarding the left-right discrimination of specific body parts (Long & Looft, 1972).

Because the ability to discriminate left-right body parts on another person requires mirror-image perception, it generally develops later than the ability to label one's own body parts. Close to perfect accuracy is achieved in general discrimination by age 10 (Williams, 1983). Due to the cognitive nature of kinesthesis, children as young as 5 years of age can be taught to correctly discriminate left-right body parts.

From a theoretical perspective, *directionality* is the motoric expression of laterality and the perception of spatial orientation. Along with the understanding that the body has two distinct sides that work independently and in unison, the identification of the dimensions of external space are prerequisite to purposeful movement in the environment. In the discussion of spatial orientation we learned that by 3 to 4 years of age children, although frequently confused, are aware of the most basic spatial dualisms (also referred to as *locatives*) such as over/under, high/low, front/back, in/out, top/bottom, and vertical/horizontal. It was also noted that children appear to master spatial dimensions in a fairly orderly sequence: from the vertical to the horizontal, to the diagonal or oblique. At this age children still lack the ability to effectively perceive mirror images and oblique lines and diagonals. Although by age 8 most children have the basic ability to effectively perceive the spatial orientation of objects, some individuals continue to have problems with right-left or mirror-image reversals.

The perception of objects and spatial dimensions are basic prerequisites to moving purposefully within the environment, but the ability to discriminate and move correctly in response to verbal cues concerning spatial dimensions is a much more complex function. Although few studies on younger children have been conducted, evidence does suggest that the basic dimensional concepts are acquired and used in the following sequence: up-down, front-back, and finally, side (Kuczaj & Maratsos, 1975).

By 3 years of age a high percentage of children can position an object in front of or behind their bodies, but a great many of them experience difficulty placing objects in front or behind another object. By a year later (age 4), most children can effectively identify front and back on objects external to the body and can perform the more difficult task of placing an object "to the side." It is important to understand that positional relationships between objects and oneself, using the self as a reference point, appear to develop prior to positional interrelationships among objects in space. In other words, in most instances children must become aware of the concept in relation to themselves before they can detect the same qualities in objects external to the body.

Remember that although most 5-year-olds realize that left and right are distinct and opposite sides of the body, it is not until about the age of 8 that children respond accurately to questions regarding the left-right discrimination of their specific body parts. Refinement of directionality is not complete until this perceptual awareness has been acquired.

Along with their findings on left-right discrimination, data from Long and Looft (1972) also suggest that a linear trend exists in the development of directionality from ages 6 to 12. The researchers found that at age 6 children are aware of the various dimensional aspects of "up-down" and are able to imitate many movements made by another person. Seven-year-olds exhibit some of the first overt uses of objective localization and directionality by describing objects as between or in the middle of two other objects. By age 8 children begin to relate their understanding of left-right to other objects and move their bodies based on that information (e.g., the ability to say "The boundary line is on my right"). This type of action reveals the first signs of true movement-oriented directionality.

Approximately a year later (age 9), children exhibit more frequent use of objective localization and will base their answers on objective directional relationships without reference to self (e.g., "The ball was hit to the left of the second baseman"). Refinement continues to about age 12, when directionality approximates adult behavior. Some mature characteristics are present as evidenced when the child begins to use a natural reference system (e.g., "The sun rises in the east").

Vestibular Awareness   The successful performance of virtually all motor skills depends on the individual's ability to establish and maintain *equilibrium* (balance). The general description for this component is called **vestibular awareness.** The newborn must establish equilibrium before any of the early developmental milestones of maintaining head position, sitting, and standing can be achieved. Remember that the vestibular apparatus is the part of the kinesthetic modality that provides the individual with information about the body's relationship to gravity and its head position. In a general sense balance, which depends primarily on information from the vestibular apparatus, refers to the ability to maintain equilibrium in relation to the force of gravity. Regardless of how the term is defined, it is clear that balance involves the successful integration of several anatomical and neurological functions (e.g., the skeletal, muscular, sensory, and motor systems).

Typically balance is subdivided into three types: postural, static, and dynamic. *Postural balance* refers to the relatively unconscious level of functioning of basic reflexes that enables one to maintain upright posture, hold the head erect, sit, and stand. *Static balance* is the ability to maintain a desired body posture when the body is relatively stationary. Common testing procedures include standing on one foot and balancing on a stabilometer or balance board. *Dynamic balance* refers to the ability of the body to maintain and control posture while in motion. This type of equilibrium is employed in numerous motor skill activities and is usually evaluated based on the ability to walk on balance beams of different widths and heights. Although these definitions may appear somewhat simple and relatively clear, balance tasks are quite complex in nature and the assessment of this ability at any age is related to the specific task used to measure it. Thus balance is better described as a set of specific characteristics or components that describe the task to be performed rather than as a single, unitary ability.

Although the exact relationship of age to balance ability is unclear, most investigators indicate that balance performance generally improves with age into the adolescent years (Ulrich & Ulrich, 1985; Williams, 1983). However, specificity of the task appears to be a major determinant in the exact pattern of change. For example, improvement for a group of children on some tasks may be significant and steady over a number of years; performance on other tasks may change very little, not at all, or may even decline. A review of several studies also indicates that across a wide range of ages and tasks there are few, if any, differences between males and females in general balance performance. However, there have been observations of female superiority on specific static balance tasks such as the stabilometer (DeOreo & Wade, 1971; Winterhalter, 1974). Table 6.2 shows selected balance abilities of young children in approximate developmental order.

A large number of investigators have utilized the task of walking across a balance beam to observe balancing ability. DeOreo (1974, 1975) identified two distinct developmental movement patterns used by young children (3 to 5 years old): the mark-time (shuffle-step) pattern and the alternate-step pattern. The task was to walk forward and backward on beams of varying widths.

The researcher found that approximately one-fourth of the 3-year-olds were unable to walk across the beam at all. Younger children more often demonstrated the less mature mark-time pattern, keeping one foot in front at all times and simply shuffling across the beam. More 5-year-olds used the more mature alternate-step pattern like that used in regular walking. As one would expect, as the width on the beam decreased the task became more difficult for all the children.

The ability to maintain and control equilibrium is widely acknowledged as a fundamental motor component. It is included in almost all motor assessment

Table 6.2 Selected Balance Abilities

Type of Balance	Task	Approximate Age
Static	Balances on one foot (3–4 sec)	3 yr
	Balances on one foot (10 sec)	4 yr
	Supports body in basic inverted position	6 yr
Dynamic	Walks on straight line (1 in.)	3 yr
	Walks on beam using alternating steps (4 in. wide)	3 yr
	Walks on circular pattern (1-in. line)	4 yr
	Walks on beam using alternating steps (2–3 in. wide)	4½ yr
	Hops proficiently (traveling)	6 yr

batteries, especially those that focus on the abilities of young children. In fact, some evidence suggests strongly that the balance performance of young children is more closely related to maturational level than to chronological age (Erbaugh, 1984). Other indications are that balance ability is a good predictor of overall fundamental gross motor skill performance in children 3 to 5 years of age (Ulrich & Ulrich, 1985). Further discussion of balance ability is presented in the chapter on adult motor performance.

Rhythmic (Temporal) Awareness All motor activities consist of spatial-temporal characteristics (i.e., movement takes place in both space and time). Coincident timing, one aspect of temporal awareness that has already been discussed, involves the ability to gauge one's position relative to the changing position of an object or another person. Another aspect of the internal awareness of time is *rhythmic awareness*. As with other kinesthetic perceptions, considerable interplay exists among the various aspects of temporal awareness.

Rhythmic awareness refers to creating or maintaining a temporal pattern within a set of movements. The temporal pattern may be initiated from within by the individual or may be matched with some external pattern. Even young infants seem to be born with the tendency to make rhythmic movements with parts of their bodies. These movements appear within a predictable time during the first year and seem to be triggered by an internal awareness (drummer) rather than in reaction to any observable external stimulation (Thelen, 1979). Examples of rhythmic awareness are found in such activities as keeping time to music, tapping the hands or feet to stay in rhythm with a sound or light, reproducing a pattern from memory, and creating a rhythmic beat or pattern without external stimulation. Two major differences between this aspect of temporal awareness and coincident timing are evident. Coincident timing occurs primarily in unpredictable situations (i.e., with moving stimuli) and requires visual information for effective completion.

Very little information is currently available on the ability of an individual to initiate rhythmic movements without an external stimulus. The absence of criteria by which to make judgments and the strong possibility that most rhythmic responses are not original but are derived from past experiences or memory combine to make this a difficult question to answer experimentally. However, there have been several reports related to the ability to keep time. One study of particular interest observed the ability of children (2 to 5 years) and adults to keep time to piano music played at several different tempos when walking and beating time with their hands (Jersild & Bienstock, 1935). The accuracy rate of the 5-year-olds about doubled that of the 2-year-olds (which was about 21%), and the adults' values were almost double those of the 5-year-olds. These data suggest that children between 2 and 5 years of age improve considerably in their ability to keep time to a rhythmical stimulus and that periods of improvement continue up to adulthood. Another interesting note is that better accuracy scores were recorded at faster rather than slower tempos. This type of behavior has also been observed in coincident timing tasks. One could speculate from these observations that the body's rhythmic mechanism responds more effectively to an external stimulus in which the tempo falls within a range of quite fast and quite slow.

Keogh and Sugden (1985) concluded from their review of this topic that the ability to keep time to an external stimulus improves significantly around age 6 and continues at a rather steady rate up to adolescence or early adulthood. Although very little information is currently available, it has also been suggested that young females may do somewhat better than males in keeping time. In more recent years, researchers have investigated both temporal accuracy and amplitude of movement (spatial accuracy). For example, a study in this area might look at the accuracy of individuals in moving their hands or bodies within a designated pattern at a prescribed tempo or beat. Smoll (1974) found steady improvements in this type of motor behavior for children 5 to 11 years of age. Cratty (1986) notes that studies indicate that a plateau effect occurs at about the age of 10 and that sex differences are usually not apparent.

AUDITORY PERCEPTION

■ *Objective 6.9*

Although generally not as vital to motor behavior as visual and kinesthetic information, the importance of auditory cues are obvious. Very few sport activities are learned or carried out without auditory information and verbal cues. As with all the perceptual modalities, auditory perception derives from a rather complex system that involves the ability to detect, discriminate, associate, and interpret auditory stimuli. In a general educational learning context, commonly addressed aspects of auditory perception include auditory localization, auditory discrimination, and auditory figure-ground perception. What follows is a brief discussion of the basic structures and functions associated with auditory perception and the general developmental characteristics that take place to maturity.

Structure and Function of the Auditory System

The portion of the ear primarily responsible for sound detection is the *cochlea,* which is located in the inner ear. It is the inner ear that receives auditory sensations and transmits them to the brain for interpretation. Of the major parts of the hearing apparatus that includes the external, middle, and inner ears, the inner ear develops first. Its essential gross parts are well defined by the second and third prenatal month. At birth, the growth of the bony structures of the ear are near completion and change little thereafter. The middle ear, which collects sound waves and processes them to the inner ear, is practically of adult size at birth, although the drum membrane is usually smaller and more oblique (Lowrey, 1986). The external ear (canal) consists primarily of cartilage at birth and is much shorter than in the adult. At birth, both the external canal and eustachian tube are shorter than the adult structures.

An infant's sense of hearing is functional prior to birth (Acredolo & Hake, 1982; Birnholz & Benacerraf, 1983). Immediately following birth, however, hearing is usually impaired because the auditory canals are filled with fluid. Normal hearing function is usually restored within a few days and hearing improves rapidly. Young infants have been observed at birth to turn their eyes in the direction of the source of sound, even though they seem unable to locate the exact position from where the stimulus originated. Although the ability is unrefined, newborns are able to make crude discriminations among different pitches. Eisenburg (1976) notes that low-pitch sounds have a calming effect on newborns, whereas high pitches tend to elicit stress reactions. Shortly after birth, the neonate can differentiate sounds of different duration, which is critical to processing spoken language. After four months most new sounds (as opposed to ambient sounds) stimulate searching activities in the infant, an aspect of auditory localization. These activities may include head turning, reaching, and movement toward the sound. Some evidence also suggests that by 6 months of age, infants are almost as sensitive to sounds as adults (Berg & Smith, 1983).

The ability to localize distant auditory stimuli occurs around 12 months and continues to improve so that by age 3, children can localize, or determine, the general direction of sounds quite effectively. The ability to discriminate auditory sounds also continues into childhood and with significant levels of improvement by 8 to 10 years of age and refinement extending to at least the early teens. Little information is available on the developmental characteristics of auditory figure-ground perception, other than the observation that some children have more difficulty than others separating irrelevant stimuli (background) from the primary task (figure). As with the other aspects of auditory perception, auditory figure-ground perception is closely associated with the ability to selectively attend to the stimuli in the environment.

In summary, although mastery of basic auditory skills is evident by the age of 3 years, refinement continues steadily until approximately 13 years of age when near adult levels are reached.

Development of Auditory Perception

Tactile perception (touch) refers to the ability to detect and interpret sensory information cutaneously (of or on the skin). Most textbooks give very little if any attention to tactile perception, perhaps because relatively little information is available on its developmental characteristics and because most aspects of this ability are not closely related to the study of motor behavior. Most information on tactile

TACTILE PERCEPTION

■ *Objective 6.10*

perception discusses the individual's ability to perceive pain, light, and temperature differences. However, from a motor behavior perspective, several considerations warrant at least a brief treatment. As we learned in an earlier section, the first phase of motor behavior is the reflexive phase. Several of these initial movements (e.g., grasping, rooting, Babinski-foot reflexes) are elicited through touch stimulation and dependent on the infant's ability to detect tactile sensations. In conjunction with kinesthetic perception, often referred to as the tactile-kinesthetic system, the individual receives vital information in a movement context in regard to the position of the body in space. For example, a competitive gymnast walking across a balance beam relies primarily on tactile input through the feet and depends relatively little on visual or auditory information. Individuals who are blind depend heavily on tactile input to get feedback for organizing their movements.

Many different neural receptors contribute to tactile perception—receptors for heat, cold, pain, pressure, and vibration. Tactile sensations are picked up by numerous receptors and are transmitted to different areas of the brain for interpretation. With growth, increasing numbers of receptors are found in the skin over a wider area and in closer proximity.

The first responses to touch are elicited in the facial area of the fetus, specifically the lips. The general trend of sensitivity to tactile stimulation during the prenatal stage follows a cephalocaudal progression (i.e., head, trunk, limb). The first prenatal reflex, elicited by tactile stimulation around the mouth, is opposite-side neck flexion at about seven and a half weeks after conception.

Newborns are sensitive to mild and strong tactile stimulation. The most sensitive parts at this time are the lips, eyelashes, soles of the feet, skin of the forehead, and the mucous membrane of the nose. The available information suggests that the development of touch discrimination and localization, as determined by a one-point tactile stimulus, is well developed by 5 years of age (Ayres, 1978; Williams, Temple, & Bateman, 1979). The ability to discriminate and localize more than one point on the skin placed either simultaneously or sequentially gets more refined up to the age of $7\frac{1}{2}$ years. Ayres (1966, 1978) found that only 50% of 5-year-olds sampled were able to consistently perceive multiple-point contacts. By 8 years of age, children are able to consistently reproduce designs that have been drawn on one of their body parts such as the back of their hand. Interestingly, evidence also suggests that infant and adult females tend to be more sensitive to touch than males (Pick & Pick, 1970).

One of the most-researched aspects of tactile perception studies the ability to recognize objects (e.g., letters, shapes) through tactile manipulation. Observations indicate that young children generally go from tactile to visual recognition of objects, suggesting that tactile perception may develop before the ability to visually identify objects (Van Duyne, 1973). In the context of learning, this type of behavior is quite evident in infants and their first responses to something hot, cold, or sharp. The infant learns by touch, places this in memory, and from that time is able to recognize the characteristics of similar items by sight (e.g., fire, a red stove burner). After the age of 4, however, visual recognition of objects is consistently superior to the use of tactile manipulation.

The discussion thus far has focused on the perceptual modalities as independent systems. The description of developmental characteristics and improvements within individual sensory systems is referred to as *intrasensory development*. The perceptual and perceptual-motor processes, however, frequently involve the simultaneous use of more than one system (i.e., *intersensory*). This process, known as **perceptual integration,** includes the translation of information from one sensory modality into another. This type of functioning is commonplace in motor skill settings. For example, a skill such as hitting a baseball requires the interaction of visual, tactile, and kinesthetic information. The normal academic task of learning to read aloud requires vision and auditory integration.

It is generally acknowledged that intersensory perceptual integration is at least partially functional at birth and improves with increasing age beyond the childhood years. Evidence suggests that there are three basic levels of intersensory integration (partially from Williams, 1983). The first level involves the automatic integration of basic sensory information as a function of the subcortical brain. This level is inherent to nervous system processing and is functional at birth or shortly thereafter.

The second level involves the recognition of a specific stimulus or the features of a stimulus as the same or equivalent when they are presented to two different perceptual modalities. For example, after being allowed to manipulate an object tactually (without use of vision), a child can recognize the object as being the same item when permitted to use the visual modality. This level of functioning is often referred to as *cross-modal equivalence.* Although serious methodological and design limitations have kept researchers from adequately describing the mechanisms involved, evidence clearly indicates that newborns can recognize that information gathered in one sensory modality is equivalent to that collected in another (Meltzoff & Moore, 1983).

The third level of intersensory integration, frequently described as the *cross-modal concept,* involves the sophisticated transfer of concepts across perceptual modalities. This function requires the interrelating of certain cognitive dimensions and may be used to solve problems that are associated with dissimilar yet related sensory input that has been presented through two different modalities. An example is the ability to determine whether an object seen and an object heard are associated with the same general category of objects (e.g., animals). This type of behavior is usually found in children 5 years of age and older. Unfortunately, much of the evidence regarding the development of this process is contradictory and ambiguous and thus offers little support for a meaningful discussion. One of the difficulties in interpreting any improvements in intersensory development is isolating improvements in intrasensory functioning, which may help improve perceptual integration. Researchers have also found it very difficult to explain how intersensory functioning affects the development of motor behavior. Despite these and other related problems, perceptual integration generally appears to increase with age through childhood.

PERCEPTUAL INTEGRATION

■ *Objective 6.11*

Visual-Auditory Integration

Although quite unrefined, visual and auditory perceptual abilities appear to be integrated at birth. A rather common observation upon which many research studies have been developed is that infants will orient their eyes in the direction from which sounds originate. Muir and Field (1979) found that even newborns exhibit this behavior in a rudimentary way. A tape-recorder rattle sounded from one of two speakers placed on each side of infants only 2 to 4 days old. Approximately 75% of the time, the infants turned their heads in the direction of the sound. By 4 months of age, infants will look toward the right person when their mother or father talks to them (Spelke & Owsley, 1979). Other studies also indicate that by 4 months infants can link auditory and visual information according to synchrony. When shown a film of a bouncing toy monkey on one screen and a bouncing kangaroo on another, infants preferred to observe the film of the animal that bounced in rhythm with the sound track (Spelke, 1979). Thus, they apparently were able to identify the way in which the visual information matched the auditory stimuli. In a somewhat more sophisticated study (Wagner et al., 1981), the researchers found what has been suggested are quite remarkable abilities for young children 9 to 13 months of age. The children were presented with pairs of visual and auditory stimuli that matched each other abstractly. For example, a visual pair of a broken line and a continuous line was matched with an auditory pair of a pulsing tone and a continuous tone. The researchers determined that the infants were able to perceive similarity (i.e., match) in three of the eight stimuli sets, which was considered quite remarkable because infants are typically not exposed to these kinds of events.

In general, visual-auditory integration abilities are quite functional by 4 months and continue to improve markedly up to approximately 12 years.

Visual-Kinesthetic Integration

As with visual-auditory integration, the visual and kinesthetic perceptual modalities appear to have some integrative characteristics in the very early stages of infancy. In fact, the abilities associated with visual-kinesthetic integration seem to become refined before those of visual-auditory integration. Information on this topic usually includes mention of tactile integration due to its interrelation with kinesthetic awareness.

The research findings on this aspect of perceptual integration are relatively consistent; most sources note an improvement with increasing age through childhood up to about 11 years of age. Some of the first indications of visual-kinesthetic integration (cross-modal equivalence) have been observed in children as young as 2 to 3 weeks of age. Meltzoff and Moore (1977) found that babies at this age were able to imitate the mouth and tongue movements of adults. This is quite impressive considering that they could not see (and probably never had) their own "matched" movement. In this instance, the infants translated visual input into an equivalent body movement, exhibiting kinesthesis.

By 6 months of age most children exhibit the ability to explore objects with one perceptual modality and transfer that information to the other perceptual system (Acredolo & Hake, 1982; Ruff, 1980). For example, in one study babies who were allowed to explore objects tactually later chose those objects after viewing several objects without tactile manipulation (Bryant et al., 1972). This type of cross-modal equivalence appears to be nearly mature by the age of 5 (Birch & Lefford, 1967; Blank & Bridger, 1974). Evidence also suggests that, among the visual, tactile, and kinesthetic types of perception, visual-tactile integration abilities are the most advanced in 5-year-olds. Hence, the refinement of visual-kinesthetic integration appears, in general, to lag behind visual-tactile integration.

Unfortunately, the evidence regarding the development of cross-modal concept and visual-kinesthetic integration in older children is contradictory. In most studies of visual-kinesthetic integration, the individual is first presented with a visual or kinesthetic stimulus (e.g., passively guided through a maze, line, or pattern) and then is asked to identify or reproduce the event in either the same or a different modality. Although most sources agree that individuals 5 to 11 years become steadily better at integrating visual-kinesthetic information and making cross-modal concept judgments, the reason for the improvement is not clear. It has been mentioned that one limiting factor in making accurate judgments is the inability to isolate intrasensory improvements and the possible effect this has on the upgrading of cross-modal functioning. The interweaving process that occurs between the visual and kinesthetic modalities is still somewhat of a mystery, as is the way in which intersensory integration affects the development of motor behavior.

Compared to the amount of information on visual-kinesthetic abilities, data on auditory-kinesthetic integration are limited. However, the available information suggests that there is a natural compatibility between the auditory and tactile kinesthetic modalities. In a frequently used testing procedure, the tester tells the name of an object or shape to the child, who is then asked to select the item tactually from a number of items to create an auditory-tactile/kinesthetic match. Consistent development differences have been reported between individuals 5, 6, and 8 years old in this aspect of sensory integration (Temple et al., 1979; Williams et al., 1978). The evidence also suggests that older children are markedly less variable as a group than younger children. Therefore, although only scanty information on a small age range is available, evidence does indicate that auditory-kinesthetic/tactile integration improves during childhood.

Auditory-Kinesthetic Integration

It is a fact of life that as most people begin their middle-age years, the perceptual systems generally become less responsive to stimulation. However, most of these changes do not significantly interfere with everyday functions until the later adult years. Aside from the basic performance decrements that are associated with the visual and auditory perceptual modalities, very little is known or clearly understood about the effects of aging on the perceptual processes. This is especially evident with the various aspects of kinesthetic perception.

CHANGES WITH AGING

■ *Objective 6.12*

A loss of visual abilities is one of the most noticeable changes in perceptual functioning as the body passes its peak performance level. Changes in visual acuity

Visual Perception

begin to be noticeable for most people between 40 to 50 years. Although there are individuals in their seventies and eighties who have 20/20 acuity, only about 25% of 70-year-olds and 10% of 80-year-olds exhibit this level of clarity. Weale (1975) notes that by age 85, individuals experience an 80% loss of acuity from what it was in their forties. The loss of acuity is especially severe at low levels of illumination and for moving objects.

Closely related to loss of acuity with age is the condition known as *presbyopia*. This condition is characterized by the reduction in the elasticity of the lens of the eye to the point where the lens can no longer change its curvature sufficiently to allow accommodation for nearsightedness (close vision). By the age of 60, this condition can greatly affect the eye's ability to focus on objects at close distance (Whitbourne, 1985). As the eye ages, the lens also becomes thicker and less transparent, thus reducing the transmission of light through the lens.

In addition to reducing the amount of light, age-related changes in the density of the lens also have the effect of scattering light rays before they reach the retina, which may produce a blurred retinal image. Changes in corneal tissue may also lead to the greater scattering of light rays, resulting in a blurring effect on vision. Another age effect on the cornea that influences the eye's optical function is a change in its curvature. Adults over 60 years of age tend to have flatter corneas, which can affect the degree of astigmatism they experience.

Sensitivity to light is important to optical functioning and a major consideration with the elderly. As a result of increased density and yellowing of the lens, the amount of light that enters the eye is reduced; thus older adults are less able to see in low levels of illumination (Carter, 1982). Closely associated with this is the ability of the eyes to adapt to dim light; sensitivity decreases after 20 years of age and is particularly evident after age 60. Another form of light sensitivity that appears to affect older individuals is glare, especially after age 60. This is also attributed in part to age changes in the lens that affect the scattering of light over the retinal surface. This, along with problems related to deteriorating acuity, has some very practical implications for older individuals who attempt to move around and perform tasks in reduced light conditions. One obvious suggestion for minimizing the effects of these changes would be to increase the light levels in environments where older adults live and participate in leisure activities.

The information related to aging and its effects on depth perception is quite limited. Current evidence suggests little or no change up to 60+ years. Another area of limited information is the effects of aging on field of vision. As noted earlier in the discussion of visual perception, lateral field of vision reaches a peak at approximately 35 years of age, after which it gradually decreases until about 60 years of age, whereupon the decline continues at a much more rapid rate. In addition to associative changes within the visual modality, age-related changes in facial structure (i.e., upper eyelid, orbital eye sockets) have also been linked to the decrease in field of vision (Shephard, 1978). Another important note related to motor behavior is the observation that older persons require a significantly longer time to process visual information; Williams (1990) estimates that older individuals are about 33% slower in processing than younger persons.

With aging comes a gradual loss of hearing that begins around the midthirties and continues to progress at least into the eighties (Whitbourne, 1985). The most common type of age-related hearing loss is associated with the gradual deterioration and hardening of the auditory nerve cells. The result of this condition is that the sensitivity to tones of high frequency is impaired earlier and more severely than the loss of sensitivity to low-frequency tones. The term used to describe this kind of hearing loss is *presbycusis.* **Auditory Perception**

Beginning at about age 40, there are marked sex differences, with females having much better hearing than men at higher frequency levels. There is some suggestion that hearing loss is due in part to environmental noise stress, especially in male-dominated occupations. Hearing loss at higher frequencies becomes more apparent after the age of 50, especially in males. For age groups over 60, hearing loss progresses to affect the lower frequencies, so that hearing in the elderly is generally impaired for a wide range of tones.

Most of the research on kinesthetic perception is grouped under the heading of somesthetic senses. You will recall that the word *somatosensory* refers to sensations from the body, which in this context also includes tactile perception. Except for functioning of the vestibular system (balance) and tactile perception, very little is actually known about the effects of aging on the kinesthetic receptors and the individual's sense of body position. As with the other perceptual modalities, it might be expected that as people approach middle age they would have diminished somesthetic function, resulting in a reduced awareness of bodily orientation, movement, and touch. However, current research findings on age differences in somesthetic sensitivity have not provided clear-cut evidence of this parallel. Although structural and functional age effects have been found, no overall deterioration across modalities in the somatic domain are evident (Whitbourne, 1985). This should be considered a preliminary evaluation of the somesthetic domain, however, since documentation of many of these areas is sparse or virtually absent. **Kinesthetic Perception**

Perhaps the most abundant information about age effects on somesthetic sensitivity is in the area of tactile perception. Some evidence suggests that elderly persons frequently suffer a loss in sensitivity to touch; however, the degree of loss varies widely among individuals and is different for various parts of the body. For example, Skre (1972) examined sensitivity to vibration in people 64 to 73 years of age and found a 40% impairment in the lower parts of their bodies but only a 5% impairment in the upper parts. Caution should be taken in interpreting results such as these, for impairments observed might be related to factors other than age, such as disease, injuries, or circulatory problems.

The primary age-related factors responsible for diminished touch sensitivity have been fairly well established. For example, there are cross-sectional age reductions in the number of touch receptors (i.e., Pacinian corpuscles and Meissner's corpuscles) in the skin.

Malfunctioning of the vestibular system as evidenced by decreased performance on balance tasks, dizziness, and vertigo (Toglia, 1975) are common

Table 6.3 Summary of Selected Perceptual Changes with Aging

Perceptual Characteristic		
Visual Perception	*Auditory Perception*	*Kinesthetic Perception (Somesthetic)*

Acuity (decreases)
Sensitivity to light (decreases)
Depth perception (little change)
Visual information processing
  (decreases)
	Sensitivity to high and low
	  frequencies (decreases)
		Touch sensitivity (decreases)
		Sense of body position and movement
		  (no major change)
		Weight discrimination (decreases)
		Balance (decreases)

characteristics among older adults. From approximately 40 years of age a gradual deterioration of vestibular nerve cells takes place. Sensory cell loss may begin as early as the fifth decade in the semicircular canals and may be quite marked in the saccule and utricle as well after the age of 70 (Rosenhall, 1973). As in the case of the cochlear nerve fibers, evidence suggests that the loss of fibers is due to the accumulation of bony material around the opening through which the nerve fibers pass (Krmpotic-Nemanic, 1969). Apparently, as the holes become smaller the nerve fibers compress and gradually degenerate. Although balance performance and the symptomatic indications of vertigo and dizziness point to vestibular malfunction in older adults, the underlying mechanisms responsible for these responses have not been clearly defined.

As noted earlier, currently little is known about sense of body position (kinesthesis) in the elderly. Although older persons generally show a decrease in the ability to discriminate weight (e.g., held in hands while blindfolded), there are some indications that no major changes in active joint-movement (body positioning) sensation occurs (Meeuwsen, Tesi, & Goggin, 1992; Williams, 1990). For example, Meeuwsen and colleagues reported no significant differences in young and older adults in their ability to blindly compare upper limb body position by location and extent (degree of change). More important, however, the researchers suggest that age-related declines in the elderly cannot be generalized to all perceptual systems.

Table 6.3 provides a summary of selected perceptual changes due to aging.

THEORETICAL VIEWPOINTS

■ *Objective 6.13*

Although the role of nature-nurture in the perceptual development of children represents one of the classic debates in developmental psychology, most contemporary theorists view the process from an interactive perspective. That is, they believe that changes in the ways human beings perceive the world reflect a fundamental interplay between nature (maturation) and nurture (experience and learning within their cultural context).

Of the various theoretical viewpoints that have received attention in the general field of developmental psychology, two approaches appear to be more dominant; these are the *constructivist approach* and the *ecological approach.*

Often associated with Piaget and several information-processing psychologists (e.g., Hochberg, 1978; Neisser, 1967), the **constructivist approach** may be considered the more traditional approach in comparison with the ecological approach. The basic tenet of this approach is that individuals make a "construction" out of the sensory input from their eyes plus information retrieved from their memories. In essence, this construction is a representation of the surrounding that is built in the individual's mind. The constructivist viewpoint also suggests that many changes in perception reflect alterations in constructive activities and/or changes in long-term memory on which constructive events are based. In another associated view, referred to as *enrichment theory,* our stored knowledge "enriches" the meaning of sensory experience (learning). Two individuals with different experiences may view an identical object or event, but have different perceptions of it.

Constructivist Approach

The constructivist view contends that individuals begin life with minimal perceptual capacities then build them up (enriches) with years of experience. The **ecological approach** as represented by the works of Gibson (Gibson, 1969; Gibson & Spelke, 1983) takes issue with the idea that perception is constructive. It proposes that individuals can directly perceive information that exists in their surroundings and need not build representations of it within their minds. Part of the justification for this view is the fact that infants have some perceptual capacities even at birth. According to this approach, infants are gifted with powerful perceptual capacities from birth and with experience become increasingly aware of their surroundings resulting in greater perceptual refinement. Gibson's theory of perception includes three interesting features: exploring, detecting invariances, and recognizing affordances. Some refer to this approach as *differentiation theory,* suggesting that as people develop they attend more and more closely to the distinctive features of the environment.

Ecological Approach

Whereas Piaget thought that infants construct their perception of objects by acting on them, the ecological approach views information as part of the object and that infants merely need to seek and they find. With the first step, the infant attends to the environment by *exploring* its features and focusing on some specific action or object. Infants perceive events and objects by detecting *invariants,* those actions or things that look or sound the same every time they appear. It is acknowledged that these abilities are innate and improve with maturation and experience. An example of detecting invariants is evident in the visual size constancy that is exhibited by young infants. Recognizing *affordances* refers to learning about the properties of objects, realizing what can be done with them, and understanding what can be expected from them. For example, a balance board affords support to a person but not to an elephant; a floor must be level, solid, and not slippery to be passable. Affordances are, in essence, the product of sensory input and reasoning (i.e., information processing). The presence of this type of behavior with young infants is well documented.

Ecological theorists, like constructivists, acknowledge that perception develops with maturation. It is assumed that as perceptual processes mature, individuals become more efficient at discovering invariants and affordances.

Movement-Based Approaches

Several theories proposed during the 1950s and into the early 1970s received considerable attention in the educational and research communities (Barsch, 1965; Delacato, 1959, 1966; Kephart, 1971). Underlying most of these theories was the assumption of a strong direct link between perceptual-motor functioning and cognitive functioning. Deficiencies in cognitive performance, such as reading and spelling, were attributed primarily to poor sensory integration and perceptual shortcomings were seen to be remediable through perceptual-motor training activities requiring those perceptual judgments. For example, some theorists believed that participation in activities involving eye-hand coordination, laterality, directionality, and balance could enhance cognitive and motor functions. Other theories proposed that it was critical that the neurological organization of the brain exhibit a dominant hemisphere. That is, if an individual was right-eyed, -handed, and -footed, he or she was considered to be neurologically "wired" to facilitate intellectual development. A proposed method of "rewiring" an individual was to have him or her perform movement patterning activities such as crawling and creeping. Another assumption was that lack of hemispheric dominance was the possible result of not having experienced (or performed improperly) early movement-patterned events.

Although some proponents of these theories are still around, little general support from the scientific community can be found (Salvia & Ysseldyke, 1991; Seefeldt, 1974; Williams, 1983). Convincing evidence of a true link between perceptual-motor skills and cognition has not been presented. Most studies have reported little or no relationship among perceptual-motor development, perception, and intellectual performance. Most of the reviews of the effectiveness of motor-based programs on the development of cognitive abilities have concluded that there is a lack of scientific support for perceptual-motor intervention.

Kavale and Matson (1983) conducted a sophisticated study of 180 scientific reports designed to research the efficacy of perceptual-motor programs. The researchers concluded that perceptual-motor intervention appears to be ineffective and noted that, compared with other forms of educational intervention, the effects of such programs are negligible. Further, Salvia and Ysseldyke (1991) state that "There is a tremendous lack of empirical evidence to support the claim that specific perceptual-motor training facilitates the acquisition of academic skills or improves the chances of academic success. Perceptual-motor training will improve *perceptual-motor* functioning" (p. 322). This in itself is a reasonable goal in motor development programming, especially during the childhood years.

SUMMARY

Both perception and the perceptual-motor process are related to information processing, which involves monitoring and interpreting sensory information, and deciding on and organizing a motor response. Of the six perceptual modalities, the two that provide the most relevance to the study of motor behavior are vision and kinesthesis. Visual perception, considered to be the dominant modality, undergoes several structural and functional changes across the life span. Although the visual system is relatively well developed at birth, every structure in the system undergoes some postnatal change. Significant structural differences between the newborn and adult are accompanied by functional changes in visual acuity, perceptual constancy, the perception of spatial orientation, figure-ground perception, whole-part perception, depth perception, field of vision, the perception of movement, and visual-motor coordination.

Kinesthetic perception, often referred to as the sixth sense, encompasses the awareness of movement and body position. It involves the ability to discriminate positions and movements of the body based on information derived from four sensory receptors located within the body. The various aspects of kinesthetic perception may be described in regard to its physiological and more applied characteristics. In physiological terms, the ability to discriminate kinesthetic information is known as kinesthetic acuity; kinesthetic memory refers to the ability to reproduce movements after a pause in time. Kinesthetic acuity normally improves faster and earlier than kinesthetic memory.

Five interrelated movement awareness components show developmental change: body awareness, spatial awareness, directional awareness, vestibular awareness, and temporal awareness. Each of these components theoretically involves information relative to knowledge of the body and its spatial-temporal characteristics (i.e., position and movement).

Although they are not generally considered as vital to motor behavior as visual and kinesthetic perception, the auditory and tactile modalities provide important cues to successful movement responses.

Though discussions of perceptual development usually focus on the various modalities as independent systems (denoting intrasensory change), the actual perceptual-motor process frequently involves the simultaneous use of more than one system (perceptual integration). Intersensory integration is at least partially functional at birth and improves with increasing age beyond the childhood years. The most frequently utilized forms of integration include the visual-auditory, visual-kinesthetic, and auditory-kinesthetic.

Although most people experience some form of regression in perceptual function beginning in middle age, most changes do not significantly affect everyday activities until old age. Most noted among the forms of regression are losses in visual acuity, sensitivity to light, hearing, and a loss of balance.

Two theories concerning the importance of perceptual experiences to human development have received considerable attention: the constructivist approach and the ecological approach. During the late 1950s and early 1970s, several movement-based perceptual-motor programs were also proposed to enhance academic success, but they have received little support from the scientific community.

SUGGESTED READINGS

Birren, J. E., & Schaie, K. W. (Eds.). (1990). *Handbook of the psychology of aging.* San Diego, CA: Academic Press.

Bornstein, M. H. (1992). Perception across the life span. In M. H. Bornstein & M. E. Lamb (Eds.), *Developmental psychology: An advanced textbook* (3rd ed.). Hillsdale, NJ: Erlbaum.

Sekuler, R., & Blake, R. (1994). *Perception* (3rd ed.). New York: Alfred A. Knopf.

Whitbourne, S. K. (1985). *The aging body: Physiological changes and psychological consequences.* New York: Springer-Verlag.

Williams, H. G. (1983). *Perceptual and motor development.* Englewood Cliffs, NJ: Prentice-Hall.

Williams, H. G. (1990). Aging and eye-hand coordination. In C. Bard, M. Fleury, & L. Hay (Eds.), *Development of eye-hand coordination.* Columbia, SC: University of South Carolina Press.

Information Processing and Motor Control

KEY TERMS

attention
memory
processing speed
reaction time
movement time
response time

programming
schema
motor program
coordinative structures
dynamical systems
neuronal group selection

Information processing is involved in virtually all forms of human behavior. Whether from the perspective of cognitive psychology or the acquisition and performance of a motor skill, the ability to process information is one of the major contemporary issues in human development.

■ *Objective 7.1*

Information processing is an approach to understanding human development and behavior that became established during the 1970s with the mass popularity of computers. The roots of the information-processing approach are in the fields of computer science (manipulation of symbols), communications theory (information coding and channel capacity), and linguistics (language). As long ago as 1949 psychologists compared human processing to basic computer functions in which information is encoded, stored, transformed, retrieved, and acted upon (Shannon & Weaver, 1949). During these early stages of information theory, mathematical models were developed to make quantitative predictions about human behavior. Of the numerous traditional and contemporary theories of human development, information processing now dominates the field of cognitive psychology.

■ *Objective 7.2*

One of the more popular approaches to studying the execution of a motor response has been to conceptualize the process in terms of an information-processing model. Motor developmentalists are concerned with the way individuals utilize information in perceiving, making decisions, and organizing activity in relationship to the demands placed on them by the environment. Four basic components have been identified in the simplified model of information processing in figure 7.1: sensory (afferent) input, reception of information through the various perceptual modalities, interpretation and decision making (processing), and an overt motor (efferent) response.

The last chapter focused on the developmental aspects of the perceptual modalities in regard to receiving, monitoring, and interpreting sensory data. This chapter will concentrate on the developmental characteristics associated with the ability to process information: attention, memory, processing speed, and programming strategies.

As individuals progress toward adulthood they are able to process more information in a shorter period of time. This concept, by itself, is relatively simple; yet the structural and functional aspects of information processing to which it refers are quite complex and in several instances remain in the "little black box" of the unknown or highly theoretical.

ATTENTION

■ *Objective 7.3*

Attention is usually considered the core of information-processing models. And there is convincing evidence that this aspect of information processing changes across the life span. About a century ago James, one of the earliest and most renowned experimental psychologists, described attention as the "focalization, concentration, of consciousness" (James, 1890). Although several definitions have been proposed since James's work, two features inherent in his statement

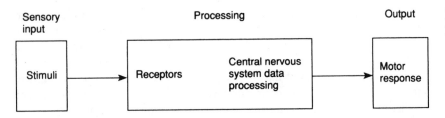

Figure 7.1
The basic information-
processing model.

are still relatively prominent today. First is the notion that attention is *limited* (an individual capacity exists) so that an individual can attend to only one thing at a time, or think only one thought at a time. From a motor behavior perspective, this statement suggests that an individual's capacity could be exceeded if too much activity were attempted. Second, attention is apparently performed in a *serial* manner, suggesting that individuals attend to one thing and then another and that it is quite difficult (sometimes impossible) to combine certain activities (Schmidt, 1988).

While attention is one of the key components in information-processing models, it is also a strong factor in the successful performance of a motor task. Three basic concepts associated with attention and motor behavior have been identified: (a) attention involves alertness and preparation of the motor system to produce a response; (b) attention is related to individuals' limited capacity to process information; and (c) the successful performance of a motor task requires selective attention, the ability to select and attend to meaningful information. The following pages will discuss the functional and developmental aspects associated with these concepts, giving particular emphasis to capacity and selective attention. For detailed discussions of attention theory see Schmidt (1988) and Magill (1993).

Alertness

Alertness and preparation of the perceptual-motor system prior to any initiated motor task is a vital aspect of attention and a successful movement response. These conditions are evident in the softball batter preparing for the incoming pitch and the sprinter in the starting block before the gun. The level of perceptual alertness and physiological preparation immediately preceding the pitch or the starter's signal are critical to the outcome of these events.

A large portion of the research on this aspect of attention has dealt with individuals' ability to exhibit alertness and response preparation by assessing their ability to detect either a visual or auditory signal in a predetermined fashion. Researchers have typically used reaction time (RT) experiments to test these abilities. *Reaction time* is the interval of time between the onset of a stimulus and the initiation of a response. It is important to note that reaction time does not include the physical movement; movement time combined with reaction time is the *response time*.

Reaction time is typically measured in the laboratory by presenting an individual with a visual (light), auditory (buzzer), or tactile (shock) stimulus and asking him or her to move some part of the body (e.g., to lift a finger off a button) in recognition of the signal, which stops an interval timer.

In the case of the sprinter, RT is the period of time from the moment the auditory stimulus (gun shot) was received by the auditory receptors to the point just prior to the first motor response. The RT phase includes attention (alertness) to the stimulus, motor system preparation (body position and muscle tension), reception and transmission of the stimulus, interpretation of the information, and finally, organization of a motor program with which to execute the response. The number of functional characteristics performed within the general alertness and preparation stages are numerous and sophisticated.

From a developmental and life-span perspective, alertness, as measured by RT, decreases (improves) with age into the twenties, then remains relatively stable until about age 60. Evidence based on RT studies also infers that the preparation required for a rapid movement response depends on the complexity of the response that must be made. It should also be noted that, in experimental studies, the use of a warning or ready signal has been shown to induce faster RTs and more accurate responses to the "go" signal. Apparently, the warning signal alerts the individual to prepare himself for the actual stimulus. This condition may also be viewed as a form of anticipation, which has strong implications for alertness, motor preparation, and successful motor performance. Successful athletes, such as the softball batter, learn to anticipate upcoming conditions by being alert.

Divided Attention

Another aspect of attention having strong implications for motor performance is the concept of limitations in the capacity to handle information from the environment. This aspect of attention represents the concept of *divided attention*. Basically, if a specific movement activity requires attention, then some (or all) of an individual's limited capacity must be allocated to its performance or control. Because capacity is believed to be limited, interference will occur if another activity requires these resources. Interference could result in (a) a loss of speed or quality in the performance of one of the activities; (b) both activities could be affected; or (c) the second activity could be ignored.

Part of the operational definition of attention generally includes the notion of interference. However, it is possible to perform two tasks simultaneously as long as the total amount of processing capacity is not exceeded. The amount of the capacity resources required for any task or tasks depends on the activity and the individual. Examples of divided attention are abundant in everyday life as individuals try to handle several cognitive tasks at once, sometimes unsuccessfully. The act of driving a car while conducting a conversation may not be difficult if both conditions are relatively light and undemanding. However, if the road conditions deteriorate or the conversation turns to something complex or emotional, processing capacity may become overloaded and performances may suffer. Numerous studies of the processing capacity of attention have used the "dual-task paradigm" in which subjects are observed as they perform two tasks simultaneously.

Keogh and Sugden (1985), in reviewing the literature on this topic, described a frequently used method of applying the dual-task paradigm. Under this method, individuals are presented a primary task and secondary task separately and then are asked to perform the tasks simultaneously. Presumably when the tasks are executed simultaneously, any differences between the events indicate where and how individuals allocate their capacity resources. In their review, the authors concluded that younger children allocate their available attention somewhat equally between two tasks, whereas older children appear to recognize the significance of the primary task, attend to it, and perform better on it.

Guttentag (1984) also found a developmental trend using the dual-task paradigm. In this study, children 7 to 11 years of age tapped a telegraph key as fast as possible in one condition and then again while simultaneously rehearsing a list of words out loud. The findings clearly indicated that tapping was slowed in all children by the divided attention created by the addition of the word rehearsal task; however, the detriment was less for the older children.

The dual-task paradigm experiment also has implications for selective attention, the ability to attend to relevant information in the environment. Wickens and Benel (1982) note that capacity allocation (time-sharing efficiency) improves with age and that developmental differences in attending to dual tasks are primarily due to *automation* and *attention deployment skills.* In the context of this research, automation is a possible source of age-related improvement in which the performance on one task remains the same with less attention used, thus freeing up processing resources for performing a concurrent task. Attention deployment skills are strategies used to process additional information.

Other evidence also suggests improved efficiency of capacity allocation in the dual-task paradigm is linked to age-related improvements in cognitive processes and the ability to adopt strategies (Manis, Keating, & Morrison, 1980). Manis and associates found that, from second graders to adults, the capacity allocated to process simple and complex tasks decreased. The underlying notion is that the cognitive processes involved in a task compete for limited central processing capacity. Therefore, as children grow older they become more adept at controlling the allocation of their attention (through the use of strategies) and require less of the capacity resources.

One of the much debated issues on the topic of processing ability is whether improvements are due primarily to structural increases (actual capacity) or control operations (e.g., the use of strategies). Most researchers appear to believe that age differences in processing are due primarily to qualitative changes and not structural increases (e.g., Chi, 1976). However, the critics of this view contend that increasing ability depends on structural changes.

Selective Attention

Selective attention, although closely related to alertness, response preparation, and processing capacity, is a distinguishable characteristic of attention. Selective attention is the processing of relevant information and nonprocessing of irrelevant information. A frequently used generalization is the "cocktail party phenomenon" (Cherry, 1953). Using a large gathering such as a party or sporting event to illustrate, an individual can attend selectively to a conversation with one person even though a number of other conversations and noise are taking place around them. Furthermore, if during that personal conversation someone in the crowd mentions the individual's name, attention is immediately diverted to that person in the crowd.

This phenomenon is closely related to another perceptual characteristic mentioned earlier—visual figure-ground perception. Under these conditions, the individual might also selectively attend (visually) to the relevant information in the environment (the figure) and set it apart from its background. Selective attention, however, may involve more than one sense; for example, an individual catching a fly ball must separate relevant information (the ball) from an environment that may be filled with trees, clouds, and crowd noise.

Evidence suggests that children as young as 4 months of age have impressive selective attention abilities (Bahrick, Walker, & Neisser, 1981). Although infants at this age exhibit relatively impressive skills, refinement appears to continue during the childhood years. In general, older children show better control of selective attention by processing more stimulus features that are relevant to a particular task and not processing those that are irrelevant. In contrast, younger children tend to select features of a task that stand out or are salient. And, as noted in the dual-task paradigm studies, older children are more organized and strategic in their processing abilities.

Based on observations of children viewing a visual display containing relevant and irrelevant features, Ross (1976) proposes that the development of selective attention abilities occurs in three stages. During the approximate period of 2 to 5 years of age, children most often pay exclusive attention to one stimulus in the display; this is termed the *overexclusive mode.* Observations of younger children also suggest that they are more easily distracted than older children. For example, while watching television they are more likely to play with toys, look around, and talk with other people during the viewing.

The second stage spans the years from about 6 to 11 years of age and is referred to as the *overinclusive mode.* In this stage children characteristically attend to several features in the display, some of which are irrelevant.

The final stage, *selective attention,* is attained by early adolescence when individuals become capable of attending to relevant stimuli in displays of varying complexity (simple to complex).

Ross's basic observations have been verified by others, demonstrating that older children are much more likely to ignore information that is irrelevant or that distracts from the central activity than younger children (e.g., Lane & Pearson, 1982; Smith, Kemler, & Aronfried, 1975). Also noted is the observation that when auditory and visual stimuli are presented simultaneously, children up to age 12 seem to rely heavily on auditory stimuli, to the extent that they are distracted from key visual information (Perelle, 1975).

As with the other aspects of attention, age differences in selective attention abilities are attributed primarily to experience and the refinement of operational functions (i.e., the use of strategies). Through experience adolescents and adults learn which stimuli are relevant to a particular situation and which are not. They then adopt a strategy to address the situation. For example, the experienced softball batter knows from past events at what approximate point in the pitcher's delivery the ball will be released. The batter selectively attends to the point of release and follows the ball exclusively to the anticipated point of contact. In contrast, the young, inexperienced batter's attention may be diverted to crowd noise, words from opposing players, and unfamiliar actions of the pitcher. Since the past experiences of younger children are likely to be limited, the use of selective attention strategies are also likely to be restricted. Several studies have noted that children can be taught to resist irrelevant cues and acquire selective attention strategies (e.g., Stratton, 1978a, 1978b).

MEMORY

■ *Objective 7.4*

The study of *memory* and its various systems is vital to understanding information processing and motor performance. Up to this point in the text, various discussions have focused on the developmental aspects of the way information is selected and perceived by the central nervous system in preparation for a meaningful motor response. However, at some point in information processing, data entering the system must be retained or stored for future use. There have been numerous structural and operational definitions proposed for the term *memory*. In a broad sense, **memory** refers to the retention and subsequent retrieval of information. Some operational definitions also use the term synonymously with *remembering*. In a more global treatment, Tulving (1985) describes memory as the "capacity that permits organisms to benefit from their past experiences" (p. 385).

Most evidence suggests that developmental differences in memory system effectiveness derive primarily from functional refinement (control processing strategies) rather than structural (capacity) increases. A basic knowledge of the various memory structures and their capacities is important to a more detailed understanding of developmental differences.

Memory Structures

To use a rather contemporary analogy, the operational relationship between the structural and functional components of memory is like that of a computer; the structure is similar to computer hardware, while control processes (strategies) are like software. Although numerous computerlike schematics and models have been developed and a vast array of terminology used to explain the concept of memory and its elements, several researchers support the multistore model of memory—that memory is made up of three separate memory systems: short-term sensory store, short-term memory, and long-term memory. A theoretical representation of these systems and their relationship to a motor response is shown in figure 7.2.

Short-Term Sensory Store   The *short-term sensory store* seemingly has a limitless capacity, yet it has a memory duration of less than one second. In a very brief period of time, massive amounts of information from the environment and from within the body can be presented to the short-term sensory store system. Working in cooperation with the various perceptual modalities, the short-term sensory store accepts this information without much recoding and "forgets" it very quickly as new information comes along. Evidence suggests that there is no

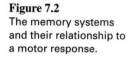

Figure 7.2
The memory systems
and their relationship to
a motor response.

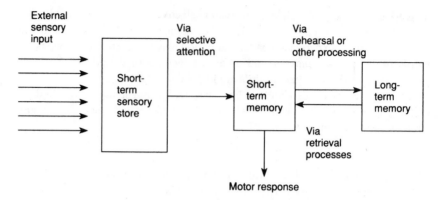

difference between young children and adults in the initial intake of information and sensory store capacity (e.g., Chi, 1976). However, if the information must be held for further processing (encoding) to the short-term memory system, the young child's recall diminishes quickly compared to the adult. An important functional (control) feature at this point in memory processing is that selective attention targets what information is relevant for further processing and allows irrelevant stimuli to be ignored.

Short-Term Memory   Short-term memory is believed to be a storage system for information derived from either of the two other memory systems (fig. 7.2). The *short-term memory* system has a small and limited capacity with a storage duration of generally not more than 60 seconds. Short-term memory has been described as a "workspace" for processing and as a processing area "for information currently (or very recently) under attentional focus" (Klatzky, 1984, p. 26).

Evidence of the way this workspace and moment of attentional focus come into play appears in numerous everyday situations. A common example is the individual who looks up a telephone number, closes the phone book, and by the time he finds the change for the call, forgets the number. Or picture being introduced to three or four people at a meeting. You could reproduce the names if you tried immediately after introductions were made, but if you conversed for a minute or two and could not rehearse the names, you may easily have lost them.

Based on these examples, it is quite evident that the major difference between short-term sensory store and the short-term memory system is in capacity. Recall that the short-term sensory store has a seemingly limitless capacity. The short-term memory system, however, has a capacity of only about seven (plus or minus two) items (Miller, 1956). For experimental purposes, the term *item* has also been used to designate the capacity of the system to hold single words or digits. Separate items grouped in some way to make larger collections is a process known as *chunking*. The capacity of individual chunks (i.e., how many pieces of information can be stored in a given chunk) is somewhere around three or four units (Allen & Crozier, 1992). This technique, along with rehearsal, significantly increases the number of items that can be retained;

however, it is unlikely that more than seven chunks or groups of items can be stored at any one time.

The capacity of short-term memory is commonly determined using the memory span technique in which an individual is presented with a list of digits or words at a given rate (usually one item per second) and asked to repeat the list. The number of items in the list is increased after each perfect recall. Despite individual differences in memory span, the general trend is for the span to grow larger with age (Dempster, 1981). This evidence suggests that memory span increases from about two digits in 2- and 3-year-olds to about five digits in 7-year-olds. And interestingly, from age 7 to adulthood the increase is only about two and a half digits (to a maximum of seven and a half digits).

Although memory span is not a pure measure of short-term memory (it may also involve long-term processes), it does suggest that short-term memory performance improves with age and that the largest changes occur between the ages of 3 and 7 years. It has also been suggested that one reason for the improvement of short-term memory with age is because control processes improve over that time span as well. However, the literature is mixed concerning whether children have a limited short-term storage capacity compared to adults (Brown et al., 1983; Case, 1985; Chi, 1976).

In some instances, researchers divide the short-term memory system into verbal and motor categories. A description of verbal short-term memory is exemplified by memory span (i.e., words, digits) determination. In contrast, short-term motor memory concerns the processing and storing of sensory information for use with movement responses. The question of whether these two entities are separate features of the memory system is yet to be resolved; speculation suggests, however, that there is one memory system that is adaptable to the unique requirements of verbal and motor skills (Magill, 1993). Unfortunately the literature is almost void of developmental studies related to short-term motor memory.

Long-Term Memory   When items are rehearsed, they are transferred in some manner from short-term memory storage to long-term storage where they can be held on a more permanent basis. Both the storage duration and capacity of *long-term memory* are seemingly unlimited. Storage can span hours, days, and with several items, years. Perhaps the most often cited examples of long-term motor memory are the ability to ride a bicycle and swim after several years without any intervening practice. One thing that is relatively clear is that long-term memory depends on what control processes and strategies people use when they are learning and remembering information. This topic will be the focus of the next major section of this chapter.

The multistore memory model is an important contribution to the understanding of memory structure, but other models have also attracted consideration. The *levels-of-processing model* does not view memory as being set in separate memory systems as in the multistore approach (Craik & Lockhart, 1972). Instead this model emphasizes the functional aspects of memory by proposing a hierarchy of processing stages. Remembering is related to how deeply the information is processed.

The hierarchy begins with shallow sensory processing and progresses to deeper more abstract and semantic levels of operation. Basically the strength of the memory depends on how deeply the information was processed. Information processed only at the shallow level is not well remembered. This basic model has been updated by one of its originators (Craik, 1979); depth is no longer viewed as a simple continuum from shallow to deep levels with no consideration of the characteristics of the information (e.g., physical, phonemic, semantic).

Another model that attempts to address more of the functional aspects of memory is the *component functional model* of memory (Baddeley, 1986; Baddeley & Hitch, 1974). This model places emphasis not so much on where the information is (as in the other two models) but on what an individual actively does with the information. The assumed need that this model addresses is to provide a description of the structural components of memory but at the same time emphasize the role of what the individual does with the information. To do this, the originators of the component functional model have organized memory into *working memory* and *long-term memory.* Working memory is similar to the sensory store and short-term systems. The structure is described as a passive place where information is stored and a workspace for integrating information coming from and going to long-term storage. In this workspace, critical cognitive functions such as problem solving and decision making are stimulated. Long-term memory is characterized in a similar manner as in the multistore model.

Memory Functions   Although the three memory models presented seemingly offer different perspectives of memory, all three models consider control processes to be vital components in the overall process. As noted in the introduction of this chapter, most evidence suggests that developmental differences in effective usage of the memory system are due primarily to functional refinement (control processing strategies) rather than structural (capacity) increases. This section will address the primary control processes involved in memory from a developmental perspective.

The identification and importance of control processes in memory functions are emphasized in the Atkinson and Shiffrin (1971) multistore model (fig. 7.3). The developers of this well-known model stress the importance of control processes in receiving, storing, retrieving, and providing information for formulating motor or verbal responses. Due to the interactive nature of the structural and functional (control processes) aspects of memory, the following discussion addresses selected processing abilities within the context of short-term and long-term storage structures.

Early Processing Abilities   The identification and extent of processing strategies used by infants is difficult to determine. However, there are indications that primitive forms of memory are present shortly after birth. Evidence of this phenomenon has been gathered primarily through observing *habituation.* A stimulus is presented to an infant and its responsiveness is monitored (e.g., time and quality

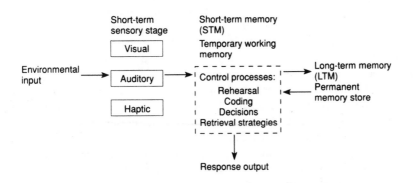

Figure 7.3
The Atkinson and
Shiffrin multistore model
depicting control
processes.
"The Control of Short-Term
Memory" by Richard C. Atkinson
and Richard M. Shiffrin.
Copyright © 1971 by Scientific
American, Inc. All rights
reserved.

of gaze). If the stimulus is repeated, reaction declines and habituation is said to have occurred; but if another stimulus is substituted, responsiveness generally increases.

Theoretically the renewed responsiveness to a new stimulus (referred to as *dishabituation*) can be viewed as evidence for memory (Fagan, 1984). Given adequate familiarization and relatively simple stimuli, memory in the form of dishabituation has been observed in infants and newborns (e.g., Gibson & Spelke, 1983). For this type of behavior to occur, it appears that even newborns actively attend to the environment and exhibit the ability to encode information from it.

Dishabituation may also be viewed as a characteristic behavior of *recognition* memory, theoretically the simplest form of remembering. In contrast, *recall* memory is a more difficult process because information must be retrieved from memory with no prompts from the environment. That is, no person or item is present to trigger memory, as in recognition.

In general, the available evidence suggests that adultlike "conscious" recall memory emerges in the second six months of life (Mandler, 1983; Olson & Strauss, 1984; Schachter & Moscovitch, 1984). This form of memory is more evident by age 2, as exhibited on memory span tests described in the earlier discussion of short-term memory. Despite the rather impressive memory abilities of infants during the first year of life, considerable refinement through the development of processing strategies is yet to occur.

Short-Term Memory Abilities Evidence of short-term memory ability as shown by memory span has been discussed within the context of short-term capacity. Memory span (recall memory) improves markedly up to early adolescence, with the largest changes occurring from about 3 to 7 years. Due to the rapid rate at which memory span and similar tests are presented, the use of active learning strategies is generally not encouraged by the nature of the task unless the individual is experienced. Evidence suggests that older children and adults perform better on such tasks because they are faster at perceiving the test items in the form of letters, words, and digits. Therefore, more time and, perhaps, capacity is available for retaining items after the point of perception (Case, Kurland, & Goldberg,

1982). This type of evidence would suggest that improvements in memory span during childhood years are related to the speed of perceptually identifying stimuli, which also has implications for their ability to selectively attend.

Another suggestion is that there may be age differences in the coding process and strategies used to encode information. *Coding* is the process of organizing a stimulus input into an acceptable form for storage. To successfully transfer meaningful information from the short-term sensory store to short-term memory, data must be coded. For example, if the initial presentation was visual, a subsequent encoding may transform the information to a verbal code. In this instance, instead of transforming the information into a verbal code (as employed by the older person), the young child may use a visual code that is more difficult to rehearse. When young children are prompted to use overt verbalization in rehearsing memory tasks, their performance improves (Hagen, Hargrave, & Ross, 1973; Kingsley & Hagen, 1969).

While a number of research reports suggest that the inability to encode information is a major stumbling block for young children, more recent data indicate that this is not the problem. Rather, it appears that the younger child has a poor understanding of how to search for and recall the information that has been encoded (Pressley & MacFadyen, 1983; Sullivan, 1982). It is hoped that future research will clarify this issue. For now it may be assumed that compared to older individuals, the young child generally has greater difficulty in coding, searching, and recalling information.

There is little reason to believe that these characteristics would be different in a motor domain. The formation of a motor program to execute a movement or series of movements is based on schemata similar to those used to produce verbal responses such as the memory span test. An important function of motor programming is to search and recall from short-term and long-term storage the past movement experiences associated with the immediate task. For example, in tossing a ball to a target, a miss to the right would be stored in memory. On the next attempt, the individual makes an adjustment based on that stored information. It should be quite clear that without short-term memory to recall prior performance, improvement would be unlikely and the benefits of practice minimal. Although inferences have been made from general short-term memory studies, unfortunately the developmental literature on short-term motor memory is quite sparse.

A common method for testing short-term memory within a motor context is the use of an arm-positioning task in which the subject is asked to reproduce the location and distance of a movement. The general findings suggest that individuals are most successful in recalling location and that, in comparison, distance information seems to decay rather quickly.

An interesting study (Thomas et al., 1983) produced similar results in a field environment. The short-term memory of children (age 4 to 12) was studied by asking them to reproduce location and distance after jogging through a predesigned course. The researchers suggested that all the children encoded location before distance information. They also found that the ability to recall distance improves with age as does the apparent use of processing strategies such as rehearsal. Other studies of short-term motor memory have also noted that one of the characteristic features of immature performers is a lack of effective use of control operations such as rehearsal (e.g., Sugden, 1980).

Although few studies specifically address the developmental nature of short-term motor memory, it appears as though the effective use of processing strategies such as rehearsal are associated with better performance and that the acquisition of these abilities seems to be a developmental process (Keogh & Sugden, 1985). These findings, although quite general in view of the several processes related to short-term memory, complement the evidence reported on primarily verbal-oriented tasks.

Long-Term Memory Abilities   Though a number of questions pertaining to memory are unanswered, one issue that has been made clear is that long-term memory depends on what people do (i.e., what control processes and strategies they use) when they are learning and remembering information. Much of what occurs during these tasks can be classified under one of the controlled processes of coding (encoding), rehearsal, decision making (organizational processing), and retrieval.

Coding (encoding) has been described as an important control process for transforming information from sensory store to short-term memory. This process is also important for placing data into the long-term memory system as a result of *encoding* and *recoding*. Recoding involves searching in the short-term memory for two or more items that can be combined (grouped by similarity) and then reentered with a new code into long-term storage. For example, a basketball player experiences the results of several different types of shots from the corner of the court during a practice session. The long-term memory of the player combines this knowledge and recodes it as information about shooting from that section of the court.

This type of information gathering also has strong implications for the formulation of motor programs. Apparently children below the age of about 8 years exhibit weak encoding and recoding ability compared to that of young adults (Chi, 1976; Ghatala, Carbonari, & Bobele, 1980; Lindberg, 1980). The use of mental imagery has been shown to enhance the coding ability of both younger (if the presentation is slow) and older children (Pressley & Levin, 1977). Another strategy related to recoding is the ability to *group* information for coding purposes. Evidence suggests that adults use this strategy quite well but that young children do not. In general, as children get older they become more efficient at coding information; younger children usually need more instruction at a relatively slow pace.

As children become older, they begin to engage in more formal memory strategies. One of the most common of these strategies is *rehearsal,* the act of repeatedly going over material for the purpose of enhancing retention. Although evidence suggests that children as young as 5 and 6 years of age are able to rehearse and memorize information successfully if they are aided (Ornstein et al., 1985), they generally do not begin to independently rehearse on memory tasks until approximately 7 or 8 years of age. The most common type of rehearsal with young children is rote repetition of material, whereas adolescents and adults usually modify their rehearsal to accommodate the structure of the material. By age 10 rehearsal is practiced regularly by most children (Flavell, Beach, & Chinsky, 1966).

When the learner rehearses by actively grouping items together into high-order units (or chunks) of information, a form of *organizational processing* occurs. For example, in a long list the words *red, blue,* and *yellow* might be grouped together. The organization of material in long-term storage determines the sequence in which the items are recalled. It would be difficult to recall the months of the year alphabetically because it is likely that they have been organized sequentially in time. As with the ability to rehearse, children 10 and 11 years old group information rather well, but younger individuals are still relatively weak in their use of this strategy. However, evidence also suggests that compared to fourth graders (9 and 10 years old) adults tend to organize information in ways that make retrieval much easier (Liberty & Ornstein, 1973).

Retrieval processes pertain to finding information located in long-term storage. The process requires both the ability to search the memory and decide that the appropriate information has been retrieved. Differences between children and adults are evident in the retrieval process (e.g., Kobasigawa, 1974). The developmental pattern for retrieving ability is similar to that of rehearsal, that is 10- and 11-year-old children are likely to use elaborate strategies to retrieve information from long-term storage but 5- and 6-year-old children are not (Kreutzer, Leonard, & Flavell, 1975). Apparently, older children are better at conducting a thorough and systematic search of the memory store and selecting appropriate material.

In regard to general long-term memory performance and effective use of the control processes, most development appears to occur between the ages of 5 to 12 years with younger children exhibiting more deficiencies than older children (Brown et al., 1983). More complex strategies such as note taking and summarizing to remember lengthy textual materials are usually not found until the adolescent years (Brown & Smiley, 1978).

Due perhaps to methodological problems associated with studies of memory over long periods of time, little information about long-term motor memory, particularly in children, is available. General observations suggest that for well-learned *continuous motor skills* (i.e., with no recognizable beginning or end) such as riding a bike and swimming, long-term motor memory loss is minimal. Studies of adults appear to support this observation. For example, Fleishman and Parker (1962) found no appreciable loss in motor performance on a tracking task after periods of 9, 14, and 24 months. Other studies using different continuous tasks have shown similar effects (Meyers, 1967; Ryan, 1962, 1965).

In regard to *discrete motor tasks* (i.e., skills with a recognizable beginning and ending) such as kicking and throwing, however, evidence suggests that forgetting can be considerable. For example, Neumann and Ammons (1957), using a stimulus-matching task, tested adult retention levels after periods of 1 minute, 20 minutes, 2 days, 7 weeks, and 1 year. The researchers found some loss in performance after only 20 minutes, with losses increasing as more time passed. After 1 year, performance indicated an almost complete loss of memory.

Speculation as to why there is a difference between continuous skill and discrete task recall has produced a number of hypotheses. Schmidt (1988) suggests that discrete skills generally have more verbal-cognitive components and these characteristics are more quickly forgotten than motor components. In a review of discrete tasks that are highly motor in nature, Schmidt proposes that retention (loss) is similar to that found in the Neumann and Ammons study, suggesting that there is more to the difference than the "motorness" of the task.

Not a great deal is known about how individuals code information for long-term storage so that it might be retrieved several years later (e.g., to ride a bicycle or swim). One possible explanation from a motor behavior perspective is that the specific patterns of a movement are not stored, but rather a motor program is formulated based on schemata that have been placed in long-term storage. The functional relationship between long-term memory and the use of schema has broader implications for motor performance; schema theory is discussed later in this chapter.

Knowledge and Metamemory   It is a widely accepted notion that the knowledge an individual possesses probably contributes in some manner to memory abilities. In an interesting study that compared adults and children on memory for chessboard displays, it was found that children had significantly better recall than adults (Chi, 1978). Initially this result seems unusual; but in fact, the reason was that the children were skilled chess players, whereas the adults were novices. It is suggested that due to the children's advanced knowledge base, they used "chunking" strategy to be more proficient at grouping items of information. It could also be assumed that because of their familiarity with the game, their processing was more efficient. Evidence such as this provides rather strong confirmation of the power of the effects of knowledge on long-term memory.

Metamemory may be defined as the knowledge or intuitions that people have about memory and themselves as memorizers. Or, more simply perhaps, it may be described as knowledge of one's own memory. An important facet of metamemory is knowledge of how to "monitor" memory during the course of learning, such as in the case when you know when you have studied enough to be sure you can accurately recall some particular information. In support of the contention that children are aware that memory is affected by knowledge, Kreutzer, Leonard, and Flavell (1975) found that as children grew older they studied differently and, as a result, were better able to remember. Even kindergarteners can verbalize about memory being affected by learning practices (e.g., study time). Metamemory has been shown to improve with age (Cavanaugh & Borkowski, 1980), but it is not clear how it is related to improvements in memory itself.

Processing speed is simply defined as the rate of speed at which information is processed. Although not all cognitive or motor tasks require information to be processed at a fast rate, processing speed may account for many performance differences between children and adults. In a movement setting, there are numerous

PROCESSING SPEED AND MOVEMENT TIME

■ *Objective 7.5*

instances when the ability to recognize a stimulus and process information quickly is critical to an effective motor response (e.g., catching a ball or playing Ping-Pong or badminton). Successful motor performance is based on a combination of perceptual recognition (attention), speed of memory functions, and neuromuscular response time.

Reaction time (RT) is the basic measure of processing speed. As noted in earlier discussions of perception and attention, **reaction time** refers to the interval of time between the onset of a stimulus (e.g., light, buzzer, shock) and the initiation of a motor response (e.g., lifting a finger off a button). When the individual being tested is asked to respond to only one stimulus, the task is referred to as *simple reaction time. Choice reaction time* tasks are more complex and involve a greater information load because more than one signal requires discrimination in a more unpredictable setting. For example, a red light may mean move one way, blue another, and yellow another. A well-known concept related to processing speed, information load, and complexity is *Hick's Law:* Reaction time increases linearly as information load increases. That is, as the complexity of the processing task increases, so does the time it takes to process that information.

The time required to complete motor response is referred to as **movement time.** Movement time combined with the reaction time is known as **response time.** Reaction time is usually calculated as the amount of information (or "bits") processed per second.

A comprehensive review on the topic of age differences in performance on speeded tasks suggests that differences are large and remarkably consistent (Kail, 1991). Speed of processing, as determined by reaction time, improves with age up to the early adult years and then remains relatively stable until about age 60 (Hodgkins, 1962; Wilkinson & Allison, 1989). The period of greatest improvement appears to be between the ages of 6 and 15 (fig. 7.4).

More specifically, Hale (1990) found that on a simple reaction time task, 10-year-olds were about twice as slow compared to young adults, whereas 12-year-olds were approximately 1.5 times slower. However, the processing speed of a group of 15-year-olds was about the same as the young adults. Adding to this, Surwillo (1977) estimates that it takes a 5-year-old nearly three times as long as a 17-year-old to process one bit of information. This estimate was based on differences between simple and choice reaction times. It has been suggested that the slower processing speed of children is due primarily to central processing factors, such as attention and speed of memory processes (Elliot, 1972).

It was alluded to in the earlier discussion of memory storage that children and adults differ in their ability to process information into memory. It is also apparent that there is a difference in the speed with which these memory functions are performed. Along with the greater difficulty they have in searching for and retrieving information, children appear to take longer than older individuals to encode information into memory (Chi, 1977a, 1977b). Chi found in her studies that 5-year-olds require more than twice as much time as adults to identify familiar

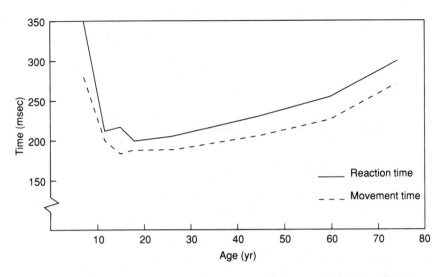

Figure 7.4
Reaction time and
movement time as a
function of age.

stimuli (pictures of classmates), suggesting that adults encode information about twice as fast as children. Apparently this efficiency allows more time for further processing in short-term memory.

Most discussions of reaction time also include movement time, particularly within the context of motor behavior. The close or, in most conditions, inseparable association between the two phases of a movement response time has been discussed (see fig. 7.4). Movement time is usually defined as the interval from the initiation of the movement (or the end of RT) to the completion of the movement.

A well-known law of motor behavior that is closely associated with movement is *Fitts's Law* (Fitts, 1954). This law is also referred to as the *speed-accuracy trade-off*, which means simply that when performers attempt to do something more quickly, they typically do it less accurately. That is, under most movement conditions, as speed increases, accuracy decreases. Exceptions are those conditions where moving too slowly is not conducive to motor efficiency that can affect accuracy (e.g., striking a fast-moving object).

Other factors such as task complexity and movement distance can also affect an individual's movement time. Movement time is often described as the capacity of the motor system to use an amount of information per second. Within this definition, capacity is expressed as the index of difficulty divided by the movement time and is evidenced when movement times become faster (lower) for a specific information load. Evidence suggests that capacity increases with increasing age up to adulthood.

According to a review by Keogh and Sugden (1985) on reciprocal and discrete tapping tasks with varying levels of difficulty, movement time is generally faster for adults at any level; and older children have faster responses than younger children. Movement times across all age groups slow as difficulty increases.

PROGRAM-MING

■ *Objective 7.6*

Perhaps the most sophisticated mental operation is that of programming. In a broad sense, **programming** can be defined as cognitive processing that results in the formulation of a thought, cognitive expression, or motor program. Depending on the task at hand, several factors are involved in these functional operations. For example, attention, perceptual awareness, and information stored in the short-term and long-term memory systems all influence programming. Recall from the general information-processing model that individuals collect information from the external and internal environments through the various perceptual modalities, interpret that information, decide whether a motor response is appropriate, and if so, generate that response. Obviously it is also possible to generate a thought or motor program in the mind without perceptual information, but in most cases, an overt expression requires additional environmental information.

A major psychological determinant in the ability to program is *cognition.* This term generally refers to any mental process that results in the generation or obtaining of knowledge. It is also known as the means by which individuals can think. Cognition is an integral part of perceiving, recognizing, conceiving (i.e., generating information), decision making, reasoning, and varying any of the perceptual-conceptual processes. One of the most popular theories of cognitive development was described in chapter 1 in reference to Piaget's work. Some of the theorist's original ideas can be linked to contemporary notions regarding verbal and motor skill learning theory. The reader may find it helpful to refer back to chapter 1 for a more detailed review of Piaget's work.

In the context of this discussion, three of Piaget's ideas are applicable. Piaget postulates that humans characteristically systematize and organize mental processes and the content of knowledge into coherent systems; this is referred to as *organization.* It was also his belief that individuals use *adaptation;* that is, people use cognitive processes to change experiences into a cognitive form that can be used later in addressing new situations. The third notion, the formulation of **schema,** has direct relevance to programming.

Piaget was not the originator of the term (Bartlett, 1932; Head, 1926) but he did popularize the concept of schema as important to a comprehensive theory of cognitive development. In Piaget's framework, organization and adaptation combine to produce cognitive structures called *schema.* Schema was defined as a number of ideas or concepts organized into a coherent plan (program), or outline, that indicates the relationship between concepts. The definition of schema that perhaps best fits contemporary motor learning theory is described by Schmidt (1988) in reference to Bartlett's (1932) thoughts: "For Bartlett, the schema is an abstract memory representation for events or skilled actions, thought of as a rule, concept, or generalization" (p. 483).

Schmidt (1975, 1988), a renowned motor behavior scientist and the originator of the schema theory of motor skill learning, notes that at the heart of schema is the notion of the *generalized motor program.* Schmidt describes the **motor program** as a memory representation of a class of actions that is responsible for producing a unique pattern of motor activity if the program is executed.

The emphasis and application of the motor program concept is the core of schema theory.

Schema theory offers an explanation of the way individuals learn and perform a seemingly endless variety of movements. The theory suggests that the motor programs stored in memory are not specific records of the movements to be performed; rather, they are a set of general rules, concepts, and relationships (schemas) to guide performance in keeping with the concept of a generalized motor program. Basically individuals store past movement experiences in memory. Storage of "movement elements" and the relationship of these elements to each other is called *movement schema*. Individuals call up the schema for programming or, in a sense, they "piece together" desired movements.

Schema Theory

Schema theory suggests an explanation for two characteristics of human performance that previously existing theories had difficulty in clarifying (e.g., Adams, 1971). First, individuals rarely repeat a set of movements precisely in the same manner. If a separate program were required for each movement variation performed, our storage capabilities would be quickly surpassed. Second, individuals are capable of performing movements to fit seemingly novel situations. An example of the theory in practice is the performance of an individual playing shortstop in baseball or guard in basketball. The shortstop can field a ball from numerous positions, many of which are unpracticed, and still get the ball to first base, just as the basketball player can shoot successfully from almost any position on the court. Schema theory treats motor programs in much the same manner as concepts are negotiated in verbal learning. The motor program begins with the cognitive domain and perception of incoming information. The individual who has practiced throwing far, hard, soft, or short has good cognitive sense of what may be in the middle.

According to Schmidt (1975), goal-directed movement leads to storage of (a) initial conditions, or the state of the internal and external conditions; (b) response specifications (e.g., release angle, muscular force, arm speed, etc); (c) sensory consequences of the movement (performance feedback); and (d) outcome or result of the movement. These four sources of information combine to produce a motor schema that is diverse in conception and scope. Ultimately, though individuals may forget the specific instances of a movement, they retain the general schema that enables them to repeat a movement or perform a new variation of it.

Schmidt's theory holds that there are two schema: recall and recognition. The recall schema is the relationship, developed from previous experience, between the response specifications and the actual outcome (knowledge of results). To produce a novel response, the subject enters recall schema with the desired outcome and the initial conditions, and the schema rule generates the specifications for that response. Once the response specifications have been determined, the response can be executed by means of a motor program.

Recognition schema is the relationship, built up over past experience, between the sensory consequences and actual outcomes. For each response the initial conditions are considered, and the sensory consequences and actual outcome are coupled to build this relationship. During a response, the subject designates the desired outcome and by means of the recognition schema predicts the expected consequences of the response.

The motor schema (concept) for a general skill area (e.g., throwing, jumping) is bounded by dimensions related to space, time, and force. Each dimension represents a continuum that may, depending on experience, be very limited or quite diverse. Theoretically, the greater the variety of experiences produced by the individual, the more diverse the schema becomes and the greater the capacity to move. The motor schema enables the individual to select the appropriate level from each dimension to program a task that may be known or novel. In shooting a basketball, for example, the child calls upon a program consisting of a relationship among distance the ball has to travel, required muscular force, arm speed, and angle release—all of which may change from one attempt to the next.

In schema theory each performance is another instance of the relationship among the dimensions of the motor program. This is true in both closed and open skills. In closed skills the goal and the environmental conditions are relatively constant (e.g., bowling, bean bag toss to a fixed target) and the development of the schematic rule leads to more and more accuracy. In open skills the environmental conditions are in constant flux (e.g., soccer, basketball, tennis); the development of the schematic rule in relation to initial conditions enables the performer to generate novel responses. For schema theory, the essential difference between open and closed skills is the unpredictability of the environment in the open skill situation. The motor response is generated in the same way for both open and closed skills; but the response for open skills is based on the individual's best estimate of the changing environment.

A major prediction of schema theory is that increasing variability in practice on a given task will result in increased transfer to a novel task of the same movement class. The strength of the schema, whether it be recall or recognition, is a positive function of the amount and variability of practice of responses within a movement class. The theory predicts that stronger schemata should (a) produce less error in the initial trials of a novel response governed by the schema and (b) result in an increased rate of learning of a novel rapid response in the absence of knowledge of results.

Most studies have tested these predictions by manipulating the variability of practice factor. Based on the variability of practice hypothesis, it is predicted that subjects exposed to a high variability of practice regime but who never experience the test condition will perform as well or better than subjects who practice only on the test condition relevant to a final observation on that test condition. In addition, it is predicted that the high-variability group will outperform the specificity group on transfer to novel tasks. Numerous studies supporting the variability hypothesis have been generated using young children and adults as subjects and a gross motor skill as the task (e.g., Carson & Wiegand, 1979; Kelso & Norman, 1978; Kerr & Booth, 1977, 1978; Moxley, 1979; Wulf & Schmidt,

1988). The general findings are that children seem to have schemas and can use them to organize movements in slightly different movement situations.

For example, Carson and Wiegand (1979) found that young children (ages 3 to 5) with more variable practice experience had greater overall success when throwing a beanbag of a new weight at a target on the floor and at a relocated target attached to a wall. The variable-practice group also maintained its performance level in all conditions after a period of two weeks, in contrast with a loss in performance by the control, low-variability, and specific-practice groups. The findings of Carson and Weigand demonstrated that young children can establish a movement rule or relationship, use it, and retain it for later use.

Evidence from the developmental psychology literature also supports the presence of schemata in children and adults. Schemata for events are referred to as *scripts;* evidence suggests these may emerge by the end of the first year (Nelson, 1977; Schank & Abelson, 1976) and that scripts are clearly evident by the time children start school (e.g., Furman & Walden, 1990). Light and Anderson (1983) suggest that scripts are highly similar in young adults and the elderly and that both groups appear to make inferences (a form of deep processing) based on their scripts.

According to a review by Shapiro and Schmidt (1982), one conclusion that emerges from the literature when comparing studies using adults and children is that children's motor skills are apparently more easily affected by variability in practice. This observation suggests that the recall schemata of children are easily developed, whereas adults may have already developed schemata for the relatively simple tasks used in the investigations. The authors also note that various rival theories (e.g., Adams, 1971) cannot explain how performance on variations of a task leads to essentially the same transfer performance in adults as practicing the criterion task itself. Nor can they explain why varied practice for children was consistently more effective than practicing the criterion task itself.

These findings can easily be explained by the schema theory. Each trial, whether it be the first one experienced or a novel criterion task, is perceived as novel; that is, subjects presumably prepare the movement on each trial as if it were the first. Because all responses under this view are novel, then an effective way to prepare for the criterion task is with varied practice. This finding has important implications for the structure of practice sessions, particularly with children for whom varied practice is most effective (Gabbard, 1984; Kerr, 1982).

Variability in practice is predictively more effective for children than for adults simply because young individuals have considerably more to learn. Typically, children learning to read are taught to recognize letters, then parts of words, then complete words, and finally sentences. Children studying mathematics learn to solve problems after they have grasped the basic functions of numbers and signs. When a child learns to play a musical instrument, he or she studies the scale before attempting a song. In physical education or a youth sport setting, however, children are frequently taught games and dances before they are able to perform prerequisite skills. Generally, schema theory supports variability in practice and problem solving (within the same class of movements and rules) during early years rather than instruction in specific sport skills.

Developmental Theories of Motor Control

■ *Objective 7.7*

Although it is generally accepted that the control of motor behavior is under the direction of some type of generalized motor program, the question of what the program actually controls and how control develops has been one of the most active areas of contemporary motor development research. Under the heading of *developmental biodynamics,* as discussed in chapter 1 (Lockman & Thelen, 1993), this general approach to the study of perception and action is based on the general hypothesis that motor coordination and control emerge from continual and intimate interactions between the nervous system and the periphery, the limbs and body segments (in essence, the brain and body). From this multifaceted approach, three ideas have emerged that would appear to have exciting implications for better understanding the processes and mechanisms involved in motor development: *coordinative structures, dynamical systems,* and *neuronal group selection.*

Coordinative Structures   The general consensus in past years was that the individual muscles were given instructions by the commands produced in the motor program. The ideas of Bernstein (1967) challenged this notion, arguing that there are too many independent operations for the motor program to control them all simultaneously. Researchers in more recent years have supported Bernstein's initial concern (Greene, 1972; Turvey, 1977), finding it difficult to consider that the nearly 800 muscles and 100 mobile joints of the body are under the control of the central nervous system in all the possible "degrees of freedom" (i.e., controlled action of an independent joint movement). Bernstein suggested that motor programs, rather than controlling individual muscles, control groupings of muscles instead. These groupings of muscles with associated joints have been termed **coordinative structures,** dynamic structures, and *synergies.* The commands generated by the motor program are said to be directed toward the specific coordinative structure that, as a group, is constrained to act as a single functional unit. A simplified analogy would be the actions of a puppeteer who controls several "degrees of freedom" (e.g., the action of the whole arm of the puppet) with a single finger.

Although the actual nature of how the motor program controls movement patterns is not yet clear, the existence of coordinative structures has been verified (e.g., Kelso et al., 1984; Kelso, Southard, & Goodman, 1979) and has stimulated considerable inquiry in developmental research, much of which has been associated with the dynamical systems perspective.

Dynamical Systems   In chapter 1, we saw that the **dynamical systems** perspective seeks to provide an understanding of "how" movement and control emerges or unfolds developmentally. Based on highly complex principles from theoretical physics, theoretical mathematics, and ecological psychology, this theory proposes that qualitative changes in motor behavior emerge out of the naturally developing dynamic properties of the motor system and coordinative structures (Kugler, Kelso, & Turvey, 1982, 1980).

The following features of dynamical systems should be noted (as interpreted from Thelen & Ulrich, 1991):

1. Movement appears to emerge from *self-organizing properties* of the body. Individuals are composed of several complex and cooperative systems (e.g., perceptual, postural, muscular, skeletal), with each developing at different rates; thus the intricacy of coordinated movement. Even the simplest skill requires a cooperative effort with each system providing critical input. For example, muscular strength of the legs and postural control are critical controllers for walking. Supposedly, for each skill there is a *rate controller* that organizes the system or systems needed to execute the task.

2. Another assumption is that movement is determined not only by muscular forces but also by mechanical interactions (e.g., body segments, joint reaction forces). The contribution of these forces may be influenced by speed of movement, body position, length and mass of body segments, and the intentions of the movement. Research in this area seeks to examine the developmental characteristics associated with the generation and apportionment of forces and the timing-based control system.

3. In regard to the question of continuity or discontinuity, the dynamical systems perspective suggests that in the transition from old movement patterns to new ones, disruptions (discontinuities) occur in performance. The major challenge is to account for these characteristics from processes that are inherently continuous.

4. In addition to the biological contexts of dynamical systems, the nature of the environment and demands of the task influence development.

Using the dynamical systems perspective in motor behavior, recent inquiries have begun to unfold the developmental picture of "how" interlimb coordination emerges in such early motor tasks as kicking (Thelen, 1985), stepping patterns (Thelen & Ulrich, 1991; Thelen, Ulrich, & Jensen, 1989), reaching (Thelen et al., 1993), hopping (Roberton & Halverson, 1988), and independent walking (Clark & Phillips, 1993). Specifics from some of these studies will be described in chapter 8.

This line of developmental research shows great promise for providing a more comprehensive understanding of motor control and performance across the life span. It is hoped that future studies will complete the picture of lifelong development by observing the effects of aging on the coordinative structures of older adults. Since the presence of dedifferentiation and other neurological losses are assumed sometime during late adulthood, it is likely that there is some accompanying effect on the coordinative structures. For a more detailed discussion of dynamical systems theory refer to Thelen and Ulrich (1991) and the special feature in *Child Development* introduced by Lockman and Thelen (1993) listed in the Suggested Readings.

Neuronal Group Selection   The dynamical systems perspective provides a glimpse into self-organizing properties of developing and mature motor systems, but by itself it does not identify specific underlying neural mechanisms. The theory of **neuronal group selection** (Edelman, 1989, 1987) holds such promise and is tied to the notion that the emergence of coordinated movement is closely tied to the growth of the musculoskeletal system and development of the brain. One of the key issues is how changes in brain circuitry-controlling synergies (coordinative structures) become matched to developmental changes in the musculoskeletal system and the environment of the organism.

According to Sporns & Edelman (1993), during the early stages of development, neuronal circuits are not precisely "wired" to execute specific skills. Instead the brain contains variant circuits (structural variability) with dynamic properties. That is, those circuits selectively form *neuronal groups,* localized collections of hundreds to thousands of interconnected neurons. Neuronal groups are arranged in the brain in neural maps, representing the body surface. Though they are separated in regions of the cortex, they are anatomically linked through long-range neural connections, thus allowing for global mapping. Theoretically these selected neuronal groups share functional properties in a temporary manner to accommodate the desired task.

In essence, the theory of neuronal group selection accounts for the brain's organization of synergies as functional units of motor control (coordinative structures). One of its merits is that it accounts for the spontaneous adaptability of coordinative structure in response to biomechanical and environmental changes. Therefore, as with the dynamical systems perspective, neuronal group selection supports the contextualist approach to the study of motor development.

A commonality among all three theoretical notions is the suggestion that the brain and body are not "prewired" for skilled movements; rather, they have amazing adaptable "self-organizing" properties that adjust for biological and environmental contexts.

CHANGES WITH AGING

■ *Objective 7.8*

Attention

It was noted in the discussion of the effects of aging on brain structures that not only does a significant loss of neurons occur, but other possibly debilitating changes occur within neurons themselves (e.g., senile plaques, vacuoles, tangles). However, even with these conditions, it is believed that the brain may only lose a small portion of its ability to function.

Although not all sources are in agreement, older adults generally have more difficulty than younger individuals in dividing their attention between two tasks. One of the more frequently noted explanations for this decrement revolves around the possibility of reduced amounts of processing capacity (Craik & Simon, 1980; Wright, 1981). These findings are similar to some of the research described earlier on divided attention in children, promoting speculation that processing capacity is a life-span trend—that is, capacity increases during childhood, that peak capacity is reached in young adulthood, and then declines in old age (Craik & Simon, 1980; Hasher & Zacks, 1979).

An alternative view is that the total amount of capacity stays constant across age but that young and old require more capacity to perform elementary functions. Thus, less space would be available for higher-level processing operations (Flavell, 1984; Kail, 1984). Although these views are somewhat different, both suggest that capacity (i.e., total or space capacity) is affected by age and that the old and young exhibit similar characteristics.

Short-term sensory capacity changes little across time, but if the information must be held for further processing (encoded) to the short-term memory system, the recall of the young child will diminish quickly compared to the adult. This also appears to be the case in older individuals. That is, elderly individuals are more likely to lose information before it can be transferred to short-term memory (Poon, 1985).

Sensory Store and Short-Term Memory

In general, short-term memory declines only slightly during most of adulthood, with greater deficits appearing in older persons. As evidenced by performance on digit span and working memory tasks, short-term memory does not change appreciably during most of adulthood, although there may be a decline after the age of 60 (e.g., Botwinick & Storandt, 1974; Dobbs & Rule, 1989; Fozard, 1980). The study by Botwinick and Storandt found that memory span remained relatively constant from the twenties through the fifties, then dropped only about one item following those years. However, greater comparative deficits between young and older adults have been found on other tasks, such as math problems and the linear ordering of items (Foos, 1989).

Explanations for the age-related changes in short-term memory are mixed. Some sources contend that the difference is a result of information-processing abilities growing less flexible with age, a loss of information during reorganization (encoding), or a diminished ability to process the encoded information (Perlmutter, 1983; Walsh, 1983). To the encoding deficit explanations has been added the hypothesis that the information may be encoded but cannot be retrieved (Smith, 1980). In addition to this processing deficit explanation, the hypothesis that older adults have a storage capacity deficit compared to young persons has also been supported by research (Foos, 1989). However, that study acknowledged that older adults may have some processing deficit as well.

Older adults often exhibit significant deficiencies on memory tasks that involve the transfer of information between short- and long-term stores. In general, older adults are less able to recall details of past events. This deficiency is likely due to the difficulty that elderly people have with both encoding and retrieving information from the long-term store (Craik, 1977; Poon, 1985). Older adults, like young children, apparently do not engage as frequently as young adults in deep elaborate encoding strategies. However, wide individual differences are normal. A study by Till (1985) found relatively small deficits on long-term memory tasks among well-educated,

Long-Term Memory

elderly individuals. Long-term memory in older adults has been shown to improve with instructions to use imagery and when plenty of time is provided.

Metamemory and Knowledge

Interestingly, older adults may use metamemory to compensate somewhat for limitations in encoding and retrieving information. A number of studies suggest that with age, adults maintain or increase their understanding about the memory system (Perlmutter & Hall, 1985). A somewhat separate but related consideration is how knowledge of materials influences memory. Just as knowledge of materials can positively affect age differences in children, it can also override age differences in adults. Barrett and Wright (1981) found that though young adults outperformed older adults on tasks of "young-familiar" words, the elderly showed superior memory with "old-familiar" words.

Although there is no doubt that a tremendous base of knowledge is accumulated throughout childhood, there is little evidence that basic knowledge shows a similar decline in late adulthood. Tests of common, factual knowledge typically show no decline from young adulthood to old age (Perlmutter, 1980). In fact, studies indicate that the ability to learn is approximately the same at the age of 80 years as it is at 12 years (Denney, Jones, & Krigel, 1979).

Processing Speed and Movement Time

One of the most consistent findings in the study of aging is a decline in processing speed as measured by reaction time and other manipulative tasks (e.g., Cerella, 1990; Dixon, Kurzman, & Friesen, 1993; Salthouse, 1985). If it takes a 20-year-old 200 milliseconds to perform a reaction time task, the same task will take a 60-year-old about 250 milliseconds, and an 80-year-old over 300 milliseconds (fig. 7.4). In contrast, a 6-year-old typically performs the task in just under 350 milliseconds. Welford (1984) estimates that older adults may decrease 20% to 50% in processing speed depending on the complexity of the task.

In regard to processing and movement time (response time) it appears on the basis of limited information that aging brings about little change in the performance of single, discrete arm/hand actions and simple repetitive tasks that can be programmed in advance (Welford, 1977; Welford, Norris, & Shock, 1969). However, much greater decrements among older persons have been found on timed (speed) tasks that involve decision making based on perceptual information and the programming of movement sequences. This is partially evidenced by large performance decreases on complex sequential tasks requiring a series of different movements, particularly when speed is a critical factor (Lupinacci et al., 1993; Welford, 1984; Williams, 1990). Overall, older adults are capable of demonstrating accurate motor performance on well-rehearsed tasks when speed is not a critical factor.

The general slowing of the body with advanced aging constitutes a universal phenomenon referred to as *psychomotor slowing*. Although the mechanisms responsible for this condition remain unexplained, Cerella (1990) contends that slowing is attributed to numerous factors associated with advanced aging (e.g., neuron loss, synaptic delay, decrease in excitability, decrease in nerve conduction velocity), which causes disruptions of connections within the neural network. Supposedly each disruption increases the time to process information. Welford

Table 7.1 Summary of Selected Information-Processing Changes with Aging

Information-Processing Characteristic	
Attention	Declines
Sensory store	Capacity remains stable; ability to "hold" for further processing diminishes
Short-term memory	Declines slightly
Encoding and retrieval abilities	Decline
Long-term memory	Declines
Metamemory and knowledge	No apparent change
Processing speed	Slows
Movement time	Slows slightly
Response time	Slows on tasks requiring that movement sequences be performed quickly
Programming	Ability to learn new motor tasks and program movement sequences (no apparent change)

(1984) suggests that although neuromuscular factors contribute to slowing with age, most of the decrement can be attributed to central mechanisms (i.e., central nervous system). The topic of psychomotor slowing will be discussed in greater detail with performance examples in chapter 11.

Programming

Welford (1980, 1984) suggested that the major limitations of older adults in the performance of a motor task are their ability to make decisions based on perceptual input and to program movement sequences. Evidence also suggests that older adults learn new motor and cognitive tasks more slowly than young individuals. One may speculate that while the schemata of older people can be complete for skills learned in past experiences, the ability of the mind to form new schemata diminishes somewhat with age.

Table 7.1 presents a summary list of selected information-processing changes with aging. A general observation is that young children and older adults have several characteristics in common.

SUMMARY

Of the numerous existing theories of human development, information processing dominates the field of cognitive psychology and motor behavior. The four basic components of the process are sensory input, reception through the various perceptual modalities, interpretation and decision making (processing), and overt motor response. Information processing can be discussed under the general topics of attention, memory, processing speed, and programming. Although much about information processing is unknown and highly theoretical, it has been generally observed that as individuals progress toward adulthood, they are able to process more information in a shorter period of time.

The core of most information-processing models is attention. The primary importance of attention to motor behavior involves alertness and the ability to select and attend to meaningful information. The developmental trend in selective attention appears to move from young children, who primarily attend exclusively to one stimulus (and are easily distracted), to older children who may attend to too much information (some of which is irrelevant), and finally, to the early adolescent who shows the more adultlike ability to attend to relevant information in complex conditions. Age differences in this aspect of information processing are attributed primarily to experience and the refinement of strategies.

The three basic memory systems of the multistore model are short-term sensory store, short-term memory, and long-term memory. Other theoretical views include the levels-of-processing model and the component functional model. Although the three models present seemingly different views of memory, all three consider control processes (functional abilities) to be vital components in the overall process.

Evidence that primitive forms of memory are present in the newborn is found in the observation of habituation. In addition to age differences in some aspects of storage capacity, differences between young children and adults can also be found in coding, searching and retrieving, and recalling information. Other functional abilities that have developmental implications are rehearsal and chunking. An individual's knowledge or intuition about memory and themselves as memorizers is known as *metamemory*. Although this function improves with age, its link to improvements in memory itself is not clear.

The rate at which information is processed, as reflected by an individual's reaction time, improves with age up to the early adult years, then remains relatively stable until later adulthood. Though not all motor tasks require that information be processed at a fast rate, processing speed may account for many age-related performance differences. This appears to be especially evident when comparing young adults to young children and the aged.

Programming is considered one of the most sophisticated operations in information processing. With regard to the development of a theoretical base for motor programming, the concept of a *schema* has received considerable attention in recent years as a result of the schema theory of motor skill learning. According to schema theory, individuals store past movement experiences in memory. The storage of movement elements and their relationship to each other is referred to as *movement schema*.

An individual, in a sense, calls up the schema to program desired motor responses. The motor programs stored in memory are not specific records of the movement to be performed; rather, they are a set of general rules, concepts, and relationships that guide programming of a generalized motor program. A major prediction of schema theory is that practicing a variety of movement outcomes within the same general class will provide a diverse set of experiences upon which to build schema.

The notion of some form of motor program has been generally accepted, but it does not provide an explanation of the link between brain on body (motor control). Three aspects of the developmental biodynamics perspective offer some insight: *coordinative structures, dynamical systems,* and *neuronal group selection*. A commonality among these ideas is suggestion that the brain and body are not "prewired" for skills movements, rather, they have adaptable "self-organizing" properties that adjust for biological and environmental contexts.

It is fairly well accepted that changes in the brain structure with advancing age produce some degree of regression in information-processing abilities. A general observation is that the abilities of the old approximate those of children; both perform below the level of the young adult. This relationship appears, at least to some degree, in all of the information-processing characteristics with the possible exception of metamemory.

SUGGESTED READINGS

Cerella, J. (1990). Aging and information processing rate. In J. Birren & K. Schaie (Eds.), *Handbook of the psychology of aging.* San Diego, CA: Academic Press.

Ellis, H. C., & Hunt, R. R. (1983). *Fundamentals of human memory and cognition* (3rd ed.). Dubuque, IA: Wm. C. Brown.

Kail, R. (1984). *The development of memory in children.* New York: W. H. Freeman.

Lockman, J. J., & Thelen, E. (1993). Developmental biodynamics: Brain, body, behavior connections. Special feature. *Child Development, 64,* 953–1190.

Magill, R. A. (1993). *Motor learning: Concepts and application* (4th ed.). Dubuque, IA: Wm. C. Brown.

Marteniuk, R. G. (1976). *Information processing in motor skills.* New York: Holt, Rinehart and Winston.

Newell, K., & Corcos, D. (Eds.). (1993). *Variability and motor control.* Champaign, IL: Human Kinetics.

Ostrow, A. C. (Ed.). (1989). *Aging and motor behavior.* Indianapolis: Benchmark Press.

Schmidt, R. A. (1988). *Motor control and learning* (2nd ed.). Champaign, IL: Human Kinetics.

Thelen, E., & Smith, L. (1994). *A dynamic systems approach to the development of cognitive and action.* Cambridge, MA: MIT Press.

Thelen, E., & Ulrich, B. (1991). Hidden skills. *Monographs of the Society for Research in Child Development* (223), *56* (1).

Tulving, E. (1985). How many memory systems are there? *American Psychologist, 40,* 385–398.

PART ∘ *four* --

-----------------▶ *c h a p t e r s*

In the study of motor development researchers gather information related to *growth* (change in size), *development* (change in level of functioning), and *motor behavior* (performance) across the life span. The text up to this point has primarily addressed the growth and development aspects of the various systems of the human body and the factors that can affect them. Part Four builds upon this foundation by documenting observable movement characteristics that occur within the phases of motor behavior.

In essence, motor behavior is the movement expression of the human body in regard to its physical size, neurological and physiological functioning, and information-processing abilities. For the study of motor development, the information provided so far serves as the basis for understanding why characteristic movement abilities

Motor Behavior across the Life Span

within each phase are feasible and what factors of change are responsible for the transition across phases. For example, what factors are responsible for the regression of motor performance? What characteristics account for the peak performance generally found between 25 and 30 years of age? What factors account for the transition from reflexive behavior to voluntary movement? These kinds of questions are the foundation of motor development inquiry.

Part Four is divided into four chapters, each describing those movement characteristics associated with one or more of the phases of motor behavior. Chapter 8 describes the reflexive, stereotypic, and rudimentary behaviors that span the prenatal stage through infancy (2 years of age). The development of fundamental movement patterns and manipulative behaviors characteristic of the early childhood years (approximately ages 2 through 6) are documented in chapter 9. Chapter 10 presents information related to motor behavior during later childhood and the adolescent years (age 7 to 18), in which individuals generally exhibit sport skill abilities and rapid periods of physical growth and development. Chapter 11 of this part of the text deals with the characteristic behaviors of the adult years, which normally feature peak motor performance and the inevitable course of regression.

PHASES OF MOTOR BEHAVIOR

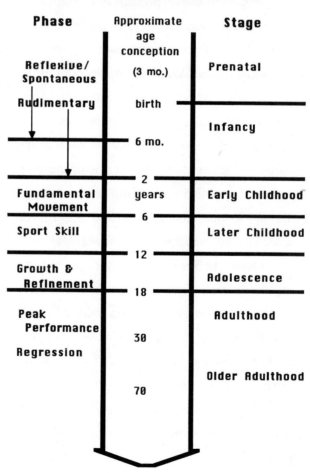

Phase	Approximate age	Stage
	conception	
Reflexive/ Spontaneous	(3 mo.)	Prenatal
Rudimentary	birth	
		Infancy
	6 mo.	
	2 years	
Fundamental Movement		Early Childhood
	6	
Sport Skill		Later Childhood
	12	
Growth & Refinement		Adolescence
	18	
Peak Performance		Adulthood
	30	
Regression		
		Older Adulthood
	70	

Early Movement Behavior

Upon completion of this chapter, you should be able to

1. Discuss reflexes in regard to definition, control, relevance, and developmental characteristics. **221**
2. Identify the three general types of reflexes and provide examples of each. **223**
3. Explain the various theoretical views of reflex behavior. **233**
4. Describe the developmental characteristics of spontaneous movements (stereotypies). **236**
5. Provide a brief overview of the rudimentary phase of motor behavior. **238**
6. Outline the developmental milestones associated with basic postural control and rudimentary locomotion. **238**
7. Describe the developmental characteristics involved in manual control during the first two years of life. **247**
8. Discuss the theoretical views associated with motor asymmetries. **251**

OBJECTIVES

---

reflexes
primitive reflexes
postural reflexes
locomotor reflexes
continuity view
spontaneous movements
rudimentary behavior

crawling
creeping
walking
prehension
manipulation
manual control
motor asymmetries

KEY TERMS

PHASES OF MOTOR BEHAVIOR

Phase	Approximate age	Stage
Reflexive/ Spontaneous	conception (3 mo.)	Prenatal
Rudimentary	birth	Infancy
	6 mo.	
Fundamental Movement	2 years	Early Childhood
Sport Skill	6	Later Childhood
Growth & Refinement	12	Adolescence
Peak Performance	18	Adulthood
Regression	30	Older Adulthood
	70	

The beginnings of human movement are those characteristic behaviors that appear during the prenatal and infancy stages of development. Coinciding with these stages that cover the first two years of life are the reflexive and rudimentary phases of motor behavior. This chapter will describe and explain the importance of reflexive and stereotypic behaviors. It will also document the development of early voluntary movement with regard to postural control, locomotion, and manual control. Evidence of early motor asymmetries and theoretical views on the topic will also be discussed.

It has already been established that motor development is not a process that begins at the time of birth but, rather, has its origins in the prenatal period. Reflexes can be elicited in the fetus as early as the second or third month after conception, and most are present by the time of birth or before. In fact, many of the superficial, cutaneous reflexes are present by the fifth or sixth fetal month. **Reflexes** are defined as involuntary movement reactions that are elicited by such forms of sensory stimuli as sound, light, touch, or body position. They are controlled primarily by the subcortical areas (lower brain centers), which are also responsible for numerous involuntary, life-sustaining processes such as breathing and heart rate.

The initial movement responses of the body are controlled in the subcortical areas due to the maturational stage of the central nervous system. As the nervous system matures, reflexes come under the command of the brain stem and midbrain rather than remain solely under spinal cord control. Finally, as the cerebral cortex (higher brain center) matures, most of these "transient" reflexes become gradually inhibited and voluntary motor behavior eventually takes over. The motor area of the cerebral cortex is generally considered to be minimally functional in the developing fetus and newborn infant, thus most movement behavior is thought to be primarily involuntary. The human organism in the first six months to one year is essentially a reflex machine that undergoes a continuous process of neuromuscular functional maturation. Of the estimated 27 major infant reflexes (Lorton & Lorton, 1984), most disappear by 6 months of age in the normal developing child and few are observable after the first year. However, some reflexes, such as coughing, blinking, and sneezing, persist throughout life.

The importance of reflex behavior is primarily associated with its role in stimulating the central nervous system and muscles, in infant survival, and in its use as a diagnostic tool for assessing neurological maturity. Until the information-processing mechanisms become mature enough to consciously formulate motor programs and execute movement responses, reflexes are one of the primary modes of stimulating the CNS and muscles. It is suggested, though somewhat speculatively, that this mode of initial stimulation is vital to optimal growth and development of the mechanisms that will eventually control voluntary motor behavior. This supposition is based partly on the fact that reflexive behaviors stimulate neural pathways, memory trace activity, and muscle tonicity. Several researchers have supported the notion that reflexes play a dominant role in the

REFLEXIVE BEHAVIOR

■ *Objective 8.1*

Importance of Reflexes in Infant Development

development of muscle tone that, in time, becomes critical to the performance of future voluntary actions (e.g., Coley, 1978; Fiorentino, 1981). In an article published in *American Scientist,* Easton (1972) comments that "normal motor coordination is based to a huge extent on the reflexes" (p. 591). Theoretical views on this topic will be discussed later in this chapter.

The presence of some reflexes offers infants certain survival advantages in regard to seeking nourishment and, perhaps to a lesser degree, protection. Two prime examples of these *primitive reflexes* are the rooting reflex and the sucking reflex. The *rooting (search) reflex* is exhibited when the newborns automatically turn their mouths toward the mother's nipple (or bottle) when it touches their cheek or lips. In the *sucking reflex,* newborns reflexively suck on anything that touches their lips. Interestingly, the neonate is not born with the voluntary capacity to ingest food. Thus the sucking reflex enables the newborn to feed by involuntary means.

While perhaps not as critical to survival as the reflexes related to feeding, certain involuntary responses have also been described as *protection mechanisms.* One example of this type of survival response is the *Moro reflex.* When a newborn is startled or begins to fall, the arms and legs fling outward, the hands open, and the fingers spread. As the reflexive action for protection continues, the arms draw close to the body, the fists clench, the eyes open, and the infant lets out a loud wail. Although some of the primitive reflexes do not readily appear to have any survival value, it has been speculated that they once served a protective function. The assumption is that certain primitive protective reflexes evolved from lower primates to human infants, as in the case of the *palmar grasp* (or *grasping reflex*). Upon tactile stimulation of the infant's palm, the hand will close reflexively around an object and grip tightly. Researchers have speculated that this reflex evolved among primates so that infants could cling to their mothers as they moved about.

Reflexes are not only important to the survival of infants but also serve as an indicator of general neurological status. The genesis of reflex behavior is well documented in regard to onset, persistence, and disappearance. Since infant reflexes occur within rather predictable age ranges, these behaviors have been used extensively by pediatricians as diagnostic tools for assessing neurological maturity. The failure of reflexes to appear, their prolonged continuation, or their uneven strength characteristics may cause the physician to suspect neurological impairment. Two of the more universal assessments of neonatal behavior are the Apgar score (Apgar, 1953) and the Brazelton scale (1984); both include reflex measures as indicators of neurological adequacy. The Apgar score assesses the newborn's well-being just after delivery. It consists of a scale of five conditions, one of which is called reflex irritability. To assess this condition, a cotton swab or small catheter is placed in the infant's nostril and the response is recorded as no response, grimace, or preferably, the reflexive response of coughing, sneezing, or crying. The Brazelton scale is first given two or three days after the child is born when the immediate stresses of delivery have passed. It may be given again on day 9 or 10 when the baby has adjusted to the home environment. The examination includes 20 reflex measures assessing the neonate's neurological adequacy. Among the reflexes measured are the rooting, sucking, Moro, and Babinski. The Moro, Babinski, and asymmetrical tonic neck reflexes are also commonly used with other infant assessment scales and during normal physical examinations. These measurements can provide highly critical information. For example, if the asymmetrical tonic neck reflex perseveres past the time that it normally disappears (about 6 months), it is an indication that cerebral palsy or other neurological impairment may be present (Lorton & Lorton, 1984).

Although most measures of reflex behavior indicate generalized neurological status, they may be used in a selective way by specialists to determine the presence of localized neurological defects of such areas as the cranial nerves, spinal roots, and spinal cord (e.g., spina bifida).

Although it may not be possible to precisely and systematically classify reflexes into neat categories, they may be generally grouped under the descriptors of primitive, postural, and locomotor.

Types of Reflexes

Primitive reflexes are primarily associated with the infant's instinct for survival and protection. The infant, upon stimulation, reflexively seeks nourishment with the rooting and sucking responses and shows indications of protecting itself with the Moro and grasping reflexes. Though it can be debated whether these primitive reflexes are actually vital to infant survival and safety, indications are that these behaviors aid the fetus in orienting itself in its prebirth environment (Milani-Comparetti, 1981). The term *primitive* may be appropriate in view of the infant's ultimate potential; however, the term may also be associated with the level of neurological control. Because most of these reflexes are mediated by the lower brain centers (e.g., spinal cord, lower brain stem) as opposed to the higher centers of control (e.g., midbrain, cerebral cortex), use of the word *primitive* in this context also seems to be appropriate. Compared with the other types of reflexes, most of the primitive responses are functional prenatally and can be elicited in utero. As the cerebral cortex matures and gains control of lower brain functions (around 3 to 4 postnatal months) many of the primitive reflexes are suppressed (Pieper, 1963).

■ *Objective 8.2*

As the higher brain centers begin to function and suppress the lower control centers, postural reactions enter the infant's repertoire. **Postural reflexes** provide the infant with the ability to react to gravitational forces and changes in equilibrium. These involuntary reactions are the mechanisms by which the infant is able to maintain an appropriate posture with a changing environment. General functions of these reflexes include coordinating movements of the head, trunk, and limbs to keep the head upright, maintaining equilibrium, rolling over, and attaining vertical postures. There is also a protective response with postural control, as evidenced with the parachute reflex. When infants are tilted forward (face down) from an upright position, they reflexively extend their arms as a protective mechanism to brace against the forward displacement.

It is not unusual to find classification methods that utilize only the primitive and postural categories. However, another group of reflexes have enough common characteristics to justify including a third category. **Locomotor reflexes** are unique in that they resemble later voluntary locomotor movements. Involuntary responses such as the reflexes associated with climbing, walking (stepping), and swimming have been described as precursors of the voluntary movements of the same name. Although these reflexes are normally suppressed by the fifth or sixth month, their voluntary counterparts do not appear until months later. Consequently there has been much debate concerning their possible linkage. This issue will arise again in the context of the discussion of the various theoretical views concerning the role of reflexes in motor development.

A fourth group of involuntary responses, known as *tendon reflexes,* should be noted in regard to their initial appearance. These reflexes are used throughout the life span to evaluate neuromuscular response (primarily contraction) and are present in standard sites of the jaw, biceps, knee, and ankle by the second day of postnatal life (Forfar & Arneil, 1992).

The following section does not attempt to describe all of the infant reflexes. Rather its purpose is to discuss those involuntary responses considered to be most appropriate to the study of motor development. For a more detailed description of specific reflexes, refer to the work of Fiorentino (1981) and Forfar and Arneil (1992). Table 8.1 offers a summary of the approximate timetables for appearance, persistence, and disappearance of selected reflexes.

Primitive Reflexes   One of the earliest involuntary responses to appear is the *sucking reflex.* This response is evident in the 4-month-old fetus as it protrudes its lips in unmistakable preparation for sucking. On occasion, babies have "sucking blisters" at birth caused by this sucking action during the prenatal state. In newborns, the sucking reflex is elicited by anything that touches its lips.

In conjunction with the sucking response, the tongue and pharynx adequately adapt to swallowing so the newborn can feed. Newborns can suck and inhale simultaneously, swallowing between breaths about three times faster than adults. After a couple of weeks, most newborns have mastered the complex function of synchronized sucking, swallowing, and breathing.

The reflexive behavior of sucking for nourishment is joined with the instinctive response to seek food using the *rooting reflex.* This response is activated by a light touch on the cheek, causing the newborn to turn its head toward the stimulus in search of nourishment. This response is also referred to as the *search reflex.* Although the sucking response appears to be more voluntary around the third month of infancy, characteristics of the reflex are still evident until 6 to 9 months of age, when feeding is totally under voluntary control. The rooting reflex follows a similar timetable, persisting until approximately the ninth month.

When the newborn is startled or begins to fall, its arms and legs extend outward, hands open, and fingers spread in the instinctive protective response called the *Moro reflex* (fig. 8.1). As noted earlier, the Moro reflex is commonly used for assessing generalized neurological status. The response can be elicited in several ways. One way is to abruptly remove support from the newborn's head by allowing it to drop sharply backward a short distance and then catching it with the

Table 8.1 Approximate Timetable of Selected Reflexes

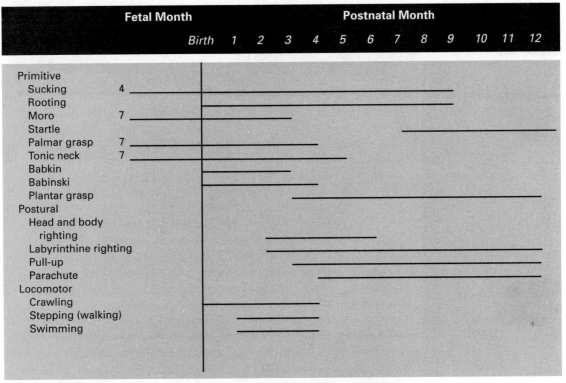

Source: Data from J. O. Forfar and G. C. Arneil, *Textbook of Paediatrics,* 2d ed. Copyright © 1978 Churchill Livingston, New York, NY.

hand. Another way to produce the reflex is to make a loud noise by clapping or slapping a table. After the arms and legs extend in response to the stimulus, the limbs return to a normal flexed position against the body. The Moro reflex crystallizes in the seventh fetal month and normally disappears by the third month of postnatal life.

Although quite similar to the Moro response, the *startle reflex* may not appear until months after the Moro reflex has disappeared (at about 7 months). Upon being startled the infant initially flexes its arms and legs rather than extending them as in the Moro reflex. The startle response is normally suppressed by the end of the first year; however, less intense indications of it may persist throughout the life span.

Grasping reflex behavior is one of the most well-known and interesting responses in the infant's involuntary movement repertoire. This reflex is evident when a small rod or someone's finger touches the newborn's palm. The fingers (excluding the thumb) fold tightly around the object with a grasping action (fig. 8.2). The reflex is present by the seventh fetal month, though in a weaker form, and gradually grows stronger for several weeks after birth. The reflex then weakens and finally is suppressed entirely by the time the infant is about 4 months old.

Figure 8.1
Moro reflex: (a) position
of rest and (b) extension
of limbs.

Courtesy of Mead Johnson
Nutritional Group.

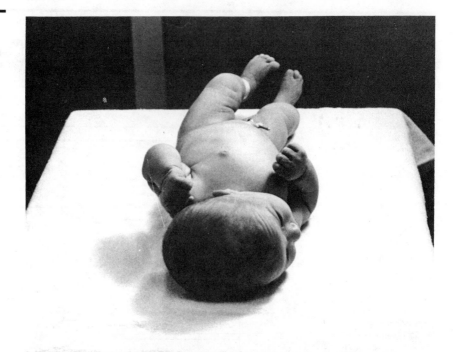

a.

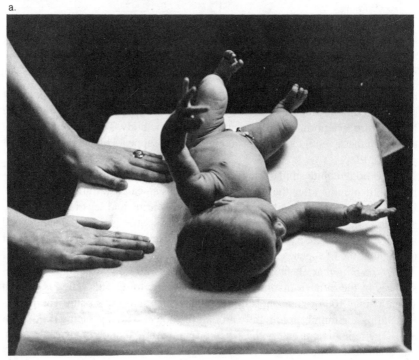

b.

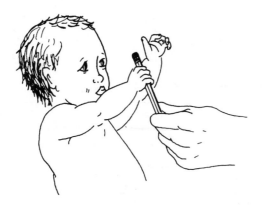

Figure 8.2
The grasping reflex.

It is around this time that the involuntary response is replaced by the initial stages of voluntary grasping, or manual control. This association will be described later in regard to the developmental progression of manual control.

Another interesting facet of involuntary grasping behavior is its strength characteristics. When pulled toward suspension while involuntarily clinging to a rod with both hands, infants are typically able to support more than 70% of their body weight (Eckert, 1987). In one of the classic studies of the grasping reflex, Richter (1934) recorded the suspension time, using both hands, of 37 newborns, aged 1 to 23 days. Large individual differences were evident, but only one infant failed to hang for any length of time. The longest suspension of 128 seconds equals that found among the better efforts of adults. It has also been noted that strength of the reflex is greater about two weeks after birth.

Two forms of the *tonic neck reflex* have been identified: asymmetric and symmetric. Both responses appear at about 7 fetal months of age and disappear by approximately 5 months of age. The asymmetric tonic neck reflex can be elicited by placing the infant in a supine position (on its back, face upward) and turning the neck so that the head is facing toward either side. In response to the stretch of the neck muscles as the neck turns, an increase of tonus in the limbs is triggered. In appearance, the infant exhibits a fencer's "on guard" position (fig. 8.3a) as the limbs extend on the side of the body in which the head is facing and flex on the opposite side.

Right-side limbs respond differently than the left-side limbs in the asymmetric pattern. In the symmetric tonic neck reflex the limbs move symmetrically, a response that can be elicited by placing the infant in a supported semistanding position and tipping the neck either forward or backward. If tipped backward, causing the neck to extend, the infant's arms will reflexively extend and the legs will flex (fig. 8.3b). If tipped forward, causing flexion of the neck, the infant's arms will also flex and the legs will extend (fig. 8.3c).

One of the more unusual infant reflexes is the *Babkin reflex,* also known as the palmar-mandibular reflex. This response can be elicited by providing pressure simultaneously to both palms, causing the infant to exhibit one or all of the following behaviors: mouth opens, eyes close, neck flexes, head tilts forward.

Figure 8.3
Tonic neck reflexes:
(a) asymmetric,
(b) symmetric when
tipped backward and
(c) symmetric when
tipped forward.

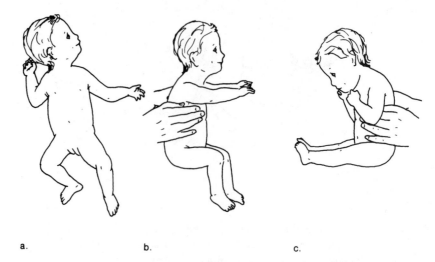

a.   b.   c.

The Babkin reflex can be elicited at birth and normally disappears by the third month.

The *Babinski* and *plantar grasp* reflexes are involuntary responses to stroke stimulation along the sole of the infant's foot (fig. 8.4). The Babinski reflex is present at birth; pressure applied to the sole of the foot causes the infant to reflexively fan out and extend its toes. This reflex is normally suppressed by the fourth month. Around this same period in the infant's life, stroking the sole will induce the plantar grasp reflex in which the toes contract or flex as if attempting to grasp the object. The plantar grasp reflex may persist through the first year of life.

Postural Reflexes   *Head and body righting* are similar reflex behaviors believed to form the basis for future voluntary rolling movements. Both involuntary responses appear around the second month and persist until about 6 months of age. The head (and neck) righting reflex can be elicited by gently turning the infant's body in either direction while in the supine position (on its back). The infant will respond reflexively by righting, or turning, its head in the same direction. In contrast, the body-righting reflex is elicited by turning the head in one direction while the infant is in a supine position. The body will reflexively right itself by turning in the same direction—first the hips and legs and then the trunk.

The *labyrinthine righting reflex* characteristically involves both righting and gravity reactions. This reflex enables the infant to maintain an upright body posture and is believed to contribute to the ability to move in a forward direction. The first evidence of this reflex is seen at around 2 months when the infant, while lying on its stomach, attempts to orient its head to gravity and look up. It is this upright head position while on the hands and knees at around 6 months that establishes the posture for forward locomotion.

The labyrinthine righting reflex can be elicited by holding the infant in an upright position and tilting its body either forward, backward, or to the side. In

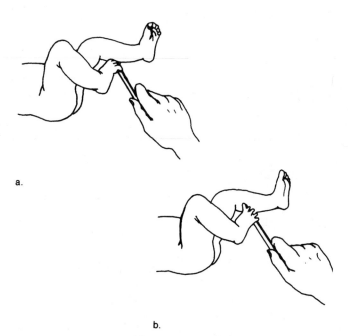

a.

b.

Figure 8.4
(a) The plantar grasp
reflex and (b) Babinski
reflex.

response, the infant will reflexively attempt to maintain an upright posture of the head by moving it in the direction opposite to that in which the trunk was moved (fig. 8.5). The head tends to maintain the original upright posture with relation to gravity. This response also detects functioning of the otolith organs housed within the labyrinth of the inner ear. The labyrinthine righting reflex generally appears around the second month of life, becomes stronger during the middle of the first year, and normally disappears by 12 months.

The *pull-up reflex* characterizes an involuntary attempt to maintain an upright position. The response is elicited by placing the infant in an upright sitting position while holding its hands and carefully tipping the infant backward or forward. In response, the infant will flex or extend its arms in an apparent effort to maintain the upright posture (fig. 8.6). This reflex usually appears around the third month and disappears by the end of the first year. It has been suggested that this involuntary response is related to the attainment of voluntary upright posture.

Also referred to as propping reflexes, the *parachute reflexes* serve as protective and supportive movements for the infant. Protective responses are evident when the infant is tipped off balance in any direction. Forward and downward parachute reflexes are present around the fourth month. The forward response is elicited by holding the infant in an upright position and tilting it toward the ground. In response, the infant reflexively extends its arms in a protective movement to break the fall (fig. 8.7a). The downward response is observed when the infant is suddenly lowered downward (from an upright position) a distance of about three feet. The infant's legs will extend and spread, and the feet will rotate outward.

- - - - - - - - - - - - -
Figure 8.5
The labyrinthine righting reflex.

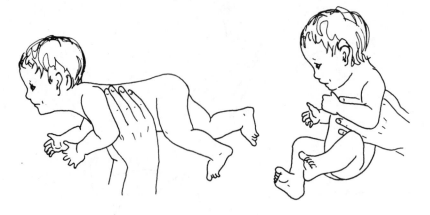

- - - - - - - - - - - - -
Figure 8.6
Pull-up reflex.

These reflexes are also seen when the infant is tilted off balance in either direction from a sitting position (fig. 8.7b). The sideward "propping" movement is usually not observable until around the sixth month, or the backward response until sometime between 10 to 12 months. In response to being tilted backward, the infant may also rotate the body to avoid falling in that direction. Until these reflexes are present, infants will make no attempt to brace or catch themselves when falling. It is not unusual for these protective responses to persist beyond the first year.

Parachute reflexes are supportive as well as protective movements. The supportive response is observable when the infant extends an arm for support

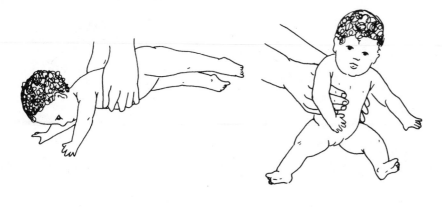

Figure 8.7
Parachute reflexes:
(a) protective reaction to
sudden downward
movement, (b) protective
reaction to being tilted
from a balanced
position, (c) propping
reaction using one arm
for support in reaching
for a toy.

a.   b.

c.

while reaching with the other hand (fig. 8.7c). It may be assumed that this supporting (propping) action enables the infant to attain a better visual perspective
of the surroundings and perform other tasks such as reaching and grasping, and
crawling.

Locomotor Reflexes   The *crawling reflex* can be observed from birth and, as
noted earlier, there is a definite similarity of movements between this response
and the voluntary pattern that appears months later. If the infant is placed in a
prone position (on the stomach) and pressure is applied to the sole of one foot or

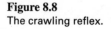

Figure 8.8
The crawling reflex.

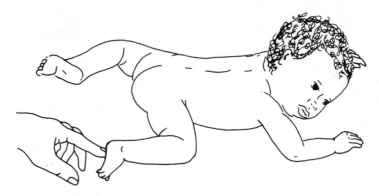

both feet alternately, the infant will reflexively crawl using its arms and legs (fig. 8.8). There is some speculation that this reflex (and others in this category) is a precursor to the voluntary crawling pattern even though there is a distinct time lag between the disappearance of this reflex and the appearance of the voluntary pattern. While the crawling reflex normally disappears by the fourth month, voluntary crawling generally is not observed until around the sixth or seventh month. The crawling reflex is considered to be important for the development of sufficient muscle tone for voluntary crawling.

The *stepping reflex* is one of the most debated involuntary responses in the infant's repertoire. The primary point of contention has been its assumed link with voluntary walking behavior. Normally present by the end of the first week after birth, this reflex can be elicited by holding the infant upright with the feet touching a flat supporting surface. Pressure on the bottom of the infant's feet will cause it to reflexively respond with crude but characteristic "walking" movements (fig. 8.9). The pattern will consist of a distinct knee lift but will not involve an arm swing or hip motions. A variation of the stepping response is the *placing reflex*. This reflex is similar in kinetics to the stepping reflex but is elicited by bringing the front of the infant's foot lightly in contact with the edge of a table. The infant responds by appearing to "place" the foot on the table. It may also appear as though the infant is lifting its foot over an object. The stepping reflex normally persists through the fourth month.

The *swimming reflex* is one of the most interesting involuntary responses seen in infants and can be elicited during the second week after birth by holding the infant horizontally in the water (with head up) or over the surface. The infant responds by moving the arms and legs rhythmically in a swimming-type movement pattern (fig. 8.10). Researchers have also noted that in the newborn, a breath-holding reflex is elicited when the infant's face is placed in the water (McGraw, 1966). However, a month or so later, infants may experience severe anxiety when the head is submerged. This involuntary response usually disappears by about the fourth month.

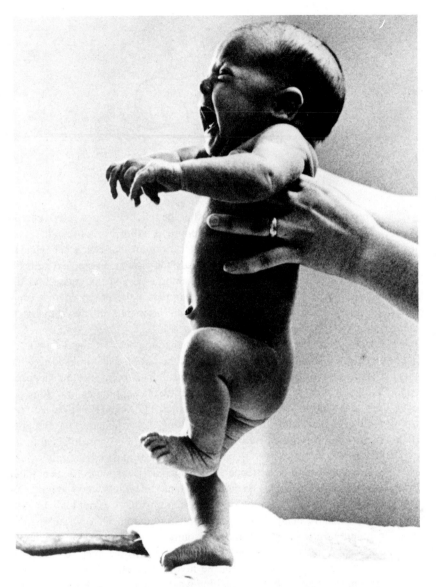

Figure 8.9
The stepping reflex.
Courtesy of Mead Johnson and
Company.

The developmental evolution of reflexive behavior into voluntary motor control can be generally described. However, an explanation of how and why these changes occur has been debated from several differing viewpoints. The relevance of the survival reflexes has stimulated some debate, especially in regard to their phylogenetic origin, but the primary issue from a motor behavior context has been related to the possible connections between involuntary responses and voluntary movement. This has been especially true of the reflexes that resemble voluntary movements, such as grasping, crawling, stepping (walking), and swimming.

Theoretical Views of Reflex Behavior and Voluntary Movement

■ *Objective 8.3*

Figure 8.10
The swimming reflex.

One prevailing point of debate is whether the link between specific reflexes and later voluntary movements that they resemble is a direct or more indirect relationship. Most viewpoints support the general notion that reflex behavior, to some degree, forms the basis for later voluntary movement. In essence, reflexes do not actually "disappear" but, rather, are suppressed and integrated into the hierarchy of controlled behavior. Other researchers have proposed that early reflexes are independent of voluntary control and thus that reflexes actually do disappear.

Tied to the theoretical view of a direct connection between the two types of responses, referred to as the **continuity view,** is the notion that stimulation of the reflex will positively affect later voluntary control. This point of view has been supported by research on the effects of stimulation on stepping and reaching reflexes. One of the more well-known series of studies (Zelazo, 1983; Zelazo, Zelazo, & Kolb, 1972) had infants "practice walking" by stimulating the stepping reflex and then compared the results with various control groups. In general, results indicated that infants whose stepping reflex was stimulated not only exhibited a prolonged phase of persistence and an increased response frequency but also began to walk voluntarily at an earlier age than the infants who had not practiced walking. One noted limitation of this research is that it is not known when the infants would have begun walking without stimulation. That is, walking depends on maturation, motivation, and environmental factors, all of which can vary considerably among individuals. Bower (1976), who subjected infants to "practice reaching" movements during the involuntary phase, obtained similar results; the emergence of voluntary reaching was accelerated, and in some cases, children experienced no disappearance phase. Results such as these have led researchers to speculate that not only is there a direct link between specific involuntary reflexes and later voluntary control, but the disappearance phase may be due to simple disuse.

Dynamical Systems Research by Thelen and associates (e.g., Thelen et al. 1993; Thelen & Bradley, 1988; Thelen & Ulrich, 1991) also supports the notion of continuity between specific reflexive movements and voluntary control. The researchers also contend that reflexes need not disappear and that they persist with adequate practice. Their view differs somewhat from Zelazo's notion that a reflex is transformed to a voluntary action by practice. They suggest instead that increased stepping frequency and the earlier onset of independent walking may simply be the result of strengthening the infant's musculature. In one set of studies, Thelen and Fisher found that the kicking and stepping reflexes involve the same neuromuscular functioning and suggest that the reflex disappears because muscle growth does not keep up with leg growth. That is, the muscle is not strong enough to lift the leg mass in the upright position. In the horizontal position the pull of gravity is less, so kicking can occur. Therefore, the researchers suggest that the stepping reflex is actually linked to later walking through spontaneous kicking. Studies involving the analysis of involuntary stepping and the walking patterns of independent walkers also report transition stages can be detected linking the initial coordinative patterns of the newborn with later independent walking (Thelen & Cooke, 1987; Thelen & Ulrich, 1991).

Perhaps the more traditional view of reflexes is that reflexes resembling later voluntary movements (primarily locomotor reflexes) must "disappear" before voluntary behavior can occur (Pontius, 1973; Wyke, 1975). Proponents of this view contend that there is a certain independence between involuntary responses and the voluntary movements that they resemble because of the natural gap of several months between them. This time lag, they suggest, indicates that there is no continuity and thus no direct link between reflexive behavior and later voluntary movement.

Milani-Comparetti and Giodoni (1967) and Molnar (1978) suggest a somewhat different but related view. They contend that certain primitive reflexes must disappear and that certain postural and locomotor reactions must appear before specific, voluntary motor skills can be attained. In their view, these postural and locomotor reflexes become "supportive" elements in the attainment of voluntary movement behavior. For example, Molnar's information suggests that only after the righting reactions appear is voluntary rolling possible. And in turn, the appearance of the righting reflexes depends on the inhibition of the primitive asymmetric tonic neck reaction (Roberton, 1984). The same research also indicates that the plantar (foot) grasp must be suppressed before voluntary walking can occur. Similar transitions can be found for other voluntary motor behaviors.

Casting some doubt on the idea that a reflex must disappear before its voluntary counterpart can appear is reported evidence that both the reflexive and voluntary patterns coexist in the infant (Touwen, 1988, 1976). In a longitudinal study of 50 infants from birth to the time of independent walking, Touwen found no indications of a sequential order of development in which reflexes appear and are suppressed by the developing voluntary activity. The results of his investigation led the researcher to conclude that support is lacking for the contention that reflexes should disappear in normal infants before particular voluntary behavior can develop.

Interestingly, one facet of research involving the study of coordinative structures has produced the speculation that simple voluntary motor tasks such as reaching or stepping movements result from a systematic relationship among muscle structures established at the spinal cord (Kelso et al., 1980; Kugler, Kelso, & Turvey, 1980). This view represents a significant change from the traditional notion that all voluntary movement has to be controlled at the higher brain centers.

The role of reflexive behavior in the later development of voluntary movement remains unclear, though each of the views presented raises questions worthy of consideration. Available evidence seems to suggest that specific reflexes play at least an indirect role in preparing the infant for later voluntary movement. Partial justification for this assertion is based on the fact that reflexes play a significant role in the regulation, strength, and distribution of infant muscularity. This supportive element has been determined by most developmentalists to be vital to the attainment of voluntary movements. This same line of reasoning may be used with other readiness elements such as postural control and balance for which specific reflexes can provide practice.

SPONTANEOUS MOVEMENTS (STEREOTYPIES)

■ *Objective 8.4*

In contrast to reflex behavior, **spontaneous movements** are stereotypic rhythmic patterns of motion that appear in the absence of any known stimuli (e.g., Hofsten & Rönnqvist, 1993; Prechtl & Hopkins, 1986; Thelen, 1979; Thelen & Bradley, 1988). Compared to rudimentary behaviors, stereotypies do not appear to serve any apparent purpose, nor can they be characterized as voluntary, goal-oriented motor behaviors. Infants seem to be absorbed in such actions as kicking, waving, rocking, and bouncing just for the sake of the activity, which characteristically accounts for as much as 40% of the infant's behavior while awake. These early patterns of coordinated movement have been observed as early as the 10th fetal week and occur with increasing frequency to peak between 6 to 10 months of age. Stereotypic behavior such as persistent rocking, waving, and other nonpurposeful spontaneous movements are considered abnormal in older children and adults.

In her pioneering efforts in this area, Thelen (1979) identified 47 distinct stereotypic movements. Of these, the most common and first noticed is rhythmical leg kicking (single leg, two leg, alternate leg). These patterns, along with various feet movements, seem to reach their peak between 6 to 8 months of age. Other commonly found stereotypies are arm waving, arm waving with an object, arm banging against a surface (with or without an object), and finger flexion. Noteworthy of these movements is that the peak frequency for hand and arm stereotypies is between $8\frac{1}{2}$ to $10\frac{1}{2}$ months. However, whereas spontaneous arm movements may appear just after birth, hand movements are typically not evident until $3\frac{1}{2}$ to $5\frac{1}{2}$ months. Stereotypic finger movements also appear in the first few months after birth but reach their peak frequency between 6 to 8 months, approximately $2\frac{1}{2}$ months before hand and arm stereotypies.

In a more recent investigation of spontaneous arm movements in neonates (3 to 5 days old), results suggested that "there is clearly more to neonatal (spontaneous) movements than thrashing and swiping or elicited reflexes" (Hofsten & Rönnqvist, 1993, p. 1057). The researchers found that these early movements are subject to several sets of organizing constraints. That is, though appearing quite

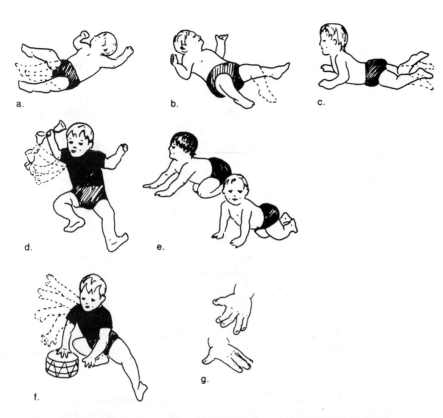

a.

b.

c.

d.

e.

f.

g.

Figure 8.11

Common stereotypies:
(a) alternate leg kicking,
(b) single leg kicking,
(c) kicking with both legs
together, (d) arm waving
with object, (e) rocking
on hands and knees,
(f) arm banging against a
surface, and (g) finger
flex.

random, they actually have an underlying temporal (timing) structuring, once again adding credence to the underlying dynamical properties of the human body.

Other characteristic stereotypies include arching the back and "rocking" when on all fours (i.e., hands and knees), rocking and bouncing when sitting, and bouncing while standing with support. These movements typically reach their peak later than the other stereotypies.

Although these movements do not appear to be goal-oriented, and their purpose is not evident, it is apparent that their appearance is quite predictable and orderly. Thus it may be suggested that they are of developmental significance. This conclusion is based on the observation that stereotypic kicking characteristically precedes and accompanies voluntary control of the legs, spontaneous finger flexion normally precedes voluntary grasping attempts, and rocking on the hands and knees precedes creeping. Thelen suggests that this type of predictive behavior reflects a specific stage process of neurological maturation controlling specific motor actions. These stages of normal maturation prepare the infant for more complex use of the neural pathways and for actions that they resemble.

Although stereotypies have not been clearly determined to be precursors of voluntary motor behavior, the fact that newborn infants can exhibit this degree of neuromuscular coordination with relatively little central nervous system activity has been somewhat of a revelation (Clark, 1988). Figure 8.11 shows seven of the more common stereotypic behaviors.

RUDIMENTARY BEHAVIOR

■ *Objective 8.5*

Rudimentary behavior occurs in that phase of motor development spanning the period from birth to approximately 2 years of age. As the infant's nervous system matures, a gradual increase in basic voluntary motor behavior occurs. During the first 6 to 10 months of the infant's life, these *initial* voluntary movement responses coexist with various reflexes and stereotypies. As more voluntary behavior is achieved, the infant develops a basic level of postural control, locomotion, and manual control. The acquisition of these basic abilities, in turn, provides the foundation for fundamental movement skill abilities.

Rudimentary movements are maturationally determined and appear in a very predictable sequence. The order of motor control generally develops in a cephalocaudal-proximodistal direction. That is, functions appear and develop earliest in the infant's head and neck, then in the shoulders and upper trunk, and later in the lower trunk and legs. The order also proceeds proximodistally from the shoulders, to the elbows, and then to control of the fingers. Progression is also characterized by the performance of gross motor before fine motor acts. This is quite evident in the development of manual control when, for example, infants initially use their whole arm to "corral" a toy before they are able to grasp it securely with a hand.

The remainder of this chapter describes the general sequence of early voluntary movement behavior in regard to the achievement of postural control, rudimentary locomotion, and manual control.

Postural Control

■ *Objective 8.6*

The ability to maintain body posture and to voluntarily move the body into a desired position underlies all motor behavior. In general, the development of these rudimentary abilities begins with control of the head and neck muscles and proceeds in a cephalocaudal direction until the infant is able to stand alone. However, before the infant can stand without support and progress to the landmark achievement of upright locomotion, a number of developmental milestones must be reached. Figure 8.12 depicts the general sequence of postural control and approximate age at which 50% to 75% of infants achieve these milestones (Bayley Scales, 1993; Denver Developmental Screening Test, 1990).

Head and Upper Trunk   Although some involuntary behavior may be evident at birth, the newborn has minimal voluntary control over the head, neck, and upper trunk. At about 1 month the infant gains enough control of its head to hold it erect momentarily when supported at the base of the neck. When prone the infant can hold its head up only for seconds; the movement is unsteady and the angle of vision is quite low.

By the second month, the infant is able to hold its head and chin steadily off a supporting surface while in the prone position. Because the head can be lifted as much as 30 degrees above the horizontal plane it allows for greater visual field. At 3 months the cephalocaudal progression is even more evident in that the infant is able to hold its head and chest up off the surface using arm

Head and upper trunk	Approximate age (mos)
Holds head erect voluntarily	1
Holds head and chin up	2
Holds chest up with arm support (prone)	3
Elevates head when supine	4–5

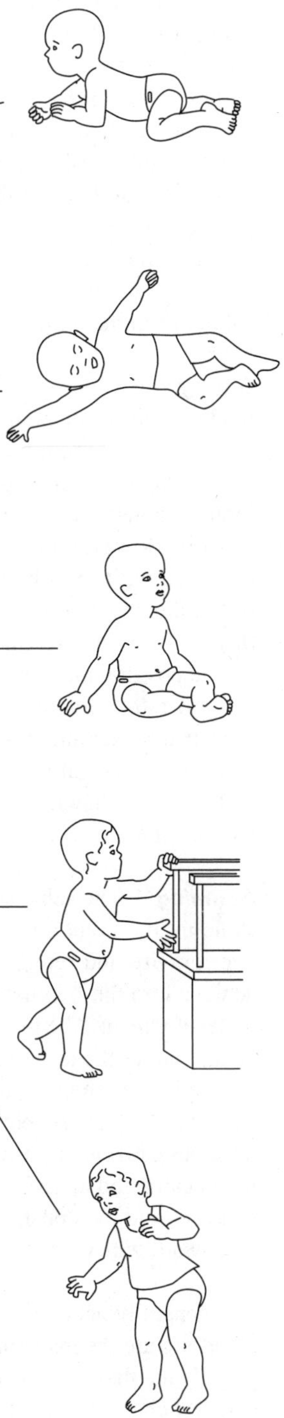

Figure 8.12
Sequence of rudimentary postural control.

Rolls (turns)	
Side to back	2
Back to side/stomach to side	3–4
Back to stomach	5–6
Stomach to back	8

Sits	
With support	3–4
Alone	5.5–6.5
Gets into sitting position	8
Sits down	9

Stands	
Holding on	6–8
Pulls self to stand	7.5–9.5
Alone	11–13

support. By the fifth month, the infant can accomplish the more difficult task of lifting the head off the surface when on its back.

Rolling (Turning)   Rolling of the body requires a certain level of control over the head, neck, and trunk. The first signs of voluntary rolling appear at about 2 months with the infant turning from its side to its back. By approximately 4 months the ability to roll from a supine position to the side and from the prone position to the side is achieved. As the infant matures and gains greater control of the hips and shoulders, more advanced abilities become evident such as rolling from the supine to prone position at around 6 months. The most difficult task, rolling from the prone to supine position, is normally achieved by 8 months.

Sitting   By approximately 3 to 4 months infants can sit with their heads relatively steady while being supported. The first attempts at sitting without support (after being placed into position) are characterized by an extreme forward lean, which may be an attempt to gain control of the lumbar region. By 6 months the infant is able to remain in the sitting position for a short time without additional support. However, before the infant can independently position itself into a sitting posture, it must have some basic body rolling ability (indicating some control of the upper body) and possess various support reactions (e.g., parachute reflexes). By 8 months infants generally can position themselves from a prone or supine position into a sitting posture and remain there without support.

A rather important early achievement is the ability to sit down from a crouched or standing position. This task requires additional postural control and flexibility, especially if the infant is attempting to position itself in a chair. The ability to sit down with reasonable control is normally achieved around 9 months of age.

Standing   The achievement of upright posture while standing is a major milestone in the infant's first year of development. With the ability to gain control over gravity and remain upright, the infant is ready to proceed with efforts to achieve upright locomotion (walking), perhaps the most monumental landmark in development. The first attempts to stand may occur quite early when the infant is held under the armpits by a parent and brought into contact with the surface. The ability to remain in the standing position while holding on to something usually occurs by the seventh or eighth month. By about 9 months of age the infant becomes capable of pushing or pulling itself to a standing position. Typically the infant can stand unassisted by the end of the first year, achieved by first getting to its knees and then pushing itself upward using the power of the legs and pulling downward with extended arms.

Rudimentary Locomotion

It is generally acknowledged that the two most basic motor skills in the human repertoire are the maintenance of upright posture and bipedal locomotion. Within roughly the first six months of life the infant develops postural control, the ability to detect changes in the body's center of gravity in relation to the base of support.

	Approximate age	
Crawling (body drag)	6–8 months	
Creeping (abdomen clear)	8–10 months	
Walking		
With support	9–10 months	
Alone (well)	12–14 months	
Backward	14–18 months	
Stairs (up/down)	2–4 years	
Perfected	4–5 years	

Figure 8.13
Sequence of rudimentary locomotor abilities.

Before infants will attempt the complex task of forward upright locomotion (walking), they will move about their surroundings in a variety of ways. As early as the third month infants have been observed to *scoot* along in a sitting position, using one leg to push the body along. However, basic forward locomotion normally begins during the second six months after birth. Although virtually all infants will develop through the orderly progression of moving forward in a prone position with crawling and creeping movement patterns, some will also use more unorthodox methods such as edging sideways like a crab and walking on all fours. After they are able to pull up to an upright standing position around 8 months, some infants may be observed to *cruise* around by moving unsteadily from spot to spot while holding on to furniture or similar supports. By the age of 13 months, most infants will have achieved the single most important milestone in motor development—independent walking.

Although walking is considered a fundamental motor skill, which is the topic of the next chapter, much of its development and refinement occurs before the rudimentary phase of motor behavior and infancy is over (fig. 8.13).

Crawling and Creeping Both *crawling* and *creeping* are well-known terms, but they are sometimes used in the developmental literature in ways that cause confusion. Some popular infant assessment batteries describe creeping as a precursor to crawling, whereas others note creeping as the advanced form of prone locomotion. By dictionary definition, crawling is slow movement by dragging the prone body along the ground; creeping is the act of moving the body along slowly and close to the ground, as on the hands and knees. Though their differences are evident, both are slow, deliberate forms of progress. Many individuals tend to perceive creeping as a slower (thus less mature) form of locomotion. However, most contemporary references to motor development and this text use crawling to designate the movement that precedes creeping.

Crawling represents the infant's first purposeful efforts at prone locomotion, which normally occurs between the ages of 6 to 8 months. Infants crawl using their arms and legs to drag the body along a surface. An important characteristic of this ability is that the abdomen remains in contact with the surface. Infants may use only their arms and legs, or a variety of arm-leg combinations to achieve forward progress. Important prerequisites to crawling are control of the muscles of the head, neck, and upper trunk. It may be speculated that for crawling and creeping, certain reflex movements (e.g., asymmetrical tonic neck, labyrinthine, body righting, crawling) provide some of the stimulus for neural and muscular development.

Creeping occurs approximately two months (between 8 to 10 months of age) after the appearance of crawling. In contrast to crawling, creeping movements involve moving in a prone position on hands and knees with the abdomen clear of the surface. Initial efforts are characterized by minimal elevation off the supporting surface and a *homolateral pattern* of movement; that is, the leg and arm of one side move together, alternating with the limbs of the opposite side. Gradually, the infant's elevation increases, as does the flexion and extension of the limbs, and the *contralateral pattern* of movement in which the right leg and left arm are used together, alternating with the left leg and right arm. The efficiency of the patterns increases with practice. Creeping appears to be a very enjoyable mode of locomotion for the infant. In fact, infants who have become independent walkers at times revert to creeping as the desired form of speedy travel. By the end of the first year, most infants have become very efficient creepers and may be observed using this pattern to move up and down stairs.

Walking   **Walking** is defined as movement by means of shifting weight from one foot to the other, with at least one foot contacting the surface at all times. Walking is perhaps the single most significant milestone in motor development, and its characteristics have probably been researched more than any other motor skill. Although some infants may be observed taking their first independent steps at 8 or 9 months, most children will not walk independently with any degree of efficiency until approximately 13 months of age. Children will usually not achieve the mature walking pattern until the age of 4 or 5 years. Before these efforts can be attempted, the infant must possess a certain level of upright postural control as evidenced by the ability to either pull up to a standing position (achieved around 8 months) or stand without assistance (achieved around 12 months). With at least some control of upright posture, infants will begin walking with the support of a parent holding their hands or by cruising, usually laterally, from point to point while holding on to furniture or other supports (achieved around 8 to 10 months).

Figure 8.14
Initial (immature)
walking pattern.

Movement Pattern Descriptions   Even when the infant walks independently, the movement pattern is usually quite immature (fig. 8.14). Balance is easily lost and falls are quite frequent in initial walking efforts. To compensate for the general sense of instability, the infant uses a relatively wide base of support and takes short steps. The leg action is inefficient as demonstrated by the short steps with limited leg and hip extension. Very little, if any, trunk rotation is evident. Contact with the ground is flat-footed with one knee locked and the other knee bent (fig. 8.14). Out-toeing is also present with minimal ankle movement and a slight pelvic tilt. Arm action is characterized by a "high guard" position in which the limbs are fixed and do not swing on each stride.

With the mature walking pattern each leg alternates between a supporting phase and swinging phase (fig. 8.15). The heel strikes the surface first as the back leg pushes off, shifting the weight to the front leg. The body leans forward slightly after the lead foot contacts the surface. The weight is then transferred from the heel to the outside of the foot, ball of the foot, and toes. The base of support is approximately shoulder-width and the feet are parallel to each other with toes pointed forward. The arms swing rhythmically in opposition to the legs; the right arm swings forward with the left leg and the left arm moves forward with the right leg.

Infants also exhibit some very characteristic patterns while walking up and down stairs. This skill is sometimes referred to as "climbing stairs," although in the truest sense, climbing involves use of the upper body. Initial efforts of walking up stairs are characterized by *marking time*. With this pattern, the infant

Figure 8.15
Mature pattern of
walking.

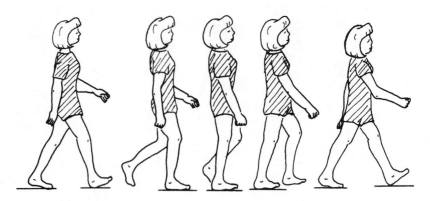

places one foot forward to the next step, then follows with the other foot to the same step. This pattern is repeated with each succeeding step. Marking time without the support of a handrail or another person's hand generally occurs between the ages of 24 and 28 months.

By the age of about 3 years, the child will progress to the more-advanced *alternate stepping* movement pattern (without support). With this pattern, the first foot moves to the first step, the other foot is placed on the second step, and the cycle is repeated. Descending stairs by marking time occurs just a few months after the child learns to walk up the stairs. However, the ability to descend using the alternating step without support does not occur until around 4 or 5 years of age, about a year (12 to 15 months) after the child learns to ascend the stairs without support.

Developmental Characteristics   The walking pattern improves gradually from initial independent walking (around 13 months) to maturity at about 4 or 5 years of age. The following general developmental trends parallel the transition from initial to mature walking:

1.   The *base of support* (dynamic base) narrows to within the approximate lateral dimensions of the trunk (fig. 8.16). This normally occurs by 4 ½ months after the onset of independent walking. In general, there is very little variation in dynamic base between age groups.

2.   *Foot contact* changes from flat-footed to the heel-toe pattern. The appearance of heel-striking is commonly achieved by the age of 19 months. "Toe stepping" is common in novice walkers who have not yet reached the more-advanced pattern.

3.   *Foot angle* changes show a decrease in the degree of toeing out. In-toeing is usually quite rare and considered abnormal.

4.   The single knee-lock pattern is abandoned for the more mature *double knee-lock* variation. This pattern involves heel-strike with knee extension followed by slight knee flexion as the body moves forward over the

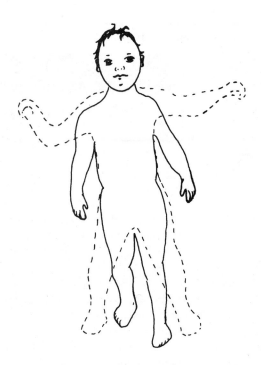

Figure 8.16
Change in arm position
as base of support
narrows.

supporting leg, then extension once again at the push-off phase. This action creates a greater range of leg motion.

5. *Pelvic rotation* increases to allow full leg motion and oppositional movement of the upper and lower body segments.. This action is normally observed in infants by 14 months of age.

6. The high-guard *arm position* (elbows flexed and abducted) is gradually lowered and oppositional arm swinging starts. The infant's first efforts at arm swing are frequently not equal and regular; both hands might swing forward in unison. A mature arm action involves movement of the opposite arm and leg together with slight movement at both the shoulder and elbow of each arm. The change in arm position usually parallels a narrowing base of support and is achieved at approximately 18 months (fig. 8.16).

7. *Step and stride length* increases, reflecting greater application of force and greater leg extension at push-off. Step and stride length nearly doubles between 1 and 7 years of age, then increases by approximately 50% by adulthood (see table 8.2). In general, step and stride length increase linearly with increasing leg length.

8. *Walking speed* increases and *step frequency* (steps per minute) decreases during the development of walking (see table 8.2). Sutherland (1984) notes that younger people normally must take more steps per unit of time to increase walking speed, due primarily to lack of neuromuscular control.

Table 8.2   Developmental Characteristics of Gait Length and Speed

Age (yr)	Step Length (cm)	Stride Length (cm)	Steps/Minute	Walking Speed (cm/s)
1	21.6	43	175.7	63.7
2	27.5	54.9	155.8	71.8
3	32.9	67.7	153.5	85.5
7	47.9	96.5	143.5	114.3
Adult	65.5	129.4	114	121.6

Source: Data from Sutherland, 1984.

Dynamical Systems Research   One of the most active lines of developmental biodynamics research has been conducted on walking using the dynamical systems approach (e.g., Clark & Phillips, 1993; Thelen & Ulrich, 1991; Thelen, Ulrich, & Jensen, 1989). As noted earlier (review chapters 1 and 7), the dynamical systems approach seeks to identify and explain the *processes* associated with motor control and of change. For example, experiments by Thelen and Ulrich have noted that some infants as young as 3 months exhibit well-coordinated, alternating stepping movements typically not common until several months later. In the experiment, infants were held by an adult over a motorized treadmill to facilitate the *stepping action* (see fig. 8.17), not a stepping "reflex" that can no longer be elicited after about 2 months. Simply stated, by holding the infant upright, balance and leg strength (subsystems), critical *rate controllers* in independent walking, were supported, therefore allowing the apparently more mature neuromuscular component to emerge. This study supports the general dynamical systems hypothesis that the development of locomotion is an *emergent* rather than prescribed (preset) process turned on in the brain. Rather, walking emerges as a consequence of sufficient development and cooperation of contributing subsystems that are quite adaptive for use in other contexts as well. As you may recall, this *functional adaptability* was also noted as a characteristic of neuronal group selection (Edelman, 1987). Other constraints, such as environmental contexts (walking surface, slope, etc.) also contribute to behavior.

Clark and Phillips (1993) more recently added to the dynamical systems literature by studying the dynamical properties of intralimb coordination in infants over the course of their first 12 months of walking. The researchers note that others have suggested that such components as body mass, limb length, and strength contribute to the dynamical system. Clark and Phillips observed that two additional control parameters, force production and balance, also play integral roles in independent walking. It is noted that the propulsive forces needed in coordinated walking are potentially destabilizing, but the infant's *self-organizing* system appears to compensate by spending more time with both feet on the ground than the more forceful adult. The novice walker begins quite slowly but after a few months produces gait parameters that are much more adultlike. The

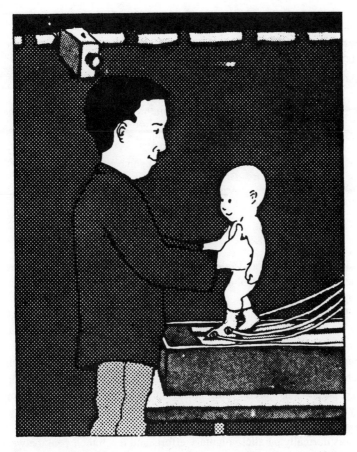

Figure 8.17
Experimental setup used
by Thelen and her
colleagues to study
dynamical components
of upright locomotion.

researchers suggest that perhaps to compensate, the infant's control parameters are being *scaled* to contribute to stability and coordination.

Clearly future research of this type will add significantly to our understanding of the mechanisms and processes that govern the development of motor behavior.

In addition to the attainment of basic postural control and upright locomotion, the first two years are characterized by several developmental milestones involving manual coordination. Three terms are commonly used to describe this area of motor development: prehension, manipulation, and manual control. **Prehension** is often used to describe initial voluntary use of the hands as characterized by basic seizing or grasping. In contrast, **manipulation** refers to skillful use of the hands, such as in stringing beads or threading a needle. Skillful manipulative behaviors are not usually observed until the middle childhood years (ages 5 to 8), even though they are a basic part of the early childhood curriculum. The term **manual control** encompasses both descriptors by referring to the developmental characteristics of hand movements. The three basic components of manual control

Manual Control

■ *Objective 8.7*

Table 8.3 General Sequence of Rudimentary Manual Control

Approximate Age (months)	Characteristics
Newborn	Reflexive and spontaneous reaching and grasping
4	Initial voluntary efforts
	Corralling (both limbs)
	Palmar grasp
5–6	Smooth arm and hand action
	One hand/pseudo thumb opposition
8	Accepts two objects
9–10	Pincer grasp (thumb opposition)
14	Adultlike reaching and grasping
18	Effective release

are reaching, grasping, and releasing. Table 8.3 shows the general sequence of rudimentary manual control and the approximate age that particular characteristics appear.

The first reaching and grasping movements appear as reflexes and spontaneous movements. Involuntary grasping usually appears in the seventh fetal month and disappears by 4 months of age. Spontaneous arm and hand (flexion, grasping) movements have also been observed prior to and during voluntary manual behavior (e.g., Hofsten & Rönnqvist, 1993; Prechtl & Hopkins, 1986; Thelen, 1979). Spontaneous "prereaching" (arm) and finger movements may appear as early as newborn to 3 months, whereas stereotypic hand movements are generally observed between $3\frac{1}{2}$ to $5\frac{1}{2}$ months. These seemingly nonpurposeful stereotypic behaviors normally parallel voluntary efforts and reach their peak between 6 and 12 months of age (Thelen, 1979). Figure 8.18 illustrates some commonly found spontaneous (stereotypic) behaviors involving arms, hands, and fingers.

By 4 to 5 months most infants make their initial efforts in voluntarily reaching and grasping objects. These first efforts are usually characterized by a "corralling" action in which both arms and hands work together to pull the object in. At this point, the infant picks up the object using the immature palmar grasp. With this form of manual control the infant grasps the object without thumb opposition and instead uses the thumb and fingers as a unit to hold the object against the palm of the hand.

During the fifth to sixth month, development normally reaches the level of reaching and grasping with one hand using *pseudo thumb opposition*. Here the thumb opposes the fingers (but not the finger tip as in true opposition) to pick up an object; there is minimal contact of the object with the palm. Infants may also be observed transferring an object from one hand to the other during this period. By 8 months infants normally exhibit the ability to receive two objects (holding one object in the opposite hand).

A major developmental milestone that usually occurs between 9 and 10 months is the ability to grasp objects using true opposition of the thumb with one

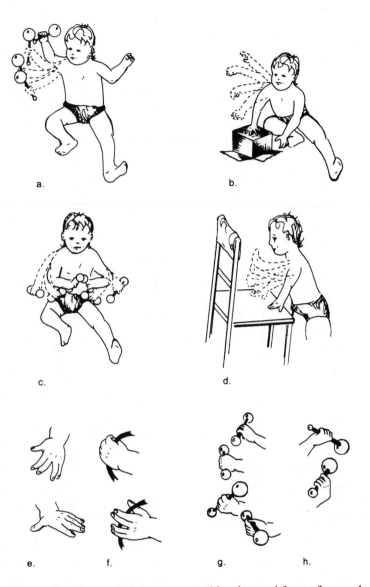

Figure 8.18
Examples of spontaneous (stereotypic) movements of the arms, hands, and fingers: (a) arm waving with object, (b) arm banging against a surface, (c) banging both arms together with object, (d) arm sway, (e) and (f) finger flexion, (g) hand rotation, and (h) hand flexion.

finger. Often referred to as the *pincer grasp,* this advanced form of manual control progresses to the level of fine motor coordination using the thumb and forefinger to manipulate objects.

To summarize, the sequence of manual control in humans moves from no thumb opposition, to pseudo-opposition, to true opposition. This developmental pattern also involves a shift in the positioning of the grasped object from the little finger side (ulnar) to the thumb side (radial). During the course of development, grasped objects are positioned to occupy a more radial and distal position next to the thumb and index forefinger rather than in the palm. Figure 8.19 illustrates the basic grasping techniques and changes in object positioning. By the

Figure 8.19
(a) Basic grasping
techniques and
(b) changes in object
positioning.

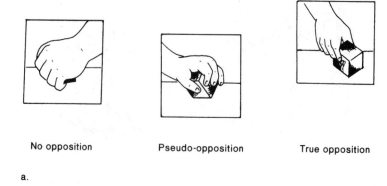

No opposition   Pseudo-opposition   True opposition

a.

Distal end

Radial side   Ulnar side

b.

time the infant is 14 months old, reaching and grasping have evolved into a smooth, coordinated adultlike action.

Landreth (1958) contends that the developmental progression of voluntary reaching and grasping includes six coordinated acts. Eckert (1987) has summarized the basic transitions that take place in each of six acts:

1.   Visual location of an object moves to the attempt to reach for the object.
2.   Simple eye-hand coordination moves to progressive independence of visual effort with ultimate expression in activities, such as piano playing and typing.
3.   Initial maximal involvement of body musculature moves to a minimum involvement and greater economy of effort.
4.   Proximal large muscle activity of the arms and shoulders moves to distal fine muscle activity of the fingers.
5.   Early crude raking movements in manipulating objects with the hands moves to the pincerlike precision of control with the opposing thumb and forefinger.
6.   Initial bilateral reaching and manipulation moves to ultimate use of the preferred hand.

Use of Visual Information   Several weeks prior to voluntary successful grasping at around 4 months, the infant projects arm and hand movements in the presence of visible objects (e.g., Hofsten, 1991). Trevarthen (1984) suggested that the prereaching movements were visually triggered. Complementing this suggestion is Piaget's (1952) initial proposal that the development of reaching is brought about by coordinated mapping of visual and manual schemas. Later literature supports these views in general, by describing the use of visual stimuli in prereaching and reaching movements as visually guided and visually elicited. The first stage, visually guided, describes the infant as learning to grasp by bringing both hand and object into view, thus guiding the hand toward the object. With the more mature technique as typically observed after 5 months, reaching becomes visually elicited. That is, the infant visually sights the object and follows with a movement, without need for sight of the hand (e.g. Bushnell, 1985). Clifton and associates (1993) conducted a relatively novel study in which they had infants attempt to reach in the dark (could not see their arm or hand) for glowing and sounding objects. Interestingly, success in the dark was similar to reaching in the light, suggesting that infant's proprioceptive abilities have been underestimated and warrant future research. However, visual information does play a vital role in reaching behavior between the ages of 4 to 7 months (Lasky, 1977; McDonnell, 1979). Beginning at around 9 months, reaching becomes less dependent on visual feedback and more automatized.

Unquestionably, *releasing* is a difficult component of manual control. The ability to control and relax the muscles in the arms and hands to provide a well-coordinated and accurate release is one of the final acquisitions in the development of manual control. Although a crude form of object release may be observed as early as 8 months when the infant opens its hand and simply allows the object to drop, accuracy of release is normally not achieved until 18 months of age.

Related to the ability to grasp and accurately release objects is the infant's perception of object weight and ability to control the amount of applied force. Prior to about 9 months of age, the application of force with the grasp appears to be unrelated to the weight of the object. Regardless of object weight, infants apply a similar force in the arm and grasping movements. By 9 months the infant normally can adjust the force used to grasp objects of varying weights, but only *after* the initial grasp of the object (Mounoud & Bower, 1974; Palmer, 1989). Not until approximately 18 months of age can the infant *anticipate* the weight of objects (after repeated presentations) and apply the appropriate amount of force. By this age most infants apparently acquire the perception that similar objects weigh more or less based upon their length. Therefore, by approximately 18 months of age, infants have acquired the general ability to reach, grasp, and release objects in a relatively well-coordinated manner.

Motor Asymmetries

One of the most evident behavioral manifestations of brain lateralization are **motor asymmetries.** That is, although the human body (and hemispheres) is symmetrical in general appearance, the paired limbs (hands, feet) and sensory organs (eyes, ears) are used in an asymmetric manner. Complementing this phenomenon is the fact that the cortex of the right hemisphere controls muscular

■ *Objective 8.8*

Right & Left Brain

activity in and receives sensory input from the left half of the body; the left hemisphere has a complementary role in conscious movement on the right side of the body (chapter 2). Other terms often used interchangeably with motor (and functional) asymmetries are *lateral preference* and *lateral (hemispheric) dominance.* Lateral dominance generally means that one side of the brain has developed to the point of establishing "dominance" for motor control; therefore the individual has an internalized lateral preference for use of one side with a specific action or task. Although lateral preference is most often identified as *right-sided* or *left-sided,* there are individuals who do not consistently favor the use of one limb or the other on a specific task; these individuals are identified as *mixed-sided.*

Theoretical Views   Perhaps the most evident and intriguing manifestation of motor behavior is that the vast majority of people are right-sided. Complementing this phenomenon is the much debated issue of whether motor asymmetries stem from innate biological factors inherited from parents (nature) or are learned behaviors influenced by environmental contexts (nurture). That is, is handedness (or footedness) learned, or is it primarily genetic based (inherited)? Despite volumes of literature dating back over a century (e.g., Broca, 1865; Brown-Sequard, 1877), no single theory explaining the nature and developmental characteristics of this phenomenon has been widely accepted. However, two general theories that have spawned considerable debate in the scientific community appear to prevail.

The first theory, proposed by Lenneberg (1967), stresses the importance of "maturational processes" between infancy and adulthood. According to this hypothesis, also referred to as the equipotentiality theory and *variant model,* the two hemispheres of the newborn are not specialized (i.e., fixed at birth) but progressively become so with age. The second theoretical perspective, referred to as the *invariant model,* suggests that functional asymmetries are in place at birth (fixed) and "remain constant" throughout the life span. Kinsbourne (1975) postulated that the neural substrates, localized at birth in the left or right hemispheres, will at some point in time subserve the specialized functions of language and visuospatial skills. This differs from the theoretical position of Lenneberg, who suggested that functions emerge from an equipotential base. In essence, the primary theoretical issue is whether functional asymmetries develop (i.e., the variant model) or remain constant (invariant) over time.

Over the years there have been numerous variations and explanations of these general models. One of the most widely accepted theoretical positions is the genetic model proposed by Annett (1978, 1985). According to the general tenets of *Annett's Right-Shift theory,* while the majority of the population has a strong right-shift genetic factor, the remaining portion of individuals may not express a dominant limb during childhood due to a weaker right-shift gene or lack of this inheritance (no right-shift factor), suggesting that environmental factors could influence the distribution of limb preference (right, left, mixed) in this portion of the population. Arguments by Corballis and Morgan (1978; see also Corballis, 1983) also suggest that limb preference is innately specified, but their contention is that in most of the population there is a *left-right maturational gradient* programming the left hemisphere to mature more rapidly than the right hemisphere, thus providing the bias toward right-sidedness in limb preference (contralateral control). Both of these positions support the variant model of development to some degree.

In essence, although functional asymmetries may very well have origins in genetic makeup and possess maturational characteristics, few theorists would deny that specific behaviors may be modified by cultural/environmental factors (i.e., experience). More specifically, while *direction* (right, left, mixed) of limb preference may be more fixed to biological foundations, the *degree* of preference could vary considerably depending on experience (e.g., McManus et al., 1988). Thus, experience may account for developing a strong or weak phenotype (functional asymmetry).

Although there have been some exceptions, research, in general has supported the variant developmental trend for handedness and footedness. This has been evidenced by a general shift toward right-sidedness with increasing age (from early childhood to adult) for hand preference (e.g., Dargent-Paré et al., 1992; McManus et al., 1988; Porac, Coren, & Duncan, 1980; Rice & Plomin, 1983) and foot preference behavior (Dargent-Paré et al., 1992; Gentry & Gabbard, 1994; Porac et al., 1980). A comprehensive review of infant motor asymmetries also suggests, in general, that there is insufficient behavioral evidence to conclude that motor asymmetries are fixed (invariant) at birth and unchanging thereafter (Provins, 1992).

In regard to gender, an interesting theory proposed by Geschwind and Galaburda (1985) (also see review by Finegan, Niccols, & Sitarenios, 1992) suggests that due to potentially higher levels of prenatal testosterone in males, neuronal growth in the left hemisphere slows, hence weakening contralateral control, suggesting less right-sidedness and more left-dominance in males. Support for this theory, however, has been quite mixed. There have been some reports that females show more right-sidedness than males (e.g., Coren, 1993; Humphrey & Humphrey, 1987); however, as with functional asymmetries in general, consistent gender differences in limb preference apparently do not exist (e.g., Coren, Porac, & Duncan, 1981; Hahn, 1987).

Early Infant Motor Asymmetries   Well before children are capable of understanding and responding reliably to conventional preference inventories, evidence of asymmetric bias may be observed. The conclusion of two recent reviews on early infant asymmetries found that the majority of newborns and young infants exhibit reflexes and spontaneous movements that are stronger and more coordinated on the "right side of the body" (Grattan et al., 1992; Provins, 1992). Types of responses favoring the right side include: spontaneous head turning, spontaneous fisting, rooting reflex, asymmetric tonic neck response, right foot lead in walking reflex, and the plantar and palmar grasps. Also worthy of note is that the asymmetry described was found only in the motor (efferent) parameters (strength and coordination) and not the sensory (afferent) parameters. This may be indicative of the developmental state of the population observed (primarily neonates). As mentioned earlier, although the relationship appears complementary, the question of whether right-biased involuntary movement behaviors in infants are the basis for right-sided voluntary responses in older persons is still being debated.

Complementing the development of motor control during childhood is the emergence of *functional asymmetries* such as handedness, footedness, and eye preference. The development of these behaviors are most evident during the early childhood years and will be discussed in the next chapter.

SUMMARY

Early movement behavior is categorized by movements associated with reflexes, spontaneous behavior, postural control, rudimentary locomotion, and manual control. Reflexes are involuntary movements that are controlled by the subcortical areas of the brain. They are primarily associated with the stimulation of the CNS and muscles, infant survival, and their usefulness as a diagnostic tool for assessing neurological maturity. The three general categories of reflexes are primitive, postural, and locomotor. Primitive reflexes are primarily associated with the infant's instinct for survival and protection; postural reflexes provide the infant with reflexive reactions to gravitational forces; locomotor reflexes resemble later voluntary locomotor movements.

Theoretical viewpoints concerning the relationship between reflex behavior and voluntary movement focus on whether the link is direct (continuity view) or more indirect (reflexes must disappear before voluntary actions emerge). Reflexes can be elicited in the young fetus. Most are suppressed by 6 months of age and only a few are observable after the first year.

Spontaneous movements are stereotypic, rhythmic patterns of motion that appear in the absence of any known stimuli and do not appear to serve any apparent purpose. Some of the more frequently observed stereotypies include: kicking, waving, rocking, and bouncing. Stereotypies are observable early in fetal life and reach a general peak between 6 to 10 months of age. Since their appearance is quite predictable and orderly, it has been suggested that they are of developmental significance.

The first two years of life are characterized by the development of several rudimentary behaviors that follow the general trend of cephalocaudal-proximodistal growth and motor control. In general, the development of postural control begins with control of the head and neck muscles and proceeds in a cephalocaudal direction until the infant is able to stand alone. Forms of early locomotion that precede independent walking are scooting, cruising, crawling, creeping, and walking with support. Even after independent walking is accomplished, other developmental changes must take place before a mature movement pattern is achieved. The notion that walking is an emergent rather than prescribed (preset) process has been confirmed by dynamical systems research. Independent locomotion emerges as a consequence of sufficient development and cooperation of several contributing subsystems.

The first signs of hand use in the form of reaching and grasping appear as reflexes and spontaneous movements during the prenatal stage. Rudimentary voluntary manual control, referred to as prehension, develops rapidly during the first two years in the form of reaching, grasping, and releasing abilities. Voluntary control begins around the fourth month with a corralling action and use of the primitive palmar grasp (without thumb opposition). By 9 to 10 months a major development is the marked ability to grasp objects using true opposition of the thumb with one finger (the pincer grasp). By approximately 18 months, infants have acquired the general ability to reach, grasp, and release objects in a relatively well-coordinated manner.

Several motor asymmetries (linked to brain lateralization) are evident during the first two years of life. In regard to its origin, the primary debate focuses on to what extent these behaviors stem from biological factors (nature) or the environment (learning).

SUGGESTED READINGS

Bard, C., Fleury, M., & Hay, L. (Eds.). (1990). *Development of eye-hand coordination across the life span.* Columbia, SC: University of South Carolina Press.

Bower, T. G. R. (1982). *Development in infancy* (2d ed.). San Francisco: Freeman.

Fiorentino, M. R. (1981). *A basis for sensorimotor development: The influence of the primitive, postural reflexes on the development and distribution of tone.* Springfield, IL: Thomas.

Forfar, J. O., & Arneil, G. C. (1992). *Textbook of paediatrics* (4th ed.). New York: Churchill Livingston. Refer to chapter on reflexes.

Lockman, J. J., & Thelen, E. (1993). Special feature on developmental biodynamics. *Child Development, 64*(4), 953–1190. This feature includes very timely articles on manual control and locomotion.

Meisami, E., & Timiras, P. S. (Eds.). (1988). *Handbook of human growth and developmental biology* (Vol. 1, pt. B). Boca Raton, FL: CRC Press. (Refer to articles by Touwen, Thelen & Bradley, and Clark).

Thelen, E., & Smith, L. (1994). *A dynamic systems approach to the development of cognition and action.* Cambridge, MA: MIT Press.

Thelen, E., & Ulrich, B. (1991). Hidden skills. *Monographs of the Society for Research in Child Development, 56*(1), serial #223.

Woollacott, M. H., Shumway-Cook, A., & Williams, H. G. (1989). The development of posture and balance control in children. In M. H. Woollacott & A. Shumway-Cook (Eds.), *Development of posture and gait across the life span.* Columbia, SC: University of South Carolina Press.

Motor Behavior during Early Childhood

KEY TERMS

locomotor skills
nonlocomotor (stability) skills
manipulative skills
fundamental movement skill
movement pattern
process characteristics
product values
composite approach
component approach
prelongitudinal screening
running
jumping
leap
vertical jump
standing long jump
hopping
galloping
sliding

skipping
throwing
catching
striking
kicking
ball bouncing
climbing
manipulation
power grip
dynamic tripod
scribbling
bimanual control
functional asymmetries

PHASES OF MOTOR BEHAVIOR

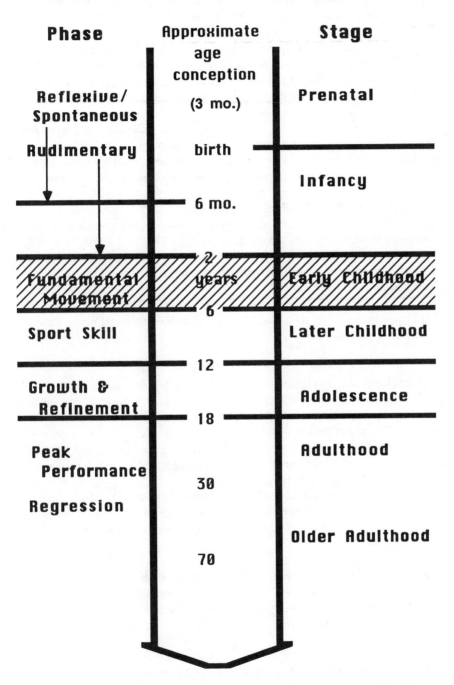

Phase	Approximate age	Stage
	conception	
Reflexive/ Spontaneous	(3 mo.)	Prenatal
Rudimentary	birth	
		Infancy
	6 mo.	
Fundamental Movement	2 years	Early Childhood
	6	
Sport Skill		Later Childhood
	12	
Growth & Refinement		Adolescence
	18	
Peak Performance		Adulthood
	30	
Regression		Older Adulthood
	70	

■ *Objective 9.1*

After the acquisition of rudimentary motor abilities, the individual's movement repertoire expands to include fundamental movement behaviors. The period during which these behaviors emerge normally spans the early childhood years (approximately ages 2 to 6 or 7), a period of landmark significance for motor development. This phase of motor behavior is sometimes referred to as the "movement foundation," based on the notion that the elements of movement behavior that develop and emerge during this period provide a substantial part of the motor skill foundation upon which more complex motor programs are formed.

Fundamental movement behaviors may be classified into three general motor skill groups: locomotor, nonlocomotor, and manipulative. **Locomotor skills** are movements that transport an individual through space from one place to another; examples are walking, running, and skipping. **Nonlocomotor (stability) skills** involve axial movements and movements of balance that are executed with minimal or no movement of the base of support; examples are bending, twisting, and swaying. **Manipulative skills** include fine motor manual movements and gross motor skills that involve the control of objects primarily with the hands and feet; examples are throwing, catching, kicking, and striking. From this foundation children begin to perform movement activities that include more advanced combinations of skills and specificity and require greater temporal and spatial accuracy.

This chapter will focus on fundamental motor skill development and will place specific emphasis on those skills in which developmental motor patterns have been established.

FUNDAMENTAL MOTOR SKILL DEVELOPMENT

Basic Terminology

■ *Objective 9.2*

The development of fundamental motor skills is perhaps the most extensively researched area of motor development because the study of basic developmental movement patterns is considered by many to form the core of the field. As early as the 1930s, researchers sought to identify the sequence and primary developmental components of basic movement patterns. In more recent years, this area of inquiry has continued to be one of activity through the efforts of several established and promising researchers.

The terminology associated with fundamental movement skills is a vocabulary that has developed over the years of research in that field of scientific inquiry. Much of the following discussion was derived from the works of Wickstrom (1983).

Fundamental Movement Skill   Also referred to as a *basic motor skill,* a **fundamental movement skill** is a common motor activity (e.g., walk, run, jump, throw) with specific movement patterns. It is also believed that these basic skills are general, in that they form the foundation for more advanced and specific movement activities.

Movement Patterns   Often used interchangeably with the term *motor pattern,* a **movement pattern** is the basic functional structure of a fundamental motor skill. The primary description involves a series of movements organized in a particular time-space sequence. The term is also used in reference to common elements that are observable in more than one motor skill. For example, several throwing and striking skills share common movement pattern elements in terms of arm, trunk, and leg action. Another term frequently associated with the identification of movement elements is *movement form*. This term is also used to describe the *process* rather than the *product* (performance) characteristics of movement.

Proficiency in movement pattern execution has been described using terms ranging from *immature* (or initial) to *mature,* and from *minimal* form to *sport skill* form. Each of these terms identifies the developmental movement pattern characteristics of children in the approximate age range of 2 to 6 years. Immature and minimal descriptors generally depict the minimal standard of a movement pattern characteristic of children 2 to 3 years of age.

With practice, by the end of the sixth year most children acquire at least some features of the *mature movement pattern.* This pattern represents the composite of the common elements used by skilled performers. Mature pattern performance is skill- rather than age-related. That is, with practice and maturity most children will acquire some, or perhaps all, mature pattern characteristics; however, others may not achieve a mature level until years later, if ever. This appears to be especially true of the overhand throwing pattern and standing long jump (with the two-foot takeoff and landing).

Mature fundamental movement patterns that have been adapted to the special requirements of a particular advanced movement activity such as pitching in baseball (from basic throwing) and running hurdles (from basic leaping) are known as *sport skill movement patterns.* These advanced skill movements retain most of the characteristics found in fundamental patterns and generally appear during the sport skill phase of motor behavior (later childhood from 6 to 12 years). Wickstrom describes the general pattern of transition from minimal to mature as a *developmental motor pattern.* More specifically, this pattern meets the minimal requirements but does not equal the features of optimal execution.

Methods for Studying Change

Several perspectives may be taken when observing and analyzing motor pattern development. Tools of assessment range from simple observation using the human eye to very sophisticated three-dimensional cinematography using computers to calculate and analyze even slight movement changes. Much of the available literature on movement pattern development is based on descriptive analysis using relatively subjective interpretation. Over the years, however, an increasing number of researchers have deemed it more appropriate to use a biomechanical approach in analyzing movement changes.

■ *Objective 9.3*

This trend has become evident in recent research activity studying coordinative structures (dynamic systems theory). *Biomechanics* is considered to be the physics of human motion, a study of the forces produced by and acting on the body. Three associated terms are *kinematics, kinetics,* and *kinesiology. Kinematics* refers to the temporal and spatial characteristics of motion. *Kinetics* is concerned with the forces that act upon (i.e., cause, modify, facilitate, or inhibit) motion. *Kinesiology* is frequently used synonymously with the term *biomechanics* or *applied* (functional) *anatomy*. The term *kinesiology* literally means the *science of motion.*

The study of movement pattern change generally includes collection of the following types of specific information: descriptive characteristics of the movements in a pattern; range of motion at each joint; muscle action; angular velocities accompanying segmental movement; timing of the movement sequences; time devoted to each phase of a pattern; and projection angles and velocities. Commonly used tools and techniques of the biomechanical approach include: anthropometry (to measure body dimensions); timing devices; optical devices (cameras, stroboscopy, videography, computerized cinematography); electrogoniometry (joint action); electromyography (electrical activity of muscle); dynamography (force production characteristics); accelerometry (acceleration); and computer modeling. For a more detailed discussion of the analysis of human movement and the study of developmental biomechanics refer to Adrian and Cooper (1989).

Much of the literature related to the development of fundamental movement patterns is based on qualitative descriptions. That is, the focus is on changes in "form" (**process characteristics**) rather than "performance" **product values.** Product values are quantifiable measures of performance such as the number of yards that a ball is thrown, running velocity in seconds, and the height of a jump in feet and inches.

■ *Objective 9.4*

Researchers have used various approaches to describe and interpret *observable* movement changes using qualitative information. Currently two practical, yet scientific, approaches have drawn considerable attention: the composite approach and the component approach. Both are, to some degree, *stage* oriented and employed primarily to identify movement pattern changes across time and describe developmental trends. The **composite approach** (Painter, 1994) describes motor pattern changes through a series of rather discrete steps (or stages) in the developmental process. Characteristic of this approach is the attempt to break down movement pattern changes into a *sequence* of stages covering all parts of the body; hence, it is also known as the *total body approach* (Seefeldt & Haubenstricker, 1982). Identified for each stage (e.g., Step 1, 2, etc.) are the actions of various primary body parts (e.g., for the skill of throwing, the arm action, trunk action, and leg action). The number of stages inevitably varies from skill to skill.

The **component approach** (e.g., Roberton, 1977, 1978) proposes that movement pattern changes must be divided into substages to gain more accurate developmental information. Under this approach, each body component (rather than the general pattern) is followed through the development process. For example, if the stage theory uses the general description of stages 1, 2, and 3 in the development of the throwing pattern, the component approach might identify five arm action stages but only three marked changes for the legs.

Each of the approaches to interpreting developmental changes includes a hierarchical structure to depict progression toward maturity. As noted, their goals are similar, but the degree of precision with which they describe the course of change varies.

The desired outcome of any hypothesized developmental sequence is that it be a valid representation of most children. As noted in chapter 1, one of the best methods of developmental research is the longitudinal design. Roberton, Williams, and Langendorfer (1980) have developed a **prelongitudinal screening** technique that accommodates such a purpose. The first step is to hypothesize a developmental sequence as a result of numerous observations for children of different ages and experience. Then data would be collected from an appropriate number and age range of subjects. Subjects' performances are evaluated and compared with the hypothesized longitudinal model, the desired outcome being that the youngest (and less-experienced) subjects exhibit the less proficient motor pattern characteristics, while the older subjects perform more-advanced (mature) patterns. To establish validity over time, subjects are retested periodically (e.g., yearly) to determine whether, for example, a child who was originally identified at Step 1 fits the model for Step 2.

The following section provides a discussion of the process characteristics associated with key fundamental movement patterns that emerge during early childhood. In addition to a summary of movement pattern and developmental characteristics for each skill, examples of component and composite sequence models are periodically included.

MOVEMENT PATTERNS (PROCESS CHARACTERISTICS)

Running

■ *Objective 9.5*

Running is a natural extension of walking and is characterized by activity in which the body is propelled into "flight" with no base of support from either leg. This is in contrast to walking, in which one foot is always in contact with the surface. Because it contains a nonsupport phase, running is a skill that is less stable than walking and therefore demands more bodily control. Most children take their first running steps at about 18 months and exhibit a true flight phase (minimum standard) between 2 and 3 years of age. By the age of 5 most children attain a reasonably skillful running form, and speed becomes a primary stimulus for performance. The general development of a skillful running pattern usually occurs without need for specific help or instruction. However, depending on the nature of the running task (e.g., sprints, long-distance running), education is usually needed to attain mastery.

Movement Pattern Descriptions   The description of the developmental pattern for running has been established primarily on running performance at maximum velocity (short sprints of 30 to 50 yards). Justification for using sprint-action analysis is that it yields relatively constant performance and the high degree of effort required produces maximum movement and acceptable reliability. During the initial (immature) stages of running, there is no observation flight phase and the base of support is relatively wide (fig. 9.1). The stride is short with

Figure 9.1
Initial running attempts
of a 15-month-old.

the feet rotated outwardly and laterally. The center of gravity is forward and foot-to-surface contact is usually flat-footed with some children exhibiting a "tiptoe" style. Arm swing is relatively rigid and more lateral and horizontal rather than vertical.

In the mature stage, the arms are bent at the elbows in approximate right angles and are swung vertically in a large arc in opposition to the legs. The recovery knee is raised high and swung forward quickly while the support leg bends slightly at contact and then extends completely and quickly through the hip, knee, and ankle. Stride length and duration of the flight phase are at their maximum. There is very little rotary movement of either the recovery knee or foot as stride length increases. Figure 9.2 shows the stages of running including the mature pattern. Wickstrom (1983) summarizes the essential characteristics of the mature running pattern as follows:

1. The trunk maintains a slight forward lean throughout the stride pattern.
2. Both arms swing through a large arc in a vertical plane and in synchronized opposition to the leg action.
3. The support foot contacts the ground approximately flat and nearly under the center of gravity.
4. The knee of the support leg bends slightly after the foot has made contact with the ground.
5. Extension of the support leg at the hip, knee, and ankle propels the body forward and upward into the nonsupport phase (*flight phase*).
6. The recovery knee swings forward quickly to a high knee raise and there is simultaneous flexion of the lower leg, bringing the heel close to the buttock.

Unfortunately, very little developmental information is available on the long-distance running pattern, even though it is considered a fundamental skill. Since almost all national physical fitness assessment batteries have been revised in recent years to include tests of distance running (rather than sprints), it is anticipated that more research in this area will follow. There are postural similarities between sprinting action and long-distance patterns, but many mechanical differences are evident. Proper mechanics and speed are primary factors in sprinting, but efficiency, pace, and endurance are paramount to the distance runner. Due to the sprinter's need for maximum forward propulsion, knee lift and rear kickup of the recovery leg are higher and the sprinter's arms move more

INITIAL

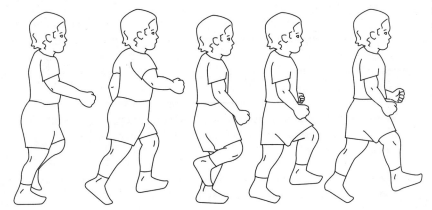

ELEMENTARY

MATURE

Figure 9.2
Stages of running
pattern.

vigorously. In contrast, the distance runner normally displays less knee flexion, uses less arm motion, contacts the ground with more of a heel-to-toe action, and displays more vertical movement. The sprinter normally relies on a combination of long stride and high frequency, whereas the distance runner exhibits a relatively short stride with low frequency and pace to conserve energy. Although the concept of pace is generally difficult for young children, the long-distance running pattern is acquired by 5 years of age.

Developmental Characteristics   Among the numerous developmental studies of the running (for speed) movement pattern, the following major trends are frequently noted:

1. Length of stride increases. This developmental trend coincides with the age-related increase in running speed.
2. Base of support narrows.
3. Relative amount of vertical movement decreases so there is greater horizontal displacement of the center of gravity.
4. Relative distance of the support foot from the center of gravity at contact decreases.
5. Hip, knee, and ankle extension during the takeoff increases.
6. Duration of the nonsupport (flight) phase increases.
7. Support knee flexion on contact increases.
8. Knee flexion in the recovery leg increases, as exhibited by the closeness of the heel to the buttock on the forward swing.
9. Height of the swing of the forward knee (recovery leg thigh) increases.
10. Lateral leg movements and out-toeing are eliminated.
11. Arm action evolves from the immature high and middle guard positions (rigid and held laterally) to elbows flexed at approximate right angles that move in opposition to the legs.

Table 9.1 presents Roberton and Halverson's (1984) component view of the running pattern.

Jumping

Jumping is a motor skill in which the body is projected into the air by means of a force generated by one or both legs and then lands on one or both feet. One of the most diverse and fundamental of all motor skills, jumping patterns range from simple leaping (a one-foot takeoff and opposite-foot landing) to hopping (jumping from one foot to the same foot rhythmically).

The primary challenges of jumping for the young child are sufficient leg strength to propel the body off the ground (for a longer period than is required in running) and the ability to maintain postural control in the air and upon landing. For these reasons, most jumping patterns are considered more difficult than walking or normal running.

Table 9.1 Developmental Sequence (Components) for Running

Leg Action Component

Step 1: The run is flat-footed with minimal flight. The swing leg is slightly abducted as it comes forward. When seen from overhead, the path of the swing leg curves out to the side during its movement forward. Foot eversion gives a toeing-out appearance to the swinging leg. The angle of the knee of the swing leg is greater than 90° during forward motion.

Step 2: The swing thigh moves forward with greater acceleration, causing 90° of maximal flexion in the knee. From the rear, the foot is no longer toed-out nor is the thigh abducted. The sideward swing of the thigh continues, however, causing the foot to cross the body midline when viewed from the rear. Flight time increases. After contact, which may still be flat-footed, the support knee flexes more as the child's weight rides over the foot.

Step 3: Foot contact is with the heel or the ball of the foot. The forward movement of the swing leg is primarily in the sagittal plane. Flexion of the thigh at the hip carries the knee higher at the end of the forward swing. The support leg moves from flexion to complete extension by takeoff.

Arm Action Component

Step 1: The arms do not participate in the running action. They are sometimes held in high guard or, more frequently, middle guard position. In high guard, the hands are held about shoulder high. Sometimes they ride even higher if the laterally rotated arms are abducted at the shoulder and the elbows flexed. In middle guard, the lateral rotation decreases, allowing the hands to be held waist high. They remain motionless, except in reaction to shifts in equilibrium.

Step 2: Spinal rotation swings the arms bilaterally to counterbalance rotation of the pelvis and swing leg. The frequently oblique plane of motion plus continual balancing adjustments give a flailing appearance to the arm action.

Step 3: Spinal rotation continues to be the prime mover of the arms. Now the elbow of the arm swinging forward begins to flex, then extend during the backward swing. The combination of rotation and elbow flexion causes the arm rotating forward to cross the body midline and the arm rotating back to abduct, swinging obliquely outward from the body.

Step 4: The humerus (upper arm) begins to drive forward and back in the sagittal plane independent of spinal rotation. The movement is in opposition to the other arm and to the leg on the same side. Elbow flexion is maintained, oscillating about a 90° angle during the forward and backward arm swings.

Source: Data from Roberton and Halverson, 1984.

Some sources document the first stage of jumping as an exaggerated step down from a higher level (stairs or box) at about 18 months, but it is generally agreed that the first true jumping movements do not occur until the child is able to project the body off the ground into flight. This form of jumping, described as a **leap,** is normally evident by approximately 2 years, when the child propels herself forward and upward into flight with one foot and lands on the other foot. Though simple leaping is achieved by the age of 2, maximum height and distance (flight) are not accomplished until approximately age 4 or 5.

Table 9.2 shows the sequence of general jumping patterns and approximate age of skillful achievement. It should be emphasized that achievement levels are for the "basic" patterns of jumping, and that these movements provide the elements for more advanced jumping variations (sport skills) commonly displayed by older children (e.g., running long jump, triple jump, high jump).

Table 9.2   General Sequence Jumping Patterns

Movement Pattern	Approximate Age of Skillful Achievement (yr)
One-foot takeoff and opposite-foot landing (simple leap)	2–2½
One-foot takeoff and two-foot landing	
Two-foot takeoff and one-foot landing	5
Two-foot takeoff and landing (vertical jump; standing long jump)	5–6
One-foot takeoff and same-foot landing (hop)	5–6

Of the several jumping variations, the patterns that have received the most attention in the developmental literature are the vertical jump, the standing long jump, and the hop.

Both the vertical jump (for height) and the standing long jump (for distance) involve a two-foot takeoff and landing, as well as distinctive preparatory, takeoff, flight, and landing phases. In the mature form of both skills, the jumper must: (a) initiate a preparatory crouch, (b) vigorously swing the arms forward and upward to initiate the action, (c) extend the legs rapidly to propel the body from the ground, (d) extend the entire body during flight, and (e) flex the hips, knees, and ankles to absorb shock upon landing.

The standing long jump pattern is slightly more difficult than the vertical style, due primarily to the angle of projection and coordination of arm action with leg movements that are required. The long jump requires that the body be propelled forward and upward. This necessitates that the center of gravity be slightly ahead of the base of support at takeoff, which may create difficulty in maintaining forward balance; there is a strong tendency for the novice to step out with one foot to avoid falling. At this angle (about 45 degrees) the jumper must also swing the legs forward under the trunk in preparation for landing, whereas in the vertical jump the legs and trunk remain relatively vertical throughout execution.

Although rhythmical hopping (one foot to the same foot repeatedly) does not require maximum effort, it is the more difficult and complex of the three forms noted. The difficulty lies primarily in the requirement of greater leg strength and better balance, as well as the ability to perform controlled, rhythmical movement. All three of the jumping patterns, however, exhibit common developmental changes in relation to increased depth of preparatory crouch and increased use and effectiveness of arm action. With few exceptions, initial jumping patterns lack effective arm action. As a general rule, leg action is considerably more advanced than arm movements in the early stages of jumping. With maturity, arm movements are used effectively to aid in takeoff propulsion and in maintaining stability through flight and upon landing.

Vertical Jump   The **vertical jump** movement pattern involves a two-foot takeoff and landing with the primary purpose of achieving maximum height. Assessment of pattern characteristics is usually conducted by observing the skill

Figure 9.3
Arm positions in novice
jumpers. The arm
position of the boy on
the left is the more
effective of the three for
attaining maximum
reach.

while the individual reaches for an overhead target that elicits maximum reach. This form of vertical jumping is often described as the jump-and-reach version. If the jumping task does not include purposeful reaching (to a target) by the arms, less mature pattern characteristics are usually displayed. Although mature process characteristics of the vertical jump have been observed in children as young as 2 years (Poe, 1976), most children do not achieve mastery until approximately age 5.

Movement Pattern Descriptions   In the performance of skillful vertical jumping, the knees are bent in a preparatory crouch and the arms are lowered with the elbows slightly flexed. As the knees straighten the arms swing upward. The body stretches and extends as far as possible into vertical flight. The landing should be on the balls of the feet, with knees flexed to absorb the force of impact. However, the novice jumper frequently displays the following immature characteristics: (a) minimal preliminary crouch, (b) arms fixed at the middle or high guard position, (c) slight forward lean at takeoff, and (d) quick flexion at hips and knees following takeoff. An important factor in accomplishing maximum reach is the position of the shoulders and nonreaching arm at peak height. Figure 9.3 illustrates three arm positions commonly observed in novice jumpers in reaching for an overhead target.

In the mature vertical jump pattern, the preparatory phase is characterized by flexion at the hips, knees, and ankles. A vigorous forward and upward lift by

Figure 9.4
Stages of the vertical
jumping pattern.

INITIAL

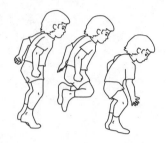

ELEMENTARY

MATURE

the arms initiates the jump, and thrust is continued by forceful extension at the hips, knees, and ankles. Just prior to the peak of the jump, the nonreaching arm is pushed downward. This movement tilts the shoulder girdle laterally and raises the hand of the reaching arm higher. Upon landing, the ankles, knees, and hips flex to absorb the shock (fig. 9.4).

Developmental Characteristics   The following general developmental trends have been noted in the vertical jump pattern:

1. A gradual increase in preparatory crouch.
2. Increased effectiveness of arm opposition.
3. Improved extension at takeoff and in flight.
4. Change from a forward flexion of the head throughout the jump to deep dorsiflexion (backward) during the entire sequence.
5. Greater extension of the trunk at the crest of the reach.

Standing Long Jump   The **standing long jump** (two-foot takeoff and landing) for maximum horizontal distance may be considered the standard fundamental jumping skill. This form of jumping has been included in numerous motor and physical fitness assessment batteries over the years, with the primary purpose of determining jumping ability in relation to maximum distance (product). Research on the early childhood years, however, has brought focus to the development of process characteristics. These components form the basis for several advanced jumping variations used by older children in a sports setting. As noted earlier, the standing long jump pattern is slightly more difficult than the jump-and-reach due primarily to the angle at which the body is projected. Although most children achieve a relatively high degree of vertical jump proficiency by age 5, mastery of the standing long jump is usually not observed until age 6.

Movement Pattern Descriptions   Skillful execution of the standing long jump requires, in general, a deep preparatory crouch and forward lean of the body that is counterbalanced by swinging the arms backward and then forcefully forward. Both feet leave the ground together with the angle of takeoff at approximately 45 degrees. The body achieves extension during flight and the landing is on both feet. However, arm, leg, and trunk action in the initial (immature) stages of jumping are quite limited. The jumping action is not initiated effectively by the arms because of their limited swing. Several ineffective arm positions may be observed in the novice jumper. Two of the most common are arms held rigidly at the side with elbows flexed and a "winging" position. At takeoff, which is usually at an angle of less than 30 degrees from vertical, the trunk is propelled in more of a vertical direction with little emphasis on the length of the jump. The preparatory crouch is limited and generally inconsistent with regard to the degree of leg flexion. Extension of the hips, legs, and ankles is incomplete at takeoff and one leg frequently precedes the other upon takeoff and upon landing. Figure 9.5

Figure 9.5
Initial patterns of the
two-foot takeoff: (top)
arms are held rigid at the
sides with elbows flexed
and (bottom) arms are
held in a "winging"
position.

shows two immature movement patterns for the standing long jump. In figure 9.5b the child displays the characteristic "winging" arm position.

With maturity and practice, greater coordination of arm and leg movements is achieved to result in a mature movement pattern. The mature pattern is characterized by a relatively deep preparatory crouch, while the arms swing backward and upward (fig. 9.6). At takeoff, both feet leave the ground together, with the thrust initiated in a horizontal direction at an angle of approximately 45 degrees. As the body moves forward, the hips, legs, and ankles extend in succession. During flight, the hips flex, bringing the thighs to a position nearly horizontal to the surface. The lower legs extend prior to landing. Upon landing on both feet, the knees bend and the body weight continues forward and downward. The arms reach forward to keep the center of gravity moving in the direction of the flight.

A component description of the developmental sequence of the standing long jump is presented in table 9.3. Keep in mind that children may move through the various levels of an action at different rates under the component perspective.

Developmental Characteristics   General developmental trends in the standing long jump movement pattern are similar to the developmental characteristics described for the vertical jump-and-reach:

1. An increase in the preparatory crouch.
2. Greater efficiency in use of the arms as displayed by an increase in the forward swing in the anteroposterior plane.
3. A decrease in the takeoff angle; theoretically, a projectile angle of 45 degrees is ideal.

Figure 9.6
Mature stages of the horizontal jumping pattern.

4. An increase in total body extension at takeoff.
5. Greater thigh flexion during flight.
6. A decrease in the angle of the leg at the instant of landing.

Hopping is a coordinated one-foot takeoff and landing on the same foot repeatedly. It is considered to be the most difficult form of basic jumping because hopping requires greater leg strength and better balance, as well as the ability to perform controlled, rhythmical movement compared to other jumping patterns. Prerequisite to hopping is the ability to balance on one foot momentarily (static balance), a skill that is not acquired until approximately 29 months of age (Eckert, 1987).

Hopping

Since hopping is a dynamic (moving) skill, repeated steps are generally not observed until after the age of 3 years. By 3 1/2 years of age most children can hop one to three steps and by the age of 5, the ability to hop 10 consecutive times is commonly seen. However, in a study of the process components of children ages 3, 4, and 5, Halverson and Williams (1985) found few children who displayed advanced levels of hopping and concluded, therefore, that the development of this skill may not reach the mature stage until the age of 6 years or older. The researchers also concluded that females are developmentally more advanced than the males. This finding appears consistently in the literature.

Movement Pattern Descriptions The early attempts at hopping are performed primarily in place with little forward movement. The general pattern is usually quite jerky with little extension and flexion of the support leg. As a result, the hop has minimal elevation and landings are flat-footed. As one would expect, arm opposition is ineffective (held in the high guard position and to the side for balance) and the nonsupport (swing) leg is often lifted high and held awkwardly to the side or in front of the body.

As the child's balance, leg strength, and coordination improve, arm and leg actions gradually transform into a smooth rhythmical mature movement pattern. Mature hopping is characterized by arm opposition that moves forward and upward in synchrony with the swing leg. The other arm moves in direct opposition to the action of the swing leg.

Table 9.3   Developmental Sequence (Components) for the Standing Long Jump

Takeoff: Leg Action Component

Step 1   One foot leads in asymmetrical takeoff.
Step 2   Both feet leave ground symmetrically, but hips or knees or both do not reach full extension by takeoff.
Step 3   Takeoff is symmetrical, with hips and knees fully extended.

Takeoff: Trunk Action Component

Step 1   Trunk is inclined forward less than 30° from vertical. Neck is hyperextended.
Step 2   Trunk leans forward less than 30°, with neck flexed or aligned with trunk at takeoff.
Step 3   Trunk is inclined forward 30° or more at takeoff, with neck flexed.
Step 4   Trunk is inclined forward 30° or more. Neck is aligned with trunk, or slightly extended.

Takeoff: Arm Action Component

Step 1   Arms move in opposition to legs or are held at side, with elbows flexed.
Step 2   Shoulders retract, arms extend backward in winging posture at takeoff.
Step 3   Arms are abducted about 90°, with elbows frequently flexed, in high or middle guard position.
Step 4   Arms flex forward and upward with minimal abduction, reaching incomplete extension overhead
     by takeoff.
Step 5   Arms flex forward, reaching full extension overhead by takeoff.

Flight and Landing: Leg Action Component

Step 1   Legs assume asymmetrical run pattern in flight, with one-footed landing.
Step 2   Legs assume asymmetrical run pattern but swing to two-footed landing.
Step 3   During flight, hips and knees flex in a synchronous fashion. Knees then extend for two-footed landing.
Step 4   During flight, flexion of both knees precedes hip flexion. As hips flex knees extend, reaching
     forward to two-footed landing.

Flight and Landing: Trunk Action Component

Step 1   Trunk maintains forward inclination of less than 30° in flight, then flexes for landing.
Step 2   Trunk corrects forward lean of 30° or more by hyperextending, then flexes for landing.
Step 3   Trunk maintains forward lean of 30° or more from takeoff to midflight, then flexes forward for landing.

Flight and Landing: Arm Action Component

Step 1   Arms move in opposition to legs as if child were running in flight and on landing.
Step 2   Shoulders retract and arms extend backward (winging) during flight and move forward
     (parachuting) during landing.
Step 3   During flight, arms assume high or middle guard positions and may move backward in windmill
     fashion. They parachute for landing.
Step 4   Arms lower or extend from flexed position overhead, reaching forward at landing.

Source: Data from Roberton and Halverson, 1984.

Longitudinal data reported by Roberton (1990) suggest that advanced, smoothly timed arm opposition is the last component of hopping to develop and sometimes takes more than 10 years to achieve. Prior to takeoff, the weight of the child is transferred along the foot to the ball before the knee and ankle extend to takeoff. The support leg reaches maximum extension on takeoff while the swing leg leads the upward-forward thrust with a pumping action. During this action, the swing leg passes behind the support leg (when viewed from the side). Upon landing, the hip, knee, and ankle joints flex to absorb the shock (fig. 9.7). Table 9.4 details the components for the developmental sequence of hopping.

Developmental Characteristics   The following major developmental trends have been observed in hopping behavior:

1. A decrease in the amount of forward body lean.
2. An increase in arm action effectiveness from stabilization to efficient arm opposition.
3. A change from leg clearance by flexion to clearance as a result of leg thrust.
4. Improvement in the use of the nonsupportive (swing) leg, from an inactive forward position to a forward-upward swing connected with takeoff.
5. An increase in range and speed of movement at the hip, knee, and ankle of the support leg.
6. Change in landing from immediate extension following knee and ankle flexion to a delay in extension while the body pivots over the foot.

Galloping, Sliding, and Skipping

Galloping, sliding, and skipping are locomotor skills that consist of a combination of basic movements. Due to their increased complexity, they are normally not mastered until around the end of the early childhood years. The relatively late appearance of these skills (around age 4) may be expected because they require a certain level of dynamic balancing ability. As children master the ability to propel their body weight onto the forward foot, these patterns begin to appear.

Galloping, which is usually the first of these three skills to emerge, combines the basic patterns of a walking step and the leap; it is an uneven rhythmical pattern. The mature movement pattern is characterized by a slight forward lean and a thrusting forward of the lead leg. The lead foot supports the body weight while the rear foot quickly closes behind the lead foot and takes the weight. During this cycle there is extension and flexion of both legs with a momentary suspension of both feet in the air. In a series of gallops, the same foot takes the lead and contacts the ground with a heel-toe action. The arms swing freely from the shoulders during execution. In comparison to sliding and skipping, opportunities for galloping and the motivation to gallop as a mature physical activity appear limited.

Figure 9.7
Mature hopping pattern.

Table 9.4 Developmental Sequence (Components) for Hopping

Leg Action Component

Step 1	*Momentary flight.* The support knee and hip quickly flex, pulling (instead of projecting) the foot from the floor. The flight is momentary. Only one or two hops can be achieved. The swing leg is lifted high and held in an inactive position to the side or in front of the body.
Step 2	*Fall and catch; swing leg inactive.* Forward lean allows minimal knee and ankle extension to help the body "fall" forward of the support foot and then quickly catch itself again. The swing leg is inactive. Repeated hops are achieved.
Step 3	*Projected takeoff; swing leg assists.* Perceptible pretakeoff extension occurs in the support leg, hip, knee, and ankle. There is little delay in changing from knee and ankle flexion on landing to takeoff extension. The swing leg now pumps up and down to assist in projection, but range is insufficient to carry it behind the support leg.
Step 4	*Projection delay; swing leg leads.* The child's weight on landing is smoothly transferred along the foot to the ball before the knee and ankle extend to takeoff. The range of the pumping action in the swing leg increases so that it passes behind the support leg when viewed from the side.

Arm Action Component

Step 1	*Bilateral inactive.* The arms are held bilaterally, usually high and out to the side, although other positions behind or in front of the body may occur. Any arm action is usually slight and not consistent.
Step 2	*Bilateral reactive.* Arms swing upward briefly and then are medially rotated at the shoulder in a winging movement prior to takeoff. This movement appears to occur in reaction to loss of balance.
Step 3	*Bilateral assist.* The arms pump up and down together, usually in front of the line of the trunk. Any downward and backward motion of the arms occurs after takeoff. The arms may move parallel to each other or be held at different levels as they move up and down.
Step 4	*Semiopposition.* The arm on the side opposite the swing leg swings forward with that leg and back as the leg moves down. The position of the other arm is variable, often staying in front or to the side.
Step 5	*Opposing assist.* The arm opposite the swing leg moves forward and upward in synchrony with the forward and upward movement of that leg. The other arm moves in the direction opposite the action of the swing leg. The range of movement in the arm action may be minimal unless the task requires speed or distance.

Source: Data from Halverson & Williams, 1985.

Table 9.5 Developmental Sequence (Composite) for Skipping

Stage 1	A deliberate step-hop pattern is employed, occasional double hop is present, there is little effective use of the arms to provide momentum, exaggerated step or leap is present during the transfer of weight from one supporting limb to the other, the total action appears segmented.
Stage 2	Rhythmical transfer of weight during the step phase, increased use of arms in providing forward and upward momentum, exaggeration of vertical component during airborne phase, i.e., while executing the hop.
Stage 3	Rhythmical transfer of weight during all phases, reduced arm action during transfer of weight phase, foot of supporting limb carried near surface during hopping phase.

Source: From V. Seefeldt and J. Haubenstricker, 1974, Developmental Sequence for Skipping. Unpublished material, Michigan State University, East Lansing, MI. Reprinted by permission of John L. Haubenstricker.

Sliding is similar to galloping except the direction of movement is sideways. When sliding to the right, the right foot moves sideward, taking the weight, and the left foot follows quickly. As with galloping, the same foot always takes the lead. The first step is usually a slow gliding movement and the second is a quick closing step. Foot contact is normally on the balls of the feet, with the weight shifted from the lead to the follow-up foot. After the follow-up foot catches the lead foot, a slight jump (momentary suspension) is displayed. At this time the body is ready for a quick change of direction or continuation in the same pathway. Arms are held in a relaxed position at approximately waist level and to the side.

Of the three locomotor skills noted, **skipping** is typically the last to be mastered (between the sixth and seventh year). The general pattern involves stepping forward on one foot, quickly hopping on the same foot, duplicating the process on the opposite foot, and so on. Complexity lies in the fact that it consists of a *step-hop pattern* that is performed in an uneven rhythm. To perform a step-hop on one foot requires additional skill in timing of sequential movements as well as a substantial degree of balance. Another aspect of its complexity is that the step-hop is performed on one foot before the weight is transferred to the other foot.

Characteristic of mature skipping is a well-coordinated and continuous cycle of weight transfer to the opposite foot and the subsequent step-hop. The trunk is held erect with the head focusing in a forward direction. The arms swing freely and in opposition to leg action. The support knee and ankle extend for takeoff (creating momentary suspension) and flex upon landing. The nonsupport leg flexes to aid elevation of the hop. The balls of the feet receive the weight of the body. A developmental sequence (composite) description for skipping is presented in table 9.5.

Since young females tend to be superior at hopping, it is not surprising that the literature has shown them to be better skippers also. For example, Haubenstricker and colleagues (1990) found that females 2 to 9 years of age were consistently more advanced than males at all ages. Males appear to be about six months to one year behind in the qualitative aspects of skipping in comparison to females. Although it is speculative, this difference could be a result of females' greater interest in and subsequent practice of the skill, as well as the fact that females have a slight edge in biological maturity for chronological age.

Throwing

Throwing is a complex manipulative skill in which one or both arms are used to thrust an object away from the body and into space. During the early childhood years, children display many different throwing patterns. According to Wickstrom (1983), there does not appear to be any definite or precise developmental order for the onset of the variations. The pattern a child uses depends on several factors, including size of the child, size of the object, and age. Of the three most common variations (overarm, sidearm, and underhand), the unilateral overarm style is the most commonly used and has been the most thoroughly studied.

The first signs of the overarm pattern appear around the age of 6 months when the child executes a crude throw involving limited use of the arm while in a sitting position. Though a wide variation in skill level may be observed at all ages, the majority of children are skillful throwers by age 6. As noted, some individuals may never acquire the mature stage of a specific motor skill. The mature overarm throwing pattern is a motor skill that has a relatively high incidence of individuals who do not reach mastery, especially among females.

Data from a longitudinal study that documented pattern changes of kindergarten children to the seventh grade support this possibility (Halverson, Roberton, & Langendorfer, 1982). By the seventh grade, 80% of the males achieved the mature level in upper arm action compared to only 29% of the females. This trend was also evident for actions of the forearm and trunk. The presence of immature throwing patterns was also reported among adult females (Leme & Shambles, 1978), thus demonstrating that not all movement patterns reach the mature level during childhood or even adolescence.

There is evidence suggesting that instruction and adequate opportunity for practice are significant factors in the development of throwing technique among children (Halverson & Roberton, 1979; Leudke, 1980). While instruction and practice have been shown to positively influence throwing technique, these factors do not appear to affect ball-throwing velocity (Halverson et al., 1977).

Movement Pattern Descriptions   Wild (1938) is recognized as having conducted one of the most definitive studies of developmental form in overarm throwing. Through cinematographic analysis of throwing form in children ages 2 to 7 years, the researcher identified four developmental stages (see table 9.6). Wickstrom (1983) notes that subsequent observations have confirmed the presence of these general stages, but it is likely that the patterns appear somewhat earlier than indicated by Wild's report. To understand the throwing action, visualize it in three phases: (a) the preparatory phase, in which movements are directed away from the line of projection, (b) the execution phase, in which movements are performed along the line of projection, and (c) follow-through, in which movement follows the release of the object.

Stage I is characterized by minimal trunk rotation and projection of the ball primarily by elbow extension. Virtually all movement is in the anterior-posterior

Table 9.6 A Summary of Wild's Four Developmental Stages of Throwing

Stage I (2–3 yr)	Stage II (3 1/2–5 yr)	Stage III (5–6 yr)	Stage IV (6 1/2 yr and older)
Throw is arm-dominated. Preparatory arm movements involve bringing the arm sideways-upward or forward-upward. Thrower faces the direction of intended throw at all times. No rotation of trunk and hips is evident. Feet remain stationary during the entire throwing act.	Body moves in horizontal plane instead of anterior-posterior plane. Throwing arm moves in a high oblique plane or horizontal plane above the shoulder. Throwing is initiated predominately by arm and elbow extension. Feet remain stationary, but rotary movement of the trunk is observable.	Forward step is unilateral to the throwing arm. Arm is prepared by swinging it obliquely upward over the shoulder with a large degree of elbow flexion. Arm follows through forward and downward and is accompanied by forward flexion of the trunk.	Forward step is taken with the contralateral leg. Trunk rotation is clearly evident. Arm is horizontally adducted in the forward swing.

Source: Data from Payne and Isaacs, 1987, as interpreted from Wild, 1938.

plane. During the entire series of movements, both feet remain stationary and the body faces toward the direction of the throw.

Stage II is distinguished primarily by movements that occur in the horizontal plane. The feet remain stationary but rotation of the pelvis and spine are evident. Greater projection force is also attained from the forward and downward follow-through.

The most noticeable occurrence at Stage III is the appearance of a step forward with the delivery. In the preparatory phase, the weight is shifted onto the left (or rear) foot while the body is rotated to the right. The arm is moved obliquely upward and behind (over) the shoulder and set in a flexed position. During delivery, the weight is transferred onto the foot on the same side as the throwing arm (ipsilateral) as the body rotates to the left and the arm projects forward. Upon completion of the follow-through, the body faces partially to the left (for a right-hander), in contrast to the facing forward as took place in the preceding stages.

In Stage IV (mature pattern), the thrower displays proper opposition by transferring weight to the foot opposite the throwing arm (contralateral). This action facilitates greater trunk rotation and, along with the horizontal adduction of the arm during the forward swing, enables the thrower to achieve maximum body leverage. Figure 9.8 depicts the mature overarm throwing pattern.

The complexity of the mature overarm throwing pattern was highlighted by Plagenhoef (1971) when he described the motion as "the properly timed coordination of accelerations and decelerations of all body segments in a sequence of action from the left foot to the right hand that produces maximum absolute velocity of the right hand." Additional clarification of the mature throwing pattern is provided by Wickstrom (1983).

Figure 9.8
Mature overarm
throwing pattern.

Preparatory Phase

1.   The body pivots to the right with the weight on the right foot; the throwing
     arm swings backward and upward.

Execution Phase

2.   The left foot strides forward in the intended direction of the throw.
3.   The hips, then the spine and shoulders, rotate counterclockwise as the
     throwing arm is retracted to the final point of its reversal.
4.   The upper arm is rotated medially and the forearm is extended with a
     whipping action.
5.   The ball is released at a point just forward of the head with the arm ex-
     tended at the elbow.

Follow-Through

6.   The movement is continued until the momentum generated in the throwing
     action is dissipated.

   A more precise developmental (component) analysis of the overarm throwing pattern has been offered by Roberton (1977, 1978, 1984) in recent years (table
9.7). Although some of the component sequences have been hypothesized from
other studies and have not been validated, this model adds significantly to our understanding of the specific developmental changes that occur. Research based on
this model has verified that development within component parts may proceed at
different rates in the same individual or at different rates in different individuals.

Developmental Characteristics   The following developmental trends have
been noted in relation to the overarm throwing pattern:

1.   A gradual shift of movement from a predominantly anterior-posterior
     plane to a horizontal plane.

Table 9.7   Developmental Sequence (Components) for Overarm Throwing

Trunk Action Component

Step 1   *No trunk action or forward-backward movements.* Only the arm is active in force production. Sometimes, the forward thrust of the arm pulls the trunk into a passive left rotation (assuming a right-handed throw), but no twist-up precedes that action. If trunk action occurs, it accompanies the forward thrust of the arm by flexing forward at the hips. Preparatory extension sometimes precedes forward hip flexion.

Step 2   *Upper trunk rotation or total trunk "block" rotation.* The spine and pelvis both rotate away from the intended line of flight and then simultaneously begin forward rotation, acting as a unit or "block." Occasionally, only the upper spine twists away, then toward the direction of force. The pelvis then remains fixed, facing the line of flight, or joins the rotary movement after forward spinal rotation has begun.

Step 3   *Differentiated rotation.* The pelvis precedes the upper spine in initiating forward rotation. The child twists away from the intended line of ball flight and then begins forward rotation with the pelvis while the upper spine is still twisting away.

Preparatory Arm Backswing Component

Step 1   *No backswing.* The ball in the hand moves directly forward to release from the arm's original position when the hand first grasped the ball.

Step 2   *Elbow and humeral flexion.* The ball moves away from the intended line of flight to a position behind or alongside the head by upward flexion of the humerus and concomitant elbow flexion.

Step 3   *Circular, upward backswing.* The ball moves away from the intended line of flight to a position behind the head via a circular overhead movement with elbow extended, or an oblique swing back, or a vertical lift from the hip.

Step 4   *Circular, downward backswing.* The ball moves away from the intended line of flight to a position behind the head via a circular, down, and back motion, which carries the hand below the waist.

Humerus (Upper Arm) Action Component during Forward Swing

Step 1   *Humerus oblique.* The humerus moves forward to ball release in a plane that intersects the trunk obliquely above or below the horizontal line of the shoulders. Occasionally, during the backswing, the humerus is placed at a right angle to the trunk, with the elbow pointing toward the target. It maintains this fixed position during the throw.

Step 2   *Humerus aligned but independent.* The humerus moves forward to ball release in a plane horizontally aligned with the shoulder, forming a right angle between humerus and trunk. By the time the shoulders (upper spine) reach front facing, the humerus (elbow) has moved independently ahead of the outline of the body (as seen from the side) via horizontal adduction at the shoulder.

Step 3   *Humerus lags.* The humerus moves forward to ball release horizontally aligned, but at the moment the shoulders (upper spine) reach front facing, the humerus remains within the outline of the body (as seen from the side). No horizontal adduction of the humerus occurs before front facing.

Table 9.7   Continued

Forearm Action Component during Forward Swing

Step 1	*No forearm lag.* The forearm and ball move steadily forward to ball release throughout the throwing action.
Step 2	*Forearm lag.* The forearm and ball appear to "lag," i.e., to remain stationary behind the child or to move down or back in relation to the child. The lagging forearm reaches its farthest point back, deepest point down, or last stationary point *before* the shoulders (upper spine) reach front facing.
Step 3	*Delayed forearm lag.* The lagging forearm delays reaching its final point of lag until the moment of front facing.

Action of the Feet

Step 1	*No step.* The child throws from the initial foot position.
Step 2	*Homolateral step.* The child steps with the foot on the same side as the throwing hand.
Step 3	*Contralateral, short step.* The child steps with the foot on the opposite side from the throwing hand.
Step 4	*Contralateral, long step.* The child steps with the opposite foot a distance of over half the child's standing height.

Source: Data from Roberton & Halverson, 1984.

2. Transition from the use of a stationary base of support to weight transfer on the same side as the throwing arm, followed by functional arm-foot opposition.

3. An increase with each successive throwing pattern in effective mechanical projection as evidenced by throwing velocity.

Catching

Catching is a fundamental, gross motor, manipulative skill that involves tracking an incoming object, stopping its momentum, and gaining control of it by use of the hands. The reception of objects with the hands is a rather complex movement pattern for children. This is due primarily to the fact that catching requires coincident timing ability. Recall that coincident timing is a form of temporal accuracy that involves timing self-movements with an object. This action includes a series of complicated perceptual judgments that can vary considerably, depending on catching conditions. Unfortunately, evidence concerning the emergence of catching skills is insufficient to provide a precise developmental description. Part of the difficulty in studying catching behavior is due to the numerous variables that influence the measurement of performance (e.g., ball size, ball speed and distance traveled, method of ball projection, angle and level of the receiver). Because of these and other conditions under which catching has been researched, it is difficult to compare and contrast various studies.

Although throwing and catching have a close functional relationship, catching proficiency normally follows that of throwing. A primitive form of catching involving trapping can be observed in 2- and 3-year-olds, but the reports of children achieving proficiency at this skill indicate that they typically range in age from 6 to 8 years. However, these figures are based on two-handed catching in relatively stable conditions; the ability to catch a ball under more complex conditions continues to develop well into the upper elementary years (age 10 to 12).

Movement Pattern Descriptions   The first attempts to stop and control a moving object occur when a ball is rolled toward a child who is seated on the floor with legs apart. Initially, the child will stop the ball by corralling and trapping it against the legs. With adequate practice, the child will develop the ability to coordinate movements of the arms with the velocity of the ball and so trap it with the hands. As the child rises from the relatively stationary sitting position, she develops the ability to chase, stop, and gain control of a moving or bouncing ball. Wickstrom (1983) notes that this series of achievements is important in the progression leading to comprehensive catching ability.

The first attempts to catch an aerial ball with two hands are relatively passive. That is, the child simply holds its arms out stiffly in front of the body, regardless of the angle and height of the incoming object. Little or no effort is made to move the body to adjust to the flight of the ball. At the point of contact with the ball, the hands exhibit very little "give"; the momentum is not attenuated. In addition, there is very little, if any, flexion at the knees to help absorb the shock of the ball's velocity.

A rather common observation in novice catchers is a negative reaction and fear of the ball. These characteristics include turning the head to the side, a slight backward bending of the trunk away from the incoming ball, and closing the eyes. Seefeldt (1972) reported these characteristics in children 4, 5, and 6 years of age but found no evidence of them in children 1 1/2 to 3 years of age. Seefeldt speculated that fear of an aerial ball is not a natural phenomenon but may be a conditioned response from earlier failures at the task. In this stage of development, success in catching is as much (or more) dependent on the accuracy of the throw as it is on the ability of the catcher. Figure 9.9 shows two variations of the immature two-handed catching pattern.

As development approaches mastery, the child displays greater coordination by adjusting the body and the hand and arm positions to accommodate the flight of the ball. In addition, the catcher learns to "give" with the object, thus allowing the momentum of the projectile to be absorbed by the body. The following general characteristics summarize the movements associated with the mature two-handed catching pattern:

1. Body is in alignment with incoming object.
2. Feet are slightly apart and parallel or in forward stride position.
3. Arms are held relaxed at sides (or in front of the body) and elbows are flexed.
4. Hands and fingers are relaxed and slightly cupped (pointing to object).
5. Eyes follow the flight of the object.
6. Hands move forward to meet approaching object.

Figure 9.9
Two variations of
immature catching
pattern. In the upper
sequence the child
displays a fear reaction.
The child in the lower
sequence exhibits
urgent-like actions,
responding to the ball
only after it has touched
his hands, then trapping
it against his chest.

7. Arms "give" upon contact to absorb the force, and fingers close around the object. Contact and closure are simultaneous with both hands.
8. Body weight is transferred from front to back.

Developmental Characteristics   Since the qualitative evidence concerning the emergence of catching skills is insufficient for a precise developmental description, only generalized observations about developmental trends and characteristics can be made:

1. Catching ability, in general, progresses from trapping (using the arms, body, and hands) to catching an object using the hands exclusively.
2. Movement pattern action improves from a relatively passive reception (e.g., simply holding the arms out and trapping after contact) to moving the body to adjust to the flight of the object.
3. Positioning of the body improves to be in line with the oncoming object.

Figure 9.10
Mature two-hand
catching pattern.

4. Arm action improves from a stiff outstretched position to a position in which the elbows are flexed and "give" occurs upon contact to absorb the momentum of the object.

5. With practice and a subsequent increase in confidence, fear reactions to catching an aerial ball decrease in most individuals.

An illustration of the mature two-hand catching pattern (with small ball) is presented in figure 9.10. Although no developmental sequence model for two-hand catching has been validated for all components, table 9.8 presents the latest update of that endeavor (Stromeyer, Williams, & Schaub-George, 1991). Table 9.9 is an example of a process rating scale for catching.

As noted earlier, the conditions under which catching behavior has been measured vary considerably. Obviously, catching a large ball that is tossed slowly in front of the body at chest level while stationary is the easier form of catching. Several factors may influence catching performance (e.g., ball velocity, trajectory angle, ball color), but one of the most consistently noted in the literature is ball size.

As may be expected, catching performance is achieved at a certain level of performance with a large ball before the same level is attained with a smaller ball. However, Wickstrom (1983) notes that mature catching, using the hands only, seems to appear earlier if the child also practices with a small ball. This practice seems to induce the child to think of using only the hands when catching a small ball, rather than arm/chest trap. When a larger ball is introduced to children after they have attained some proficiency using a small ball, they seem more likely to resort to the immature pattern (Isaacs, 1980; Payne, 1985; Wickstrom, 1983). The best catching performance seems to be elicited using a ball that can be cupped in the hands (e.g., tennis ball size) but is not so small that it requires extraordinary visual-motor control.

Striking is an action in which a part of the body or an implement is used to give impetus to an object. Depending on the striking situation, the skill can be executed using a variety of body parts (most commonly a hand, foot, or the head) and a variety of implements (e.g., paddle, racquet, bat). Striking skills can also be performed using various movement patterns, the most common being the overhand,

Striking

Table 9.8   Developmental Sequence (Components) for Catching

Preparation: Arm Component

Step 1   The arms are outstretched with elbows extended, awaiting the tossed ball.
Step 2   The arms await the ball toss with some shoulder flexion still apparent, but flexion now appears in the elbows.
Step 3   The arms await the ball in a relaxed posture at the sides of the body or slightly ahead of the body. The elbows may be flexed.

Reception: Arm Component

Step 1   The arms remain outstretched and the elbows rigid. There is little to no "give" so the ball bounces off the arms.
Step 2   The elbows flex to carry the hands upward toward the face. Initially, ball contact is primarily with the arms, and the object is trapped against the body.
Step 3   Initial contact is with the hands. If unsuccessful in using the fingers, the child may still trap the ball against the chest. The hands still move upward toward the face.
Step 4   Ball contact is made with the hands. The elbows still flex but the shoulders extend, bringing the ball down and toward the body rather than up toward the face.

Hand Component

Step 1   The palms of the hands face upward. (Rolling balls elicit a palms-down, trapping action.)
Step 2   The palms of the hands face each other.
Step 3   The palms of the hands are adjusted to the flight and size of the oncoming object. Thumbs or little fingers are placed close together, depending on the height of the flight path.

Body Component

Step 1   No adjustment of the body occurs in response to the flight path of the ball.
Step 2   The arms and trunk begin to move in relation to the ball's flight path, but the head remains erect, creating an "awkward" movement to the ball. The catcher seems to be fighting to remain balanced.
Step 3   The feet, trunk, and arms all move to adjust to the path of the oncoming ball.

Note: These sequences were hypothesized by Harper (1979), cited in Roberton & Halverson (1984), and updated by Stromeyer, Williams, & Schaub-George (1991).

sidearm, and underhand. Kicking, also considered a striking skill, is discussed separately because of its rather unique pattern characteristics.

Unfortunately, relatively little information is available on the developmental sequence of striking ability. Much like catching, a number of factors may affect striking performance including the size, weight, and length of the implement used, the physical characteristics of the object to be struck, and the speed of the incoming object (whether stationary or moving).

Young children frequently display relatively skillful striking patterns before they gain the ability to contact a moving ball with enough frequency to provide reasonable quantitative measurement. Of the various striking skills, the ones

Table 9.9 Example of a Process-Oriented Rating Scale for Catching

0 = Initial body contact; subject makes no attempt to contact the ball.
1 = Arm and/or body contact, miss: initial attempt to contact is made on the arms and/or body, and the ball is missed.
2 = Arm and/or body contact, save: initial contact is on the arms and/or body and the ball is retained.
3 = Hand contact, miss: initial contact is made by the hands, but ball is then dropped immediately or dropped following arm or body contact.
4 = Hand contact, assisted catch: initial contact is made by the hands. The ball is juggled but retained by using arms and/or body for assistance.
5 = Hand contact, clean catch: the ball is contacted and retained by the hands only. The ball may be brought into the body on the follow-through after control is gained by the hands.

Source: Data from Payne and Isaacs, 1987, as derived from Hellweg, 1972, and Isaacs, 1980.

most commonly used are the one- and two-handed sidearm striking pattern in the horizontal plane. A well-defined sidearm pattern may be evident in some children at approximately 3 years of age (Eckert, 1987), but the mature pattern is generally not achieved until age 4 or 5. The ability to consistently intercept a moving object is considerably more difficult and continues to improve through the late elementary years (ages 10 to 12). Wickstrom cautions that although there appears to be a general trend toward improvement during this period, specifics concerning the rate and nature of change are quite unclear.

Movement Pattern Descriptions Striking behavior has been observed in the form of spontaneous overarm hand banging against a surface in children as young as 10 to 12 weeks of age (Thelen, 1979). Voluntary striking appears to develop in much the same sequential order as throwing in terms of general age level of performance. In the initial stage, the child uses an overarm "chopping" action (with or without an implement) in a vertical plane. From that initial stage the child seems to progress slowly through a downward series of planes that are more horizontal (sidearm) or even underarm.

The position of the object does, however, influence the angle of approach of the arm. Most striking skills are performed with a sidearm pattern. One of the most utilized techniques for eliciting and assessing developmental sidearm striking characteristics is to require the individual to forcefully strike (with a racquet or bat) a stationary ball suspended at approximately waist level.

Initial efforts at sidearm striking are similar to immature pattern attempts at overarm throwing. That is, the individual "chops" at the ball using an overarm pattern by bending forward slightly, extending at the elbow, and using minimal trunk and leg action. The general striking action is predominantly flexion and extension of the forearm. In addition, the individual usually directly faces the object (instead of standing sideways to it) and if a step forward is taken (though none may be evident), it is taken with the foot on the same side as the striking arm (homolateral) (fig. 9.11). Therefore, as with immature throwing, arm motion occurs primarily in the anterioposterior plane.

Figure 9.11
Immature one-hand
striking.

With maturity and practice, the individual gradually lowers the striking plane to the horizontal and abandons the arm-dominated movement pattern. Although the one-arm and two-arm variations of the striking skill have their own unique characteristics, each involves the same basic sequence of three movements: the step, the turn, and the swing. Basic actions within this sequence are as follows:

1. In the preparatory phase, the feet are positioned approximately shoulder-width apart.
2. Body weight is shifted initially away from the intended hit, and then in the direction of the strike (sideways to the target) while the shoulders and arms are coiled in the opposite direction. That is, for right-handers, the weight is shifted onto the right (rear) foot. The individual "steps" with the opposite foot into the hit to apply a straight-line direction of force. The length of the step is slightly greater than one-half of the individual's height.
3. Hips and spine are rotated in rapid succession in the same direction as the weight shift.
4. The arm or arms swing around and forward (horizontally) in close succession with the other rotary movements. The swing is performed through a full range of motion to apply adequate force.
5. The wrist or wrists are cocked in preparation for the strike and uncocked just prior to contact with the object.
6. Eyes follow the flight of the ball until just before contact is made.
7. Follow-through after contact.

Figure 9.12 describes the mature one-arm striking pattern. The only additional basic characteristic unique to the two-arm striking pattern (batting) is the position of the elbow. That is, in the preparatory phase, the lead elbow is held up and out from the body with the bat held off the shoulder (fig. 9.13).

Developmental Characteristics   The following developmental trends and characteristics have been noted for the sidearm striking pattern:

1. Increased use of a forward step or a forward weight shift to initiate the pattern.

Figure 9.12
Mature one-arm striking.

Figure 9.13
Mature two-arm striking.

2. Increased range of motion in the various joints during the swing.
3. Increased hip and trunk rotation preceding the action of the arm(s) in the swing.
4. Change of wrist action from a relatively stiff position to a more distinct cock and uncock action during the swing.

Table 9.10 describes a developmental sequence (composite) for striking with a bat.

Table 9.10   Developmental Sequence (Composite) for Striking with a Bat

Stage 1	The motion is primarily posterior-anterior in direction. The movement begins with hip extension and slight spinal extension and retraction of the shoulder on the striking side of the body. The elbows flex fully. The feet remain stationary throughout the movement with the primary force coming from extension of the flexed joints.
Stage 2	The feet remain stationary or either the right or left foot may receive the weight as the body moves toward the approaching ball. The primary pattern is the unitary rotation of the hip-spinal linkage about an imaginary vertical axis. The forward movement of the bat is in a transverse plane.
Stage 3	The shift of weight to the front-supporting foot occurs in an ipsilateral pattern. The trunk rotation-derotation is decreased markedly in comparison to stage 2, and the movement of the bat is in an oblique-vertical plane instead of the transverse path as seen in stage 2.
Stage 4	The transfer of weight in rotation-derotation is in a contralateral pattern. The shift of weight to the forward foot occurs while the bat is still moving backward as the hips, spine, and shoulder girdle assume their force-producing positions. At the initiation of the forward movement the bat is kept near the body. Elbow extension and the supination-pronation of the hands do not occur until the arms and hands are well forward and ready to extend the lever in preparation to meet the ball. At contact the weight is on the forward foot.

From V. Seefeldt and J. Haubenstricker, 1974, Developmental Sequence for Striking with a Bat. Unpublished material, Michigan State University, East Lansing, MI. Reprinted by permission of John L. Haubenstricker.

Kicking

Kicking is a fundamental manipulative skill in which the foot is used to strike an object. Unfortunately, information on developmental kicking behavior has not been extensive. Although the characteristics of mature pattern variations and general trends have been reported, qualitative changes made within the various components are not well documented. Another limitation has been that kicking performance (distance, accuracy) of young children (below 4 or 5 years) is difficult to measure in quantitative terms. Most research data on the development of fundamental kicking behavior have been collected on pattern characteristics using the placekick, with a stationary ball. The placekick is considered the basic foundation upon which other skills such as kicking a moving ball, dribbling, and punting are developed. These forms of advanced kicking are usually not achieved until some degree of placekicking skill has been acquired and additional perceptual abilities developed. As with catching and striking, variations in speed and position of the ball in space can significantly alter the level of difficulty.

The ability to kick a stationary ball (with minimal form) appears around the age of 2, but it is not until 5 or 6 years that the mature pattern is achieved by most children. It should be noted that proficient kicking, like proficient throwing, may not be achieved through the natural course of childhood development.

Movement Pattern Descriptions   Before an individual can execute a true kicking action, the kicker must be able to maintain an upright posture while balancing momentarily on one foot and imparting force to an object with the other foot. This ability, which is normally present by 2 years, provides the basis for initial kicking behavior.

The immature stage of kicking is characterized by a limited range of action in the propelling leg and minimal backswing and follow-through. There is very little movement of the upper body, and the arms are held out from the sides for balance. The kicking leg frequently contacts the ball while it is deeply flexed. General leg movements are described as a "pushing" action and may be displayed with either leg. During the early stages of this skill, the child may also respond to a ball placed in front of the body by running into it.

As the child attempts to kick more forcefully, range of leg motion increases (more backswing and follow-through), forward lean of the trunk increases, and the arms elevate to aid in maintaining balance. Opposition of the arm and foot also begins to develop. Wickstrom (1983) describes the following actions as fundamental to the mature pattern of all basic kicking variations:

1. Preparatory forward step on the support leg to rotate the pelvis backward on the opposite side and to extend the thigh of the kicking leg.
2. The support foot is placed to the side and slightly behind the ball.
3. Forward pelvic rotation and swing of the kicking leg with simultaneous flexion at the hip and at the knees.
4. Vigorous extension (whipping) of the lower part of the kicking leg.
5. Momentary slowdown or cessation of thigh flexion as the lower leg whips into extension just before the foot makes contact with the ball.
6. Forward swing of the opposite arm in reaction to the vigorous action of the kicking leg (arm/leg opposition).
7. Follow-through is forward and toward the midline.

Figure 9.14 illustrates four stages of kicking behavior as described by Deach (1950).

Developmental Characteristics   The following developmental trends and characteristics have been noted for the placekicking movement pattern:

1. An increase in the range of preparatory movement at the hip and the knee of the kicking leg.
2. An increase in the range of motion in the kicking leg.
3. A tendency to start farther behind the ball and move the total body forward into the kick.
4. An increase in compensatory trunk lean and arm opposition.

Ball bouncing and dribbling are skills that propel a ball in a downward direction. It has been speculated that the development of bouncing originates as the child drops a ball, causing it to bounce, and attempts to strike the object repeatedly. As the child's control of the ball progresses, the term dribbling is used to describe

Ball Bouncing and Dribbling

Figure 9.14
Deach's four stages of
kicking behavior: (a) the
girl keeps her kicking leg
nearly straight and
exhibits minimal
coordination of rest of
her body; (b) increased
flexion of the kicking
leg (precontact) and
some arm opposition;
(c) increased preliminary
hip extension, greater
range of leg motion,
and additional body
adjustments; and (d) the
mature form of kicking
behavior.

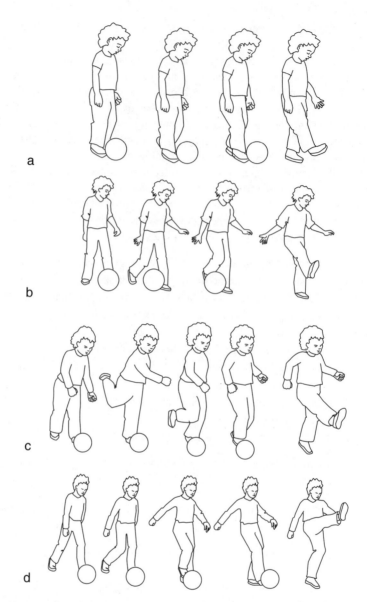

the action. Wickstrom (1983) suggests that minimal form in dribbling involves
the ability to bounce a ball three or four consecutive times.

    Although some 2-year-olds may exhibit a minimal degree of skill in two-
handed bouncing, it is not until approximately 5 or 6 years of age that the mature
pattern of one-handed dribbling in a relatively stationary position is mastered.
Dribbling proficiency while moving is much more complex than dribbling in a
stationary position and is considered a sport skill. The difficulty of this skill is attributed, in part, to the fact that the ball loses forward speed after each bounce
and a special push is necessary to maintain the desired forward speed.

Movement Pattern Descriptions   Although a specific developmental sequence has not been validated, the following general progression has been noted:

1.   Bouncing and catching.
2.   Bouncing and "slapping" on the rebound.
3.   Dribbling with the ball in control.
4.   Dribbling with the individual in control.
5.   Dribbling as a sport skill such as that utilized in basketball-type activities.

In one of the few studies of ball-handling skill among young children, Wickstrom (1980) found that children at the lowest skill level held the fingers of the striking hand close together (often hyperextended) and displayed a distinct slapping motion. Minimal elbow extension followed by a quick retraction of the hand following contact was also observed (fig. 9.15). Because of their limited eye-hand coordination, inexperienced dribblers often strike the ball in an inconsistent manner. In essence, the ball was controlling the dribbler.

Developmental Characteristics   The transition from immature to a more mature dribbling action is characterized primarily by a progression that moves from slapping to pushing the ball. To achieve the mature form of multiple controlled bounces requires greater consistency in hand position at contact and more continuous contact from the upper part of the rebound through the downward pushing action. Maximum control also necessitates that the proportionate size of the hand and ball be such that the dribbler can control the direction of the ball by placing the hands at the center of the ball's mass.

In contrast to the inexperienced dribbler, a pushing action is used to propel the ball with the elbow nearly fully extended. The dribbling arm stays extended with fingers pointed toward the ball and recontacts it after it bounces approximately two-thirds of the way up from the rebound. Upon contact, the forearm flexes and moves up with the ball; the hand remains in contact with the ball until after the downward push for the subsequent bounce. The fingers stay spread out to conform to the shape of the ball and the height of the bounce is maintained at approximately waist level (fig. 9.15).

Climbing is a fundamental locomotor skill that involves ascending and descending movement using the hands and feet. The upper limbs usually initiate primary control. Climbing, an outgrowth of creeping, is often performed before walking, especially if the opportunity to practice is made available. Unfortunately, very little information is available concerning the developmental process components used in climbing. Depending on the conditions under which climbing takes place (ladders, frames, nets, ropes, stairs), a number of movement pattern variations may be utilized. Most information available on climbing patterns (as defined in the context of upper body involvement) has been derived from ladder and stair climbing observations.

Climbing

Figure 9.15
Immature and mature
dribbling. The upper
series demonstrates the
slapping motion
characteristic of
immature dribbling. The
lower series shows a
mature dribbling action.

Figure 9.16
Marking time pattern:
(a) start and
(b) completion.

a.   b.

Movement Pattern Descriptions   Two basic movement patterns appear to predominate during ladder and stair climbing: the immature pattern of marking time
and the advanced cross-lateral pattern. During the initial stages of climbing, movement is characterized by *marking time.* That is, the child steps up or down to the appropriate level with the same foot each time. This is followed by movement of the
trailing foot to the same level (fig. 9.16). The dominant arm initiates and guides the
action, which is followed by leg movement on the same side. As with the foot action, both arms are placed on the same level before the next cycle begins.

a.   b.

Figure 9.17
Cross-lateral pattern:
(a) start and
(b) completion.

With increased maturity and practice, the child progresses to the alternate-foot ascent and descent described as the *cross-lateral pattern.* That is, rather than placing both feet on the same level, the child alternates sides and places only one foot on each level (fig. 9.17). Hand placement may vary from positioning both hands on the same level before following with leg action to the more-advanced alternating pattern (matching the leg movements). With practice and confidence, the child prefers to "reach" for the next highest level, therefore displaying the alternating cross-lateral pattern, with the arms, as well.

Developmental Characteristics   Though there are wide differences in climbing ability at every age level, children as young as 2 years have been observed to display initial climbing patterns. By 6 years the majority of children are reasonably proficient climbers and exhibit the more mature pattern characteristics. The skill of ascending an apparatus is usually achieved before an individual attempts to descend. Children are usually capable of skillfully descending ladders or stairs by 5 years of age. As one might expect, the physical characteristics of the climbing apparatus may markedly affect climbing performance (e.g., the angle, height, and distance between ladder rungs or step risers). It is not uncommon for children who normally use advanced climbing techniques when operating at low heights to revert to more immature and cautious marking time movements when attempting to climb at a significantly greater height.

In the discussion of early motor behavior several developmental milestones were noted in relation to manual control. The primitive and rudimentary behaviors of reaching, grasping, and releasing that flourish during the prenatal stage and the first two years of life represent three developmental categories: reflexes, rhythmic stereotypies, and prehensile behaviors. The grasping (palmar) reflex appears prenatally and is suppressed by the fourth postnatal month. Rhythmic

**FINE MOTOR
MANIPULATIVE
BEHAVIOR**

■ *Objective 9.6*

stereotypies appear in the arms, hands, and fingers shortly after birth and reach their peak during the second half of the first year. The peak of proficient prehensile behavior is normally displayed by 18 months of age. During this period most children are reasonably proficient in coordinating basic reaching and grasping skills, as evidenced by their ability to use a precision pincer grip and to effectively release objects.

The final stage of manual control is known as **manipulation.** Manipulation refers to skillful and refined use of the hands, such as in stringing beads, drawing, and writing. The development of several aspects of manipulation occurs during the early childhood period and shortly after (up to around 8 years of age). With an understanding of the development of manipulation skills and early motor behavior characteristics, a relatively complete picture of manual development is created.

Finger Differentiation

Although most children are capable of displaying a pincer (thumb and forefinger) grip around 9 to 10 months, it is several years before they are able to perform precision movements with individual fingers. One of the most utilized tasks to evaluate finger differentiation is the test of finger opposition. This task requires the individual to touch each finger, in order, to the thumb. Generally the task is scored under speed stress (e.g., counting the number of cycles of finger opposition within a given time). This finger opposition test is frequently used by pediatric neurologists to identify possible impairment and by researchers to investigate the developmental characteristics of the hands.

Early signs of individual finger control (differentiation) appear within the first year of life as evidenced by the appearance of stereotypic finger flexion, pseudo thumb opposition, and, as noted earlier, true thumb-to-forefinger opposition by 10 months of age. However, the ability to oppose more than one finger to the thumb does not reveal its beginnings until around age 3. Initial efforts at this age usually require that the tester provide a slow visual demonstration. Regular improvement continues in finger differentiation up until about age 8 when performance approximates that of adults (Denckla, 1973, 1974; Lefford, Birch, & Green, 1974).

Other tasks such as stringing beads, moving pegs, and picking up pennies or matchsticks and placing them in a container are also commonly used to determine level of manipulative ability. In general, studies involving these types of tasks have reported similar basic patterns of change. That is, improvement is greater during the early developmental years, with relatively small changes occurring after 8 to 10 years (e.g., Gardner & Broman, 1979; Keogh & Sugden, 1985; Schulman et al., 1969). Figure 9.18 illustrates this trend among individuals 3 to 15 years of age with the results of a test measuring the speed of moving pegs reported by Annett (1970).

Construction and Self-Help Skills

Developmental scales such as the Denver Developmental Screening Test (1990) typically include several manual construction and self-help tasks. Along with assessing the fine motor abilities used in such tasks as building a block tower, these tasks highlight the importance of manual skills in everyday personal, or self-help, tasks. The attainment of such skills as tying one's shoes, being able to "button up,"

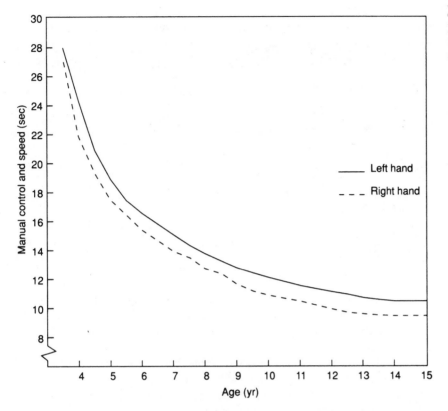

Figure 9.18
Manual control and
speed (moving pegs).

holding and drinking from a cup, and feeding oneself with a spoon are considered major developmental achievements. Table 9.11 is a list of selected construction and self-help achievements attained during the early childhood period and commonly included in developmental assessment scales.

Some of the most-investigated aspects of fine motor manual control are those techniques and processes involved in writing and drawing. These skills progress from primitive scribbling and coloring to the highly coordinated and dynamic fine motor movements used to draw figures and write in cursive. In general, the development of drawing and writing follows the proximodistal trend, with males lagging behind females at virtually all ages; due in large part to females' advanced neurological maturity (e.g., Blote & Haastern, 1989).

Around the age of 18 months, children can hold a writing or coloring implement (e.g., pencil, crayon) and by 4½ to 6 years have acquired an array of holding grips (Erhardt, 1982). Although a number of grips are applicable depending on the implement (e.g., pencil or paintbrush) and the condition (e.g., on a desk, at an upright easel), two basic developmental styles appear to be most prominent: the power grip and the tripod (fig. 9.19). From these two basic styles, several variations exist.

Drawing and Writing

Table 9.11   Selected Construction and Self-Help Achievements (ages 1–4)

Achievement	Approximate Age (mo)
Drink from cup	13
Builds tower of two cubes	14
Use spoon	15
Builds tower of four cubes	18
Builds tower of eight cubes	24
Puts on shoes	36
Cuts paper	
Buttons up (clothes)	
Feeds self (spills little)	
Pours well from pitcher	
Laces shoes	48
Brushes teeth	
Cuts paper following line	

Source: Data from Denver II, 1990, and Santrock, 1993.

The palmar grasp, also referred to as the **power grip,** is usually the initial technique used among children below the age of 3 years (e.g., Saida & Miyashita, 1979). With this style, all four fingers and the thumb are wrapped around the implement and most movement is initiated and guided by the shoulder and arm. The more advanced holding style is the tripod, in which the implement is held by the thumb, index finger, and middle finger. During initial use of this style, the child will hold the implement with the proper tripod grip, but movement will still be controlled by the arm rather than hand.

The mature version of the basic tripod is the **dynamic tripod.** At this stage, development has reached the level of using the proper grip and skillfully manipulating the implement with the wrist, fingers, and thumb. The beginnings of "dynamic" manual control in this context are usually evident by age 4 and progress to a relatively mature level by 7 years, with refinement continuing through early adolescence (Ziviani, 1983).

The development of writing ability also progresses in a proximodistal order. That is, initial movements are through use of the shoulder and arm and then through the elbow. Finally the wrist, fingers, and thumb are used to execute dynamic manipulation. From the ages of 2 to 6 years, children's writing ability increases dramatically as the hand moves closer to the tip of the implement. See the Suggested Readings (Erhardt, 1982) for a detailed description and assessment techniques for the development of prehension skills, including writing and drawing grips.

The child's first attempts at drawing usually occur by accident around the age of 15 months while spontaneously scribbling. **Scribbling** is described as a type of kinesthetic, or motor, babbling that is similar to primitive vocalizations. This form of early drawing appears to be nonpurposeful and unrelated to any visual model. By the approximate age of 5, development has progressed to the level that a definite organization in drawing and copying is evident.

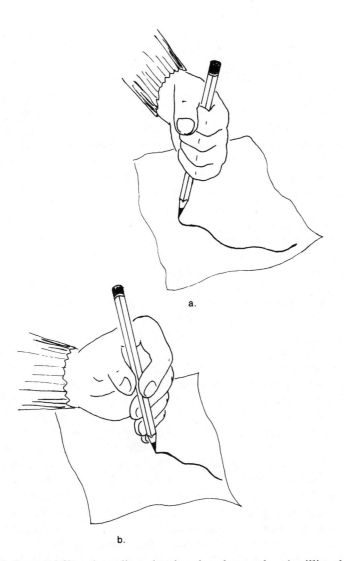

a.

b.

Figure 9.19
Basic holding grips:
(a) the immature power
grip and (b) the more
advanced tripod.

Kellogg (1969), who collected and analyzed more than 1 million drawings done by children, describes the development of drawing abilities as a four-step process made up of the scribbling stage, the combine stage, the aggregate stage, and the pictorial stage.

The *scribbling stage* is characterized by random and repetitive actions that primarily produce circular shapes, then alternating (repetitive) lines. These movements soon become less spontaneous as the child slows down in attempting to control hand movements with visual attention. Less frequently observed lines at this stage are vertical (up-and-down), horizontal (back-and-forth), and diagonal.

The *combine stage* provides evidence that the child is beginning to slow down and guide hand movements with the eyes. Characteristic of this stage is the child's effort to enclose space, as in completing a crude circle or square. This is also the period when basic geometric diagrams and combinations of diagrams are created. Common drawings at this stage include spirals, simple crosses, circles, squares, rectangles, triangles, and their combinations.

In the *aggregate stage,* the child not only combines figures but does so in multiples of three or more. This stage of development represents the ability to create more complex drawings. The final stage of basic drawing development, the *pictorial stage,* designates the child's ability to draw with more precision and complexity. As this ability develops the child begins to draw items from within their world with more detail and precision. Frequently chosen items include people (and faces), animals, transportation, and houses.

Although most children appear to follow the progression suggested by Kellogg, the determination of age-related norms is difficult because this type of skill may be influenced greatly by environmental factors such as opportunities for practice and the particular drawing conditions and availability of materials.

Most research in the area of drawing ability has been conducted in relation to copying figures. Copying tasks are standard items on most developmental assessment scales. The following discussion describes selected age-related drawing characteristics that have been noted in the literature.

By the age of 2, children can crudely draw circular, vertical, and horizontal lines that follow the appropriate direction. However, the quality of performance among children may vary considerably. Knobloch and Pasamanick (1974) suggest the following general progression and approximate age expectancies for common figure copying: circle (36 months), cross (48 months), square (54 months), triangle (60 months), and diamond (72 months). Other studies have reported similar findings. In one of the most comprehensive studies of design copying among children 5 to 11 years of age, Birch and Lefford (1967) found that tracing abilities for all forms are generally mature by 6 years. They also determined that the development of line grid drawing skills continues to be refined until around age 9 and that the more difficult skill of freehand drawing a diamond shape continues to be refined through age 11. The researchers noted that children improve and become more consistent on all three techniques with age and that a marked rate of improvement takes place between the ages of 5 to 6 years.

The task of forming letters and numbers as required in handwriting tasks presents the child with additional manual challenges. By the age of 4 years, most children are capable of printing at least a few recognizable letters but frequently position them in a disorganized manner on the page. With adequate practice, many 5-year-olds are capable of printing their first names. By 6 years most children can print the alphabet and the numbers up to 10. Lowercase letters seem to be more difficult to copy than uppercase letters, a condition that may continue through the third grade (Stennet, Smythe, & Hardy, 1972). Younger children (ages 4 and 5) typically write in large letters and numbers that are commonly $1/2$ to 2 inches high. By age 7, the height of the letter decreases to approximately $1/4$ inch. The ability to space letters remains a difficult task for many children until around 9 years of age (Cratty, 1986).

Several manipulative activities require the controlled use of the two hands, or **bimanual control,** in either a symmetrical or asymmetrical function. In *symmetrical function,* the two hands perform identical and simultaneous movements, such as those used in clapping and performing selected calisthenic exercises. The other form of bimanual control is more common to general manipulative behavior and is referred to as *asymmetrical function.* This term describes the function in which the two hands make different movements in a coordinated and complementary manner (de Schonen, 1977). This ability is evident in the performance of such tasks as cutting paper with scissors, tying shoelaces, and dealing cards. In each of these conditions, one hand is the primary manipulator, while the other functions in a complementary manner to position and stabilize the object.

Although they are not intentional (i.e., voluntary), symmetrical bilateral arm movements can be observed at birth in the form of primitive reflexes (e.g., the Moro reflex) and in the spontaneous behavior of newborns during the first month of life. The earliest patterns of bimanual activity (e.g., hand interplay, clasping) are primarily nonvisual and occur after 3 months of age. By 4 to 4 1/2 months, when crude attempts at voluntary reaching are displayed, arm actions are still basically bilateral and rather symmetrical, even though one hand contacts the object first. Although some form of asymmetrical behavior has also been noted around this age, it is generally not before the end of the first year that the complementary characteristics of bimanual asymmetrical coordination are unquestionably present (see Fagard, 1990 in the Suggested Readings). The variety and complexity of tasks that can be performed using bimanual control increase enormously during childhood, though they may be quite unrefined.

Closely associated with proficiency of asymmetrical bimanual control is hemispheric lateralization and the establishment of a consistent hand preference. Although specific age-related changes are difficult to establish, children display basic mastery of asymmetrical and symmetrical bimanual control by 6 years just after hand preference has been established. From this basic foundation develops the motor control that ultimately manifests itself in the ability to perform complex bimanual skills.

Although age-related characteristics and the nature of developmental change have not been clearly established, a general trend has been noted. Apparently children progress from an early stage of crude bimanual movements that tend to be symmetrical to the development of unilateral behavior, or establishment of a preferred limb. With the development of a preferred and nonpreferred hand evolves more highly refined bimanual control that culminates in the role differentiation of the hands that is observed in most forms of asymmetrical manipulative behavior. Refer to Fagard (1990) in the Suggested Readings for a more comprehensive review of the development of bimanual coordination in children.

Bimanual Control

FUNCTIONAL (MOTOR) ASYMMETRIES

■ *Objective 9.7*

Complementing the development of motor control, especially during early childhood, is establishment of limb (hand, foot) and eye preference. As noted in previous chapters (brain lateralization), although the human body (and hemispheres) is symmetrical in general appearance, the paired limbs (hands, feet) and sensory organs (eyes, ears) are used in an asymmetric manner. The behavioral manifestations of this phenomenon are referred to as **functional asymmetries.** For example, in tasks where only one limb can be used, such as throwing or kicking a ball, most individuals exhibit a consistent preference for the use of one limb over the other (also known as handedness and footedness).

Although lateral preference is most often identified as *right-sided* or *left-sided,* there are individuals who do not consistently favor the use of one limb or the other on a specific task; these individuals are identified as *mixed-sided. Ambidexterity* is a somewhat confusing term in a neurological context and one that has received very little attention in the developmental psychology literature. This term describes the use of either limb with somewhat equal proficiency on the same task. This characteristic does not necessarily indicate "dual" dominance of hemispheric control. It is generally believed that individuals possessing this ability with a specific task have adapted or trained their nonpreferred side to be as, or perhaps more, skillful than their dominant side for a specific skill(s). This type of behavior is quite evident in skilled (practiced) activities such as soccer, handball, and boxing that require proficiency in both limbs.

Following are brief discussions of the developmental characteristics associated with handedness, footedness, and eye preference behavior, which as noted earlier, are closely linked with the development of motor control during early childhood.

Handedness

Of all the motor asymmetries, handedness has been investigated more than all the others combined. Hand preference on tasks such as writing, throwing a ball, hammering, stacking cubes, and cutting with scissors is currently the most widely used method of determining *degree* of handedness. It is generally acknowledged that handedness (and footedness) is a trichotomous continuum ranging from left to mixed to right, rather than a dichotomy (right or left only). With a series of tasks, many individuals will not exhibit a totally consistent right or left hand preference.

Individuals generally become more right-handed with increasing age and approximately 80% to 90% of the population establish this preference (e.g., Dargent-Paré et al., 1992; McManus et al., 1988; Porac & Coren, 1981; Rice & Plomin, 1983). With increases in cortical development, more precise measures on handedness are possible, as frequently observed in the 3-month-old with voluntary hand use ("prereaching"). By the middle of the first year, these abilities become more differentiated, with most infants exhibiting a preference for the right hand for visually directed reaching. By 3 to 4 years of age children generally show differential hand preference and hand skill of a kind comparable to adult characteristics. It is generally acknowledged that at 5 years, handedness is stable in most children (Tan, 1985).

Is there a disadvantage to being left- or mixed-handed? For many years, left-handedness was associated with cognitive and language dysfunctions and general motor clumsiness because left-handed individuals were more often found to have such deficits. However, these findings and assumptions were based more on clinical samples and have not been supported by more recent investigations of normal individuals (e.g., Smith, 1983; Tan, 1985). Within the context of manual motor performance, studies of young children and adults have found that neither left- nor mixed-handed individuals exhibit inferior performance when compared to their right-handed counterparts (Gabbard, Hart, & Kanipe, 1993; Peters, 1990). In most instances, right- and left-handers perform better with the preferred limb, and those with a mixed preference exhibit no significant difference between either limb.

Another argument complementing this line of research has been the suggestion that children who exhibit consistent (left or right) hand preferences are better coordinated than their inconsistent (mixed) counterparts (Gottfried & Bathurst, 1983; Tan, 1985), the implied underlying premise being that children who are more strongly lateralized (right or left) would exhibit more-pronounced functional asymmetries in comparison with inconsistent peers. Although this hypothesis has found some support in general "field tests" of motor performance, it has not been verified through more rigorous experimental studies.

Footedness

A myriad of data are available on hand preference, but research on foot laterality is relatively limited. Compared to most handedness inventories, which consist of unimanual (e.g., writing, drawing, throwing) and bimanual (e.g., threading bolts, cutting paper, unscrewing lid) motor tasks, recommended foot preference tasks contain both stabilizing and manipulative characteristics. By operational definition, the *preferred* foot is the one used to manipulate an object or to lead out, as in kicking or jumping, while the nonpreferred limb is used as the primary stabilizer (Peters, 1988). Recommended tasks to measure footedness include: kicking a ball, figure tracing with toes while standing, and foot stamping a target.

As noted earlier, the developmental course of foot preference behavior follows a similar general developmental trend as handedness; that is, there is a significant shift toward greater right-sidedness with increasing age. In comparison to the development of handedness, however, the genesis of foot preference exhibits a few unique characteristics. During early childhood about twice as many individuals are mixed-footed (32%), compared to persons not exhibiting a preferred hand (16%). During late childhood a significant shift toward greater right-footedness with a corresponding decrease in mixed-footedness occurs, after which preferences remain relatively stable into adulthood, with approximately 80% of adults favoring the right side. The incidence of left-footedness appears relatively consistent across the life span (Gabbard, 1993; Gentry & Gabbard, 1994).

Perhaps the most intriguing theoretical question is, What accounts for the mixed-footed phenomenon? One possible explanation is linked in part to Annett's Right-Shift hypothesis, which proposes that in individuals who lack a strong right-shift factor, limb preference may be determined at random. This suggests that environmental factors such as culture (the "right-sided world phenomenon") and experience may influence *direction* and/or *degree* of preference. Considerations are that: (1) footedness is a more unbiased index of lateral specialization in being less subject to dextral social pressures than handedness, and (2) activities that require use of the feet (relative to those of the hands) are usually less complex and practiced. From these observations it seems likely that the environmental factor that may affect (e.g., strengthen) "degree" of hand preference during childhood may be less of an influence on foot laterality. This would suggest that in those individuals who do not possess a strong biological propensity for right-side control, the probability of mixed-sidedness would seem greater for the feet than for the hands.

Of the numerous questions that will challenge researchers in this relatively recent line of inquiry, one of the most interesting and basic issues is the definition of footedness itself. As noted earlier, the operational definition of footedness, as used in most inventories, describes the *preferred* limb as the one used to manipulate an object or lead out, while the other foot has the role of lending postural/stabilizing support. Previc (1991) proposes interesting arguments that are somewhat contradictory based on observations of asymmetric prenatal development of the vestibular mechanisms (stability) and left hemisphere maturation. Briefly, the researcher suggests that in most humans, antigravity extension (including postural reflexes) on the left side emerge *before* voluntary motor control (flexion-manipulation) on the right side of the body. The researcher illustrates by describing the predominant sport-skill stances in a basketball layup shot and in a golf swing; that is, for the right-hander, the left leg's antigravity muscles (extension) are used for postural support, whereas the right leg is flexed.

Selected tenets of this theory stimulate at least two explanations related to dominance of foot preference behavior. The first suggests there may be no such thing as a clearly *dominant* foot. That is, the roles of the feet are relatively complementary and equivalent, one providing postural support and the other voluntary (manipulative) action. On the other hand, it could be that in the formulation of a motor program, the first consideration in limb selection would be for postural control, meaning the left leg for most people (their dominant limb?). An additional consideration is that of *task complexity*. For example, with virtually all foot preference tasks, one leg performs a relatively simple static postural position (supposedly the nonpreferred limb) and the other limb flexes and extends (which is arguably more complex), such as in kicking. If greater complexity were shifted to stabilizing, rather than flexion/manipulation, would the individual switch limbs, thus adjusting the *preferred* limb to the complexity of the task? This is one of the questions to be challenged in future research.

Eye preference can be observed when sighting a telescope or aiming a gunlike instrument. A review of research on eyedness suggests that, as with the other asymmetries described, most individuals prefer the right side, and degrees of right-eyedness increase between childhood and maturity (e.g., Coren, Porac, & Duncan, 1981; Porac, Coren, & Duncan, 1980). The levels of right-sidedness in adults, however, are not as high as with hand and foot preference. Approximately 70% of adults prefer the right eye (Coren, 1993; Dargent-Paré et al., 1992), whereas values for the same preference in children range between 55% to 60% (Nachshon, Denno, & Aurand, 1983; Porac et al., 1980). These data also suggest that the major shift over time, complementing the increase in right-sidedness, is a decrease in the incidence of left-eyedness. Approximately 35% of children are left-eyed, compared with a value of 20% in adults. As with most functional asymmetries, there do not appear to be any differences due to gender.

Eye Preference

SUMMARY

Early childhood is characterized by fundamental movement actions that can be categorized into locomotor, nonlocomotor, and manipulative (gross and manual) behaviors. The developmental characteristics of locomotor and gross motor manipulative skills are described according to process-related movement pattern changes. Movement pattern status is frequently identified using terms that represent proficiency ranging from immature to mature, or minimal to sport skill form. The minimal characteristics are normally observed in children at ages 2 and 3. With adequate practice, most children acquire at least some features of the mature movement pattern by the end of the sixth year. However, some individuals may not achieve the mature stage until years later, if ever. This is especially true of overhand throwing and the standing long jump among females.

The study of movement pattern change is best conducted using a biomechanical approach, which identifies the sequential temporal and spatial characteristics of the motion. Two general approaches associated with describing and interpreting qualitative movement changes across time are the composite approach and component method. The composite approach describes the sequence of movement pattern changes through a series of "total body" (general pattern) stages. The component approach divides pattern changes into substages depicting specific body component changes.

A body of information is available concerning the sequential progression and developmental characteristics of most fundamental locomotor and gross motor manipulative skills. In general, running is a natural extension of walking characterized by a "flight" phase. Most children take their first running steps at about 18 months and by age 5 attain a reasonably skillful form. Among the most noted developmental changes with advancing age are increases in stride length, flight time, range of motion, extension, and running speed, paralleled by decreases in base of support and vertical movement.

Jumping includes several pattern variations distinguished by takeoff and landing on one or both feet. The first true jumping movement is a leap, which is a one-foot takeoff and landing on the other foot observed at about 2 years. Other commonly observed forms are the vertical jump (for height), the standing long jump (for distance), and the hop. In contrast to the other patterns, the hop (a one-foot takeoff and landing on the same foot repeatedly) is considered the most difficult form of basic jumping. The development of hopping may not reach the mature stage until age 6 or older.

Galloping, sliding, and skipping each consist of a combination of basic walking, stepping, leaping, and hopping skills and are normally acquired by the end of the early childhood years.

One of the most complex fundamental skills is throwing. During the early childhood years, a number of different overarm, sidearm, and underhand patterns may emerge depending on several factors. Of the most common variations, the one-arm overarm style is the most commonly used. The first signs of this pattern appear around the age of 6 months, and by 6 years the majority of children are skillful throwers. Overarm throwing has a relatively high incidence of individuals who do not reach mastery, especially among females.

Catching also involves a relatively high level of complexity for children, primarily because it requires coincident timing ability. Although a primitive form of two-handed catching (trapping) under relatively stable conditions may be observed in 2- and 3-year-olds, proficiency under similar conditions generally does not occur until 6 to 8 years of age. The ability to catch a ball under more complex conditions continues to develop well into the upper elementary years.

Striking is an action in which a part of the body or an implement is used to give impetus to an object. This skill can be performed using a variety of body parts, implements, and movement patterns. The most common forms of striking are the one- and two-handed sidearm patterns. A smooth sidearm pattern may be evident in some children by 3 years of age, but the mature pattern is generally not achieved until age 4 or 5. However, although the pattern may be well defined by those ages, the ability to consistently intercept a moving object continues to improve throughout the late elementary years.

The placekick (of a stationary ball) is considered the basic form of striking an object with a foot. From this pattern more-advanced kicking skills such as kicking a moving ball, dribbling, and punting are developed. Minimal form in this variation of kicking appears around the age of 2, and by 5 or 6 years the mature pattern is achieved by most children.

The origins of ball bouncing begin with the child dropping a ball and attempting to catch it and progress to slapping the ball on the rebound, dribbling with the ball in control, and finally to control the ball by the individual. Although minimal form may be observed in some 2-year-olds, it is not until approximately 5 or 6 years of age that the mature pattern of one-handed dribbling in a relatively stationary position is mastered. Dribbling proficiency while moving is much more complex and is considered a sport skill.

Climbing is a locomotor skill that involves moving upward or downward by using hands and feet, with the upper limbs usually initiating primary control. The two basic movement patterns that predominate in ladder and stair climbing are the immature pattern of marking time and the advanced cross-lateral pattern. There are wide differences in climbing ability at each age level beginning at about age 2, but by 6 years of age the majority of children are reasonably proficient climbers.

The final stage of fine motor manual control appears during early childhood in the form of manipulative behavior. Manipulation refers to the skillful and refined use of the hands, such as that displayed in tasks of finger differentiation, construction and self-help skills, and drawing and writing. As one would expect, the development of manual control abilities generally proceeds in a proximodistal direction.

One of the most-investigated aspects of manual control is writing and drawing. These skills progress from primitive scribbling and coloring to highly coordinated and dynamic fine movements such as cursive writing and drawing figures. The two basic grips used are the initial power grip (palmar grasp) by children younger than 3 years and the more-advanced tripod. The mature version of the tripod, the dynamic tripod, is relatively developed by age 7. The ability to copy designs (a circle being the easiest and a diamond the most difficult basic design) is generally mature by 6 years of age. However, refinement in such skills as freehand drawing may continue to at least 11 years of age. In regard to letter and number writing, most children can print the alphabet and the first 10 numbers by the age of 6 years. The more difficult task of spacing letters correctly is generally not achieved until around 9 years of age.

Bimanual control involves the use of two hands in either a symmetrical or asymmetrical function. Bimanual coordination clearly improves during childhood, with the initial actions being more symmetrical before complete role differentiation of the hands can be observed. Mastery of asymmetrical bimanual control seems to be associated with the development of hand preference. With the establishment of a preferred and nonpreferred hand, more highly refined bimanual control develops.

Closely linked to the development of manual and overall motor control during early childhood is establishment of limb and eye preference (functional asymmetries). A general developmental trend for all asymmetries is a shift toward greater right-sidedness with increasing age. Whereas adult right-sided values for handedness and footedness are close, during early childhood a significantly greater number of children express a consistent preference for right-handedness, compared to the lower limbs; thus, handedness appears to mature earlier. Right-eyedness is not as high as with hand and foot preference values at any period.

SUGGESTED READINGS

Erhardt, R. P. (1982). *Developmental hand dysfunction.* Laurel, MD: RAMSCO.

Fagard, J. (1990). The development of bimanual coordination. In C. Bard, M. Fleury, & L. Hay (Eds.), *Development of eye-hand coordination.* Columbia, SC: University of South Carolina Press.

Roberton, M. A. (1984). Changing motor patterns during childhood. In J. Thomas (Ed.), *Motor development during childhood and adolescence.* Minneapolis: Burgess.

Roberton, M. A., & Halverson, L. E. (1984). *Developing children: Their changing movement.* Philadelphia: Lea & Febiger.

Wickstrom, R. L. (1983). *Fundamental motor patterns* (3d ed.). Philadelphia: Lea & Febiger.

Motor Behavior during Later Childhood and Adolescence

OBJECTIVES

product performance
sex-related differences

youth sports
physical education

KEY TERMS

PHASES OF MOTOR BEHAVIOR

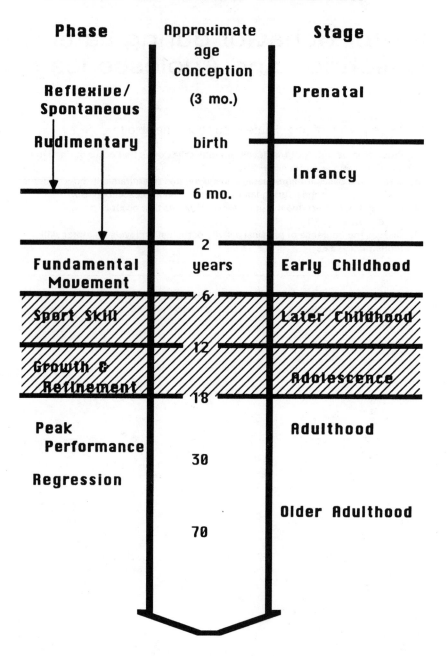

Phase	Approximate age	Stage
	conception	
Reflexive/ Spontaneous	(3 mo.)	Prenatal
Rudimentary	birth	
		Infancy
	6 mo.	
	2 years	
Fundamental Movement		Early Childhood
	6	
Sport Skill		Later Childhood
	12	
Growth & Refinement		Adolescence
	18	
Peak Performance		Adulthood
	30	
Regression		Older Adulthood
	70	

The period from later childhood through adolescence (approximately 7 to 17 years of age) is characterized by several growth and development milestones, many of which are manifested in significant improvements in motor skill performance. Around the beginning of later childhood, individuals begin to utilize with increasing frequency the fundamental movement abilities that were acquired during early childhood. At the same time, changes in physical growth, body structure, and physiological development combine to produce greater **product performance.** This period is also distinguished by the emergence of sport skill behaviors that, in essence, are advanced versions of basic skills that are the primary vehicle by which the individual's increased level of motor behavior is displayed. Along with age-related improvements in motor performance that appear in both sexes during this period, **sex-related differences** also come increasingly into play. This trend is especially apparent during the adolescent growth spurt.

The focus of this chapter will be on the description and bases for quantitative changes in motor performance exhibited during preadolescence and the adolescent years. The related issues of sex differences in motor performance and the refinement of fundamental skills as displayed by sport skill behavior will also be addressed.

■ *Objective 10.1*

Quantitative changes in motor performance tend to parallel increasing levels of experience and changes in physical growth, physiological development, and neurological functioning. Previous chapters describing these changes have established that with advancing age, most individuals (both sexes) experience increases in such characteristics as body size, muscle mass, strength, cardiorespiratory capacity, and perceptual-motor ability. As a result of these and other developmental changes, virtually all aspects of product-oriented motor performance show an upward trend beginning at the preschool level.

Differences between the sexes before puberty are minimal but do exist. As early as 3 years of age, reports suggest that males tend to outperform females on selected tasks of running, throwing, and jumping. Females, on the other hand, appear to excel in hopping, skipping, and tasks requiring fine-motor control, balance, and flexibility (Broadhead & Church, 1985; Morris et al., 1982). Although some performance differences have been attributed to biological factors such as body fat (Thomas, Nelson, & Church, 1991), favoring males, and advanced neurological development, favoring females, it also seems evident that at this age level, environmental agents are a primary determinant (Thomas & French, 1985). Several studies confirm that males, beginning at the preschool level, are generally more active than females (e.g., Baranowski et al., 1993, Branta, Painter, & Kiger, 1987; Eaton & Ennis, 1986). Complementing this is the fact that parents and teachers usually express specific sex role expectations, dictating that males and females should behave differently at play and sport. Thomas et al. (1991) note that from a physiological perspective, due to the faster maturation rate in young females prior to puberty, their overall motor performance should be better. The fact that it generally is not argues for a sociocultural explanation.

BASES FOR AGE- AND SEX-RELATED QUANTITATIVE (PRODUCT) MOTOR PERFORMANCE CHANGES

■ *Objective 10.2*

With the onset of the adolescent growth spurt (at age 10 to 12), differences in motor performance product characteristics between the sexes become increasingly evident and in favor of the male. The literature reveals a rather consistent trend showing that males continue to improve in motor performance on a variety of skills through adolescence but that females have a tendency to peak around the age of 14 and then level off (plateau) or decrease in performance (e.g., Eckert, 1987; Haubenstricker et al., 1992; Thomas, Nelson, & Church, 1991). From another perspective, females are, in essence, maturing earlier in motor performance, as evidenced by achieving adult performance levels at earlier ages than males (Haubenstricker et al., 1992).

Although cultural practices and participation levels may significantly influence the range of individual differences within and between the sexes during this period, numerous biological factors may provide an advantage to males in several motor skill activities. Earlier discussions have highlighted that the primary factors providing males with a biological advantage over females in motor performance are sex chromosomes and hormones. Most of these factors become increasingly apparent after the age of 10 years. A list of selected biological differences between the sexes is presented in table 10.1. The earlier chapters on physical growth and physiological changes would be useful to review for their discussion of the biological differences between the sexes.

In general, males tend to be larger, possess more muscle mass and less body fat, and have a greater oxygen transport capacity than their female counterparts. These characteristics give males the biological advantage in performing activities that require strength, power, and endurance. Because females possess more body fat and less muscle mass, they are at a distinct disadvantage in performing motor tasks that require lifting the body (e.g., jumping) or moving their mass against gravity (e.g., running).

Several research studies involving children and adults confirm a negative relationship between body fatness and performance on tasks requiring the vertical and horizontal movement of the body weight (e.g., Hensley, East, & Stillwell, 1982; Smoll & Schutz, 1990; Thomas et al., 1991). In essence, body fat adds to the mass of the body without contributing to force-producing capacity. Swimming is one activity for which a higher body fat value may be somewhat of an advantage. Higher values of body fat have been shown to allow individuals to swim higher in the water with less body drag. Though this factor may decrease the difference between the sexes, males still tend to outperform females in swimming events simply because they are stronger.

Structural factors providing a mechanical advantage may also favor males in sprinting, jumping, and throwing skills. Males have longer arms and wider shoulders, producing the leverage and rotation torque needed to forcefully propel objects as in throwing or striking. Longer legs and narrower hips may tend also to be advantageous in sprinting and jumping tasks. Greater proportionate hip width and angle of insertion of the femur in females has been suggested as a mechanical disadvantage in achieving maximal running speed for sprinting. On the other hand, because females generally have a lower center of gravity and wider

Table 10.1   Selected Biological Differences between the Sexes

Male	Female
Tissue	
Lower percent body fat	
Greater muscle mass (stronger)	
Heavier (greater body density)	
Anatomical structure	Greater angle of insertion of femur (more oblique)
Taller	to the hip
Wider shoulders (more rotation torque)	Lower center of gravity
Longer legs (relative to total height)	Wider hips (relative to shoulder width)
Longer forearms (more lever torque)	Shorter legs
Narrower hips relative to shoulder width	
Larger thoracic cavity (chest girth)	
Physiological functioning	
Larger heart (greater basal stroke volume and	
cardiac output)	
Greater maximal oxygen uptake (consumption)	
Faster heart rate recovery	
Greater physical working capacity	
More blood volume	
Greater number of red blood cells	
Higher mean hemoglobin values	
Greater vital capacity	
Greater ventilation volume	
Greater basal metabolic rate (active tissue)	

Source: Data from Brooks and Fahey, 1985, and Eckert, 1987.

hips, they may have the advantage in motor skills requiring a high degree of balance. Although females tend to outperform males at virtually all ages on tests of flexibility, current evidence does not suggest that biological factors are the reason for the difference.

Although biological factors such as neurological maturity, body size, anatomical structure, and physiological functioning can influence performance differences, there is also some suggestion that social and cultural agents may offer the explanation for some of the variation. This may particularly be the case when the motor task involves a high degree of skill (e.g., skipping, throwing) rather than higher levels of basic strength and cardiorespiratory endurance.

Performance of tasks that require a higher degree of skill (i.e., coordination, refinement) can be significantly influenced by practice. Although the trend appears to be changing, males and females in the United States tend to practice only those physical activities traditionally regarded as male-oriented or female-oriented, respectively. That is, females participate more often in activities that involve skipping, hopping, and perhaps flexibility and fine motor control. Males traditionally tend to spend more time in physical activities that practice running, throwing, jumping, and striking skills. Though females today participate in many more traditionally male-oriented activities (e.g., females play baseball, soccer, and football) and vice versa (e.g., males dance and play volleyball), some segments of society still adhere to the more traditional roles for the sexes. It can be speculated that this practice may discourage individuals from attaining their potential in specific motor skills.

Assuming the presence of a cultural stymie that, at least to some degree, hinders the level of practice and optimal motor skill potential among females, it is predictable that females would experience an early "plateauing effect" in their performance compared to males. Interestingly, in describing data collected during the 1950s, Eckert (1987) notes that European females continue to improve their running, throwing, and jumping skills up to 16 and 18 years of age. Their American counterparts, in contrast, displayed a leveling off between age 13 and 15. It may be assumed from this data that cultural values have an effect on motor skill performance.

Similar findings of a plateauing effect or decline among females after the age of 12 have also been reported for tests of physical fitness (Corbin, 1980; Thomas et al. 1991). In this instance, it has been speculated that females in our society are less likely than males to be encouraged and motivated to improve their fitness levels through vigorous training. Part of the problem appears to be the perpetuation of the myth that physical activity in females leads to greater masculinization and increases the risk of injury. Another explanation may be a preference for females to remain the "weaker" sex.

Although differences favoring males still exist for most tests of motor performance (especially on strength-related tasks), females in recent years appear to be narrowing that differential. This trend seems to be related to increased participation resulting from physical fitness and sports promotion, effective school and university programs, an increase in athletic scholarships and participation opportunities (through Title IX), and a gradual change in society's view of the "weaker" sex.

A word of caution should be given regarding the general interpretation of sex differences. Though clear-cut differences generally favoring males are evident, sometimes very early in development, comparisons are normally based on average performance data by group (males and females). In nearly all motor performance studies it is clear that large individual differences exist within the sexes at each age level. For example, even though the mean performance of males may be significantly greater on a specific motor skill, the performance of some females may equal or exceed that of some males. Therefore, mean data may be useful from a developmental perspective, but the true developmentalist is also sensitive to individual progress.

QUANTITATIVE (PRODUCT) CHANGES IN MOTOR PERFORMANCE

Running performance is commonly measured by requiring the individual to run as fast as possible (sprint) over a distance of 30 to 50 yards. Performance times are usually reported in tenths of a second and then, for comparative purposes, are converted to either feet or yards covered per second. The speed with which an individual can run depends significantly on the length of the stride as well as the stride rate. Stride and leg *length* generally increase with age, whereas stride *rate* remains relatively constant. Thus an increase in running velocity may not necessarily be due to an ability to move the legs more frequently but because the body is propelled farther with each stride.

    Age-related improvements in running speed can be explained in large part by the increase in body size and muscular strength. Figure 10.1 shows performance curves for males and females that were originally composed by Espenschade (1960) after reviewing the literature up to that period. Espenschade reported that velocity in both sexes increases for an average rate of just under 4 yards per second at age 4 to slightly over 6 yards per second at age 12. Males continue to increase in velocity up to age 17 (about 7 yards per second), but females begin to level off or regress slightly beginning at around age 13. A review of studies since 1960 confirms this trend for males but reports slightly higher scores for females (Haubenstricker & Seefeldt, 1986; Keogh & Sugden, 1985); these studies did not find a regression of performance in females after age 13 or 14 but identified more of a "plateauing" trend up to age 17. Keogh and Sugden found that the performance of both sexes increased 23% between the ages of 7 to 12 years. Over the next five-year period (to age 17), male performance increased steadily another 26%, whereas the females exhibited only a slight improvement during that period.

The quantitative characteristics of throwing are most often determined by measuring the maximum distance thrown, throwing velocity, and accuracy. Both sexes improve dramatically from childhood to adolescence with age on all three abilities, but sex differences (favoring males) appear quite early and are relatively large in comparison to other fundamental skills. As noted earlier, it may be speculated that much of the difference favoring males is due to their greater strength, biomechanical characteristics, and the fact that females generally exhibit less mature throwing patterns. Another consideration is the possibility that cultural and social factors provide more of a positive influence for the development of throwing among males.

    Regardless of throwing task, balls projected by the overhand throw have been measured more frequently than those of any other movement pattern. The distance that an individual can throw has received the most attention as evidenced in the developmental literature. Although sex differences are more obvious at the younger ages and become greater over time, throwing distance performance curves show trends similar to those found with running speed performance (fig. 10.2). That is, both sexes improve up to around puberty (12 to 14 years) after which males continue to improve performance, and females begin to level off and then decline in performance. Unlike running speed performance, however, recent studies do not indicate that female performances in throwing plateau rather than decline after puberty (Haubenstricker & Seefeldt, 1986). Espenschade (1960) reported from a review on the topic that throwing distances for males increased from approximately 24 feet at age 5 to about 153 feet at age 17. In contrast, females threw 14.5 feet at age 5, 75.7 feet at age 15, and then

Running Speed

■ *Objective 10.3*

Throwing

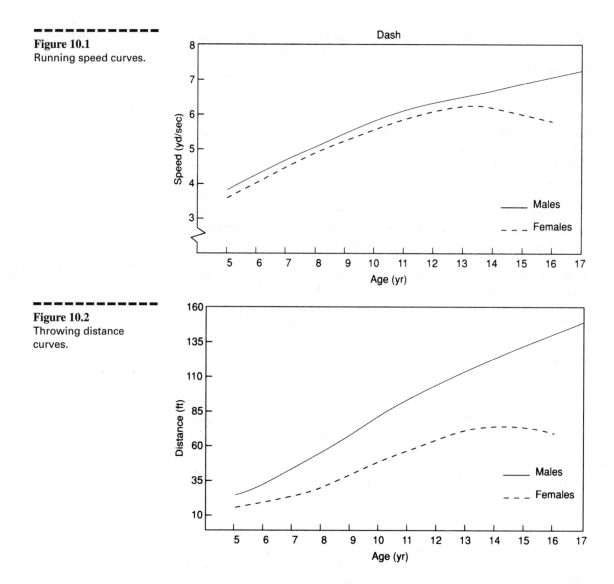

Figure 10.1
Running speed curves.

Figure 10.2
Throwing distance
curves.

declined to 74 feet at age 16 (fig. 10.3). From this data and that reported by Keogh and Sugden (1985), it is clear that both sexes display dramatic improvements in distance thrown during the preadolescent period when scores more than triple. Furthermore, by age 17 male performance more than doubles that of the female. Females and males apparently increase relatively more in the ability to throw for distance than in other fundamental skills.

Although the number of developmental studies has been somewhat limited, changes in horizontal ball velocities (speed) at various ages have been reported. Figure 10.3 shows the results of a longitudinal study of throwing velocity changes from kindergarten through the seventh grade (Halverson, Roberton,

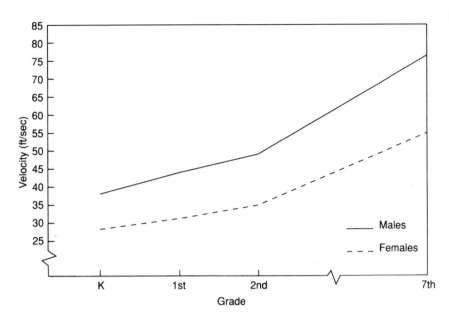

Figure 10.3
Longitudinal changes in ball throwing velocities.

& Langendorfer, 1982). In this study, females improved from an average of 29 feet per second while in kindergarten to an average of 56 feet per second by the seventh grade. As one would expect, the scores of males also improved over the study period and were significantly greater than those of the females—39 feet per second in kindergarten to 78 feet per second in the seventh grade. Like distance thrown performance, both males and females improve in throwing velocity rather dramatically through childhood, and sex differences appear early and are relatively large. Standards for skilled performers range from 70 to 80 feet per second for females and 100 to 120 feet per second for males (Adrian & Cooper, 1989).

Because testing procedures associated with throwing accuracy have varied considerably, an acceptable developmental description cannot be given. However, the general statement that individuals improve with age through childhood and that males are usually more accurate than females is supported by the literature (e.g., Hoffman, Imwold, & Koller, 1983; Keogh & Sugden, 1985). Since distance and the target area are usually confined, and the production of force not as critical (as in throwing for maximum distance and velocity), sex differences in throwing accuracy are not as marked.

Jumping

A considerable amount of developmental information has been gathered on jumping ability as measured by the standing (horizontal) long jump and vertical jump. Although data have been amassed from a multitude of studies, the general patterns of change are quite similar.

General performance curves for the standing long jump are shown in figure 10.4. Both sexes improve approximately 3 to 5 inches per year up to around age 11, with males jumping 3 to 5 inches farther than females at each age (Keogh &

Figure 10.4
Standing long jump
performance curves.

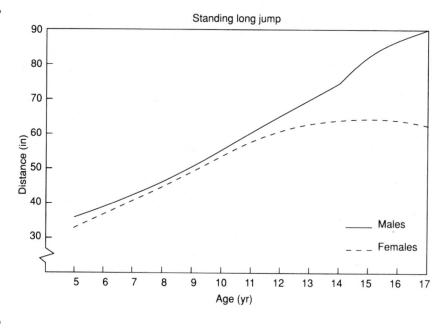

Figure 10.5
Vertical jump
performance curves.

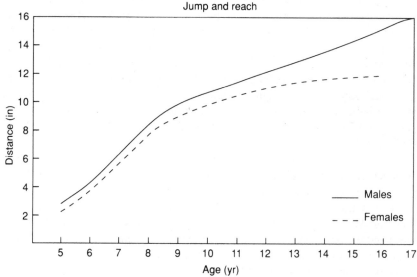

Sugden, 1985). After that period, males continue to improve to at least age 17, whereas the female's scores begin to level off, reaching a peak at around age 14. Both sexes improve approximately 45% between the ages of 7 and 12. Thereafter, females show only a 4% increase to age 17, whereas the males display an additional improvement of approximately 38%. In her review of the literature, Espenschade (1960) determined that at age 17, males achieve an average long jump of approximately 90 inches, whereas females reach a peak of about 64 inches at 14 years of age. Haubenstricker and Seefeldt's (1986) more recent view confirms the general trend across time but reports somewhat higher scores for females.

Data for the vertical jump (jump-and-reach task) provide similar performance curves to those for horizontal jumping (fig. 10.5). Males improve from approximately 2 inches at age 5 to about 17 inches at age 17. Females also show significant improvement by starting out just below the males at age 5 and reaching a peak of approximately 12 inches by age 16. As noted in figure 10.5, a plateauing of female performance begins around the age of 11.

Perhaps no other fundamental or sport skill has received the attention that has been given to kicking over the last 15 years or so, due to the increasing popularity of soccer in the United States. Unfortunately, even with such focus on the sport, little product data on kicking are available.

Kicking

The kicking performance of individuals is difficult to measure in quantitative terms prior to the age of 5 years. This is due primarily to lack of experience and lack of ability to judge ball speed and coordinate the movement pattern. After this period, however, the ability to kick a ball with force and accuracy becomes consistent enough to allow a fairly reliable measurement and assessment of ability. The quantitative characteristics of kicking are most often determined by measuring the distance that a ball is kicked, velocity, and kicking accuracy. In keeping with the previous general developmental trend noted with running, throwing, and jumping, kicking performance improves with increasing age (for both sexes) through childhood regardless of the test; and males are generally better performers than females. In addition, the differences between male and female kicking performance tend to increase as children get older.

Figures 10.6 and 10.7 show placekicking ball velocities and kicking distance characteristics for children. This information was compiled by DeOreo and Keogh (1980) from the works of Dohrmann (1965) and Williams and associates (1970). Ball velocities projected from a placekicking position using a utility ball show an almost linear increase for both sexes from ages 5 through 9 years. Yearly increases are approximately 2 to 3 feet per second for both sexes, except between the ages of 7 and 8 when increases average 5 to 6 feet. The male performance edge increases from about a 5 feet per second difference at age 5 to approximately 7 feet per second by age 9.

As one would expect, the performance data for distance kicking follows a similar trend to that of ball kicking velocity. As shown in figure 10.7, yearly increases for males are around 6 to 7 feet. Female values also increase with age but at relatively smaller increments during the earlier years (2, 5, and 7 feet, respectively). Differences between sexes range from about 4 feet at age 5 to approximately 10 feet at 8 years. In a different type of distance kick study, one that measured the soccer punting (drop kick) ability of children in grades 1 through 6, similar results were reported (Hanson, 1965). That is, distance increased with advancing age in both sexes, and males outperformed females at all ages.

Accuracy kicking is usually measured by asking the individual to kick a ball into a target area as many times as possible within a specific time, or by

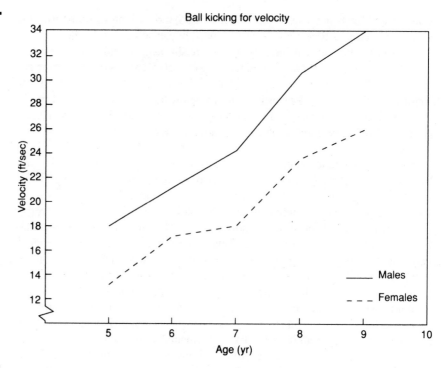

Figure 10.6
Ball velocities in place
kicking.

Sources: Data from H. Williams,
et al., "A Study of Perceptual-
Motor Characteristics of Children
in Kindergarten through Sixth
Grade." Unpublished Paper,
University of Toledo, 1970; and
P. Dohrman, "Throwing and
Kicking Ability of 8-Year-Old Boys
and Girls," in *Research Quarterly*,
35, (1965):465.

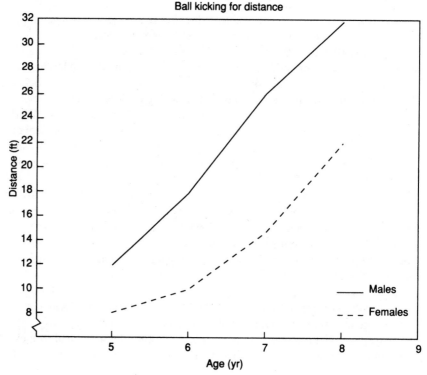

Figure 10.7
Ball kicking distance
performances.

Sources: Data from H. Williams,
et al., "A Study of Perceptual-
Motor Characteristics of Children
in Kindergarten through Sixth
Grade." Unpublished Paper,
University of Toledo, 1970; and
P. Dohrman, "Throwing and
Kicking Ability of 8-Year-Old Boys
and Girls," in *Research Quarterly*,
35, (1965):465.

attempting to hit a variety of targets from a variety of distances (Johnson, 1962; Latchaw, 1954). Studies using both of these techniques show a steady increase in performance for both sexes across the upper elementary school years (ages 9 to 12), with males being somewhat more accurate than females.

Catching

Catching is a coincident timing task that involves the complex interplay of coordinating visual information and motor behavior to a single point of interception (refer to chapter 6). The information-processing literature suggests that the ability of individuals to accurately time their movements to an interception point with a moving object improves through childhood before reaching a relatively mature level in the early teens. In the review by Williams (1983), it was noted that several developmental trends in the ability to judge the speed, direction, and interception point of a moving object can be observed between the ages of 6 and 11 years. Older children (ages 9 to 11) are generally much more accurate in judging the flight of a moving object than younger children. Though these data provide useful information with which to understand the perceptual and motor mechanisms that are prerequisite to the successful interception of objects, they do not provide specific product performance information.

Unfortunately, much of the literature concerning the quantitative characteristics of the catching ability of young children is unclear because many research studies have used scoring systems that reflect the process by which the object is intercepted as well as the actual success of the catch. It should also be noted that although much of the literature describes 6- to 8-year-olds as being relatively proficient catchers, much of this information was collected when a large ball was used and the thrower was positioned directly in front, tossing the ball slowly from about 6 feet away.

Studies that have focused primarily on the end product characteristics of catching (i.e., interception and control of the ball), under both simple and complex conditions, suggest that males and females display yearly increases in performance through the elementary years and, in some cases, beyond (e.g., Bruce, 1966; Fischman, Moore, & Steele, 1992; Seils, 1951; Williams & Breihan, 1979). These data appear to parallel the information related to the development of perceptual-motor and coincident timing abilities. In the Bruce study, children in grades 2, 4, and 6 were required to move in preparation to catch a ball that was projected from a ball-throwing machine at varied speeds and from several trajectories. Both sexes improved their catching performance with each advancing grade level. In the more recent Fischman et al. investigation, one-hand catching was studied among a sample of 5- to 12-year-olds. Subjects were tossed tennis balls at four different locations (waist height, shoulder height, above the head, out to the side) from 9 feet away and scored by the number caught. Results indicated that catching performance improved with age, with males outperforming females. One of the noted conclusions is that even 5-year-olds were able to integrate, to some degree, perception and motor response. The literature is mixed with regard to sex differences; therefore, a clear-cut statement cannot be made. Along with the general age-related trend of increased catching ability, the literature also suggests that improvement is characterized by the ability to catch increasingly smaller balls and display more sophisticated hand-catching techniques (Wickstrom, 1983).

Striking

Striking is a coincident timing skill that presents some difficulty for those who would attempt to establish a precise developmental picture of product performance. This inherent problem is that striking skills are performed in a variety of planes (overhand, sidearm, underhand) and under widely varying conditions. Along with a wide assortment of striking motions, other variables include the size and type of ball, and the striking implement. The diversity of popular sports that use striking skills include softball, tennis, hockey, golf, and volleyball. Although a limited number of studies have assessed product characteristics across the broad spectrum of striking, a clear and persistent trend of yearly increases through childhood (and perhaps beyond) has been identified. In contrast to the interception skill of catching, however, most data indicate that males are better strikers than females.

Much of the data on the product characteristics of striking have been collected using striking velocity, distance that the ball travels, and number of hits (out of a set number of trials) as the characteristics measured. Williams (1983) reported that velocity with the one-arm and two-arm sidearm motion (using a racquetball racquet) improved substantially for both sexes from 4 to 8 years. Interestingly, one-arm striking performances were slightly better than when using two arms. It has been speculated that this is due to the difficulty a young child has in effectively timing the motion with two hands.

Unfortunately, little information is available in relation to the distance that a ball travels after being struck. One would generally expect distance (force) performance trends to parallel those found in velocity scores (as in throwing and kicking); however, there has not been enough evidence gathered to make such a statement. Using the volleyball serve as the task for measuring underarm distance among first through sixth graders, Hanson (1965) found that scores increased significantly for both sexes with advancing grade level. It was also noted that males served a greater distance than females at each grade.

Perhaps the most commonly used measure of striking performance has been number of hits. Two of the most-cited studies were conducted by Seils (1951) and Johnson (1962). In both investigations, children were positioned with a bat in a typical batter's stance and asked to strike a ball attached at the end of a rope and swung over the plate. Similar results were found; both sexes increased in their striking ability at successive grade levels and, at each grade, males performed better than females.

The actual batting performance of males 7 to 15 years was investigated by Sheehan (1954) using a batter and pitcher at regulation distance from each other. Although the study was not tightly controlled (e.g., pitching speed was not always the same), the data did suggest that the batting ability of children improved substantially through the elementary grades.

Wickstrom (1983) notes that although there appears to be a general trend toward improvement through the childhood years, the characteristics associated with change in general striking ability have not been clearly identified.

We learned in the last chapter that hopping is considered to be the most difficult form of basic jumping and is an integral part of several advanced movement skills used in games, dance, and gymnastic activities. The movement presents difficulty because it requires leg strength, balance, and controlled rhythmical movement. During the childhood years, hopping is one of the fundamental skills in which sex differences favor the female. In terms of quantitative characteristics, hopping is usually measured by determining: (a) the number of consecutive hops completed before balance is lost, (b) the ability to complete a specified distance (usually 25 or 50 feet) while in balance, and (c) the time required to hop a specified distance. Available data suggest that, in general, hopping abilities, regardless of the particular type of task, improve with age through childhood and females perform better than males. Unfortunately, minimal data are available for individuals beyond the elementary years, and therefore the developmental perspective is limited.

By 3½ years of age, most children have the ability to hop repeatedly for three steps (the definition of a true hop), and by age 5 the ability to hop 10 consecutive times is normal. Keogh (1965) investigated the ability of children ages 5 to 9 to hop 50 feet without stopping or exchanging hopping legs. In his findings he reported that with advancing age, an increasing percentage of children completed the distance and that females were approximately one year ahead of the males in acquiring that ability. At age 5 about one-third of the males could not complete the distance, but only about one-fifth of the females displayed the same problem. By age 9 few individuals of either sex were unable to hop 50 feet. Hopping speed over a specified distance also improves with age during childhood. Cratty (1986) suggests that by age 5, most children can hop a distance of 50 feet in about 10 seconds and that females are approximately 3.5 seconds faster than males.

On tests of rhythmical hopping ability, improvement with age during childhood has been noted for both sexes. However, as with other tests of hopping ability, females tend to be as much as a year in advance of males. On a rhythmical test consisting of two hops on one foot followed in rhythm by two hops on the other foot, a significantly greater percentage of females than males passed at each age of 5 through 9 years (Keogh, 1968a). This type of hopping task seemed particularly difficult for young males; only 3% of 5-year-old males performed acceptably compared to 23% of the females. By age 9 both sexes displayed significant improvement in rhythmical hopping (67% of males passed compared to 87% of females).

Although ball bouncing and dribbling (i.e., three or four consecutive bounces) are two of the most widely practiced fundamental and sport skills, very little product data are available. Early research by Wellman (1937) established that an important developmental consideration in ball handling ability is the size of the ball in relation to the size of the hand. According to Wellman's research, 2-year-olds are generally able to bounce a ball 9½ inches in circumference for a distance of 1 to 3 feet using one hand. By the age of 40 months, the distance

Hopping

Ball Bouncing
and Dribbling

improves to approximately 4 to 5 feet. By 46 months, a 16 ¼-inch ball can be bounced a distance of 4 to 5 feet using both hands, and around the age of 6, the same distance can be achieved using only one hand.

In an unpublished study by Williams and Breihan (1979) as reported by Williams (1983), males and females aged 4, 6, and 8 were asked to repeatedly bounce an 8-inch ball within a 12-inch square using one hand. Results indicated that performances for both sexes improved with advancing age and that whereas males performed better than the females at ages 4 and 6, females were substantially better at 8 years. As a group, the 4-year-olds averaged less than two successful bounces within the square before control was lost. In contrast, 6-year-olds averaged four controlled bounces and the 8-year-olds slightly more than five good bounces.

This topic should be studied much more extensively to develop a more complete picture of dribbling as a sport skill. Currently no information is available that assesses dribbling performance using the right and left hands while moving.

Balance

As noted in chapter 6 under the discussion of vestibular awareness, the successful performance of virtually all motor skills depends on the individual's ability to establish and maintain equilibrium, or balance. Refer to that chapter for a review of the development of balancing abilities through adolescence.

MOTOR SKILL REFINEMENT

■ *Objective 10.4*

Sports and physical education are two of the most influential developmental factors in motor skill development and refinement during childhood and adolescence. Sport and other physical activity participation normally occurs through school programs, spontaneous play activities, and organized youth sport programs. The dramatic increase in nonschool youth sport programs has received considerable attention from developmentalists in recent years.

Youth Sport Participation

Estimates suggest that more than 30 million children, ages 6 to 16 years, are engaged in an organized program of **youth sports** (Martens, 1978). Approximately 50 different individual and team sport activities, ranging from baseball to long-distance running, have been identified.

Seefeldt and Branta (1984) suggest six explanations for the increased participation of children in youth sport programs:

1. Passage of Title IX, which provides the mandates for equal opportunities in sport at all levels for females.
2. Greater accessibility to programs due to the gradual movement of the population from rural to urban residences.
3. Changing lifestyles of adults, including greater concern for health and physical fitness, may be influencing the decisions that parents make about children's participation in physical activities that include sports.

Figure 10.8

4. An increase in the number of women in the work force. It may be speculated that this has led to an increase in the enrollment of children in activity programs as a caretaking function. It may also be that some single-parent mothers view the coach as an additional role model for their child and want to encourage that interaction.

5. Glorification of sport. The media's constant focus on the glory of sport undoubtedly leads some parents to involve their child in sport at an early age. For some parents, sport involvement is perceived as an avenue through which to increase their child's opportunities for social and financial success.

6. Declining role of public schools in providing sport programs below the high school level. This has stimulated increased involvement by municipal recreation and community agencies in providing activity programs.

Other explanations for the increase in youth sport participation include the fact that children are getting involved at an earlier age; programs for 4- and 5-year-olds are commonplace today. Increased participation in what is considered nontraditional sport activities (e.g., soccer, gymnastics, bowling, cycling, distance running) and the dramatic increase in the number of programs and events organized for handicapped individuals are two other areas of increased involvement in sport activity.

During 1984 and 1985, the U.S. Public Health Service conducted two of the most comprehensive studies of physical activity ever undertaken (National Children and Youth Fitness Study I & II, Ross & Gilbert, 1985; Ross & Pate, 1987). Both studies collected data on individuals in grades 1 through 12 (ages 6 to 18 years). The findings produced considerable insight about physical activity participation characteristics in community settings (i.e., outside of school).

Table 10.2   Top 15 Physical Activities Performed by Children (grades 1–4) in Community Settings

Males	Rank	Females
Swimming	1	Swimming
Racing/sprinting	2	Racing/sprinting
Baseball	3	Bicycling
Soccer	4	Playing on a playground
Bicycling	5	Gymnastics
Playing on a playground	6	Walking
Football	7	Basketball
Basketball	8	Climbing
Climbing	9	Hiking/backpacking
Hiking/backpacking	10	Locomotor skills
Locomotor skills	11	Playing (unspecified)
Walking	12	Jumping or skipping rope
Calisthenics/exercises	13	Roller skating
Playing (unspecified)	14	Soccer
T-ball	15	Games (unspecified)
		Ballet/square/folk dance

Among the children in grades 1 through 4 (ages 6 to 9 years), 84.3% participated in physical activity in at least one community organization. In the context of this study, community organizations were defined as parks and recreation programs, sports leagues, churches, YMCAs and YWCAs, clubs and spas, scouts, and farm clubs (e.g., 4-H). Interestingly, more than twice as many males (42.4%) as females (20.4%) participated in sports teams and leagues. Participation rates in the other community organizations were not noticeably different between the sexes. The five most frequently performed activities for both sexes were swimming, racing/sprinting, baseball/softball, bicycling, and soccer. There were differences between the sexes in most frequently performed activities, however (table 10.2). In general, and in comparison to the results for youth in grades 5 through 12, the number reporting participation in any single activity was not particularly high. The number reporting participation in the top two activities for both sexes (swimming, racing/sprinting) ranged from 28% to 43%. Only two of the remaining top 15 activities for both sexes had a participation rate higher than 20% (males: baseball 23%; soccer 21%). The findings also indicate that as the children got older their preferences changed. Baseball/softball, basketball, football, and hiking/backpacking all progressively moved up in frequency, whereas participation in locomotor activities, climbing, and playground play declined with advancing age (grade level).

Within the grades 5 through 12 group (ages 10 to 18), it was reported that the typical student spends over 80% of his or her physical activity time outside of the school setting. Participation level was highest in the summer, fell off dramatically in the fall and winter, and resumed at a more typical level in the spring. Only 18.2% of the youth did not perform at least one physical activity through a

Table 10.3 Top 15 Physical Activities Performed by Youth (grades 5–12) in Community Settings

Males	Rank	Females
Bicycling	1	Swimming
Basketball	2	Bicycling
Football (tackle)	3	Disco or popular dance
Baseball/softball	4	Roller skating
Swimming	5	Walking quickly
Weight lifting or training	6	Baseball/softball
Fishing	7	Basketball
Football (touch or flag)	8	Calisthenics/exercises
Hunting	9	Jogging (distance running)
Jogging (distance running)	10	Gymnastics-free exercise
Soccer	11	Volleyball
Lacrosse	12	Aerobic dance
King of the hill	13	Tennis
Roller skating	14	Horseback riding
Disco or popular dance		
Wrestling	15	Cheerleading/pom-pom
Karate/judo/martial arts		

community organization. In fact, according to the self-report survey results, the typical youth was exposed to 21.3 different activities (i.e., participated in each at least three times) over the course of a year.

The overall weekly participation level for the typical male in grades 5 through 12 was only 10% greater than that of the average female. The typical youth in this grade range devoted slightly over 30% of his or her activity time to five activities. In descending order these were cycling, swimming, basketball, baseball/softball, and tackle football. Just as for the results of participation levels for the lower grades, the predominance of top activities varied between males and females (table 10.3). The top five activities for males were bicycling, basketball, football (tackle), baseball/softball, and swimming. For females, the top five were swimming, bicycling, disco or popular dance, roller skating, and walking quickly. As one would expect, as most youth grew older, some activities virtually dropped out of the picture. For example, activities that were important for fifth and sixth graders such as relays, tag, jumping rope, and kickball dropped off the top 15 list by high school. Meanwhile, additional activities such as tennis assumed a higher status in the adolescent's repertoire.

In a more recent survey of sports activity of 8 to 12 years old (*Sports Illustrated for Kids,* 1992), similar findings to those described were reported. Two noteworthy changes are that bicycling, a consistent top favorite across all grades, has apparently declined in popularity, dropping out of the top five, while in-line skating has emerged as a top 10 favorite pastime.

Although involvement in youth sport programs has been found to contribute significantly to the participant's physical, psychomotor, social, and emotional development, several issues have been raised in relation to potential psychological and physical harm (e.g., Martens, 1978; Scanlan, 1984). Frequently addressed issues include the related competitive stress and anxiety, the physical risks, especially in contact and collision sports, and the effects of intense early participation. Refer to the Suggested Readings list for additional information on this topic. Selected psychological considerations will be discussed in chapter 13.

School Physical Education

Another potentially powerful source for motor skill and physical fitness development is the school **physical education** program. According to the National Children and Youth Fitness Study findings, virtually all children (97%) in grades 1 through 4 are enrolled in physical education programs of one sort or another. The typical student is provided with a structured physical education three times per week; about 36% have the privilege of participating in classes daily and the average class meets for 30 to 35 minutes. The five most common physical education class offerings in grades 1 through 4 (in order) were identified as movement experiences and body mechanics, soccer, jumping or skipping rope, gymnastics, and basketball. Other activities include throwing and catching activities, calisthenics/exercises, rhythmic activities, kickball, relays, running, and baseball/ softball. A definite trend is noted with advancing grade level, that being a shift in emphasis from movement education (i.e., fundamental skill development) toward greater participation in physical fitness activities and sports. Competitive team sports introduced by the third and fourth grades, such as softball, basketball, soccer, and volleyball, remained the core of the high school physical education program.

Data for grades 5 through 12 indicated that just over 80% of American youth are enrolled in physical education classes. Enrollment level peaked at 97% in grades 5 and 6 before tapering off to a low of approximately 50% in grades 11 and 12. Similar to the findings for the lower grades, students average just over three weekly sessions, with about one-third taking physical education daily. The typical physical education session for this group lasts about 45 minutes.

The average student at this level spends the largest portion of his/her physical education class time on the following five activities: basketball, calisthenics/ exercises, volleyball, baseball/softball, and distance running. Other predominant class activities include gymnastics, aerobic dance, swimming, hockey, weight training, and running sprints. Although several activities are performed by both sexes (e.g., basketball, volleyball), some of the top 15 activities are unique to gender. Exclusive to the female's top 15 activities are aerobic dance and gymnastics, whereas hockey, tackle football, and wrestling are restricted to the predominant 15 activities of the male only. Along with the overall trend toward competitive team sport participation, which begins in the third and fourth grade, individual and dual activities such as badminton, tennis, weight training, and swimming also become more commonplace in the high school curriculum.

Although team sport activities dominate most physical education curriculums from the fourth grade through high school, such focus has stimulated considerable debate regarding its benefits to health-related physical fitness. Top national associations have issued statements strongly suggesting the need for schools to adopt health-related physical activity goals (e.g., The American Academy of Pediatrics, 1987; American College of Sports Medicine, 1988). More important, school physical education is viewed as the only major institution that can address such needs for the largest portion of children and youth. Unfortunately, there are indications that due to financial limitations and the return to educational "basics," support for physical education programs could be decreased or lost altogether. Through such actions as the setting of national objectives for health promotion and disease prevention (Healthy Children 2000, Public Health Service, 1991), which includes physical activity and fitness as key elements, it is hoped that the importance of physical education will be reexamined.

SUMMARY

The period from later childhood through adolescence encompasses the sports skill and growth and refinement phases of motor behavior and is marked by rapid biological change, increases in product performance, sex-related differences, and the emergence of sport skill behavior. Many of the quantitative (product) changes in motor performance of both sexes are based on increasing levels of experience and a variety of characteristics associated with the adolescent growth spurt (e.g., increased body size, muscle mass, strength, cardiorespiratory capacity). Although a few sex differences favoring males are observable shortly after infancy, these differences become increasingly evident after puberty. Due primarily to the favorable effects of the growth spurt on males, increases in their product performance continue through adolescence, whereas the performance of females tends to plateau or, in some instances, decline shortly after menarche. Several biological factors may provide an advantage to the male in several motor skill activities, but cultural practices and participation levels can significantly influence wide individual differences within and between the sexes. In general, females appear to be closing the gender gap that in earlier years was quite wide. This trend seems to be related to increased participation and a change in society's view of females in sporting activities.

A general and rather consistent trend in product performance change is that both sexes improve significantly from childhood to adolescence and that males continue to improve on a variety of skills through adolescence (to at least age 17), whereas females have a tendency to peak around the age of 14, then level off or decrease in performance. More recent studies have shown females to plateau rather than regress after age 13 or 14. This has been observed in running (speed), throwing (distance, velocity), jumping (standing long jump, vertical jump), and kicking (distance, velocity, accuracy). A general trend of improvement with advancing age may also be found in catching, striking, hopping, and ball bouncing or dribbling. Clear-cut differences between the sexes are not evident in catching or ball bouncing abilities, but most data indicate that males are better strikers than females. Though very little data are available for performance beyond the elementary years, hopping has been identified as one of the fundamental skills in which sex differences favor the female.

Organized youth sport programs and school physical education programs are two of the most influential factors in motor skill development and refinement. In recent years there has been a dramatic increase in youth sport participation with a wide variety of activities being offered for both sexes. Younger children (6 to 9 years) tend to participate more frequently in swimming, racing/sprinting, baseball/softball, bicycling, and soccer, whereas older youth (10 to 18 years) generally appear to prefer cycling, swimming, basketball, baseball/softball, and tackle football.

As children and youth grow older, their activity choices change. Although numerous developmental contributions have been linked to participation in organized youth sport programs, several issues have been raised in relation to their potential to result in psychological and physical harm.

The vast majority of children and youth participate in some sort of school physical education program at least three days per week for 30 to 45 minutes per session. About one-third of the individuals are enrolled in physical education on a daily basis. Among the most common class offerings in grades 1 through 4 are movement experiences/body mechanics, soccer, jumping rope, gymnastics, and basketball. By the third or fourth grade, competitive team sports are introduced and remain the core of the high school physical education program. Along with team sport participation in the upper grades, individual (e.g., aerobic dance, calisthenics/exercises, running) and dual activities (e.g., badminton, tennis) are commonplace.

SUGGESTED READINGS

Haubenstricker, J., & Seefeldt, V. (1986). Acquisition of motor skills during childhood. In V. Seefeldt (Ed.), *Physical activity and well-being* (pp. 41–102). Reston, VA: American Alliance for Health, Physical Education, Recreation, and Dance.

Roberton, M. A. (1984). Changing motor patterns during childhood. In J. Thomas (Ed.), *Motor development during childhood and adolescence.* Minneapolis: Burgess.

Wickstrom, R. L. (1983). *Fundamental motor patterns* (3d ed.). Philadelphia: Lea & Febiger.

Youth Sports

Brown, E. W., & Branta, C. F. (Eds.). (1988). *Competitive sports for children and youth.* Champaign, IL: Human Kinetics.

Gillian, T. (1982). Answers to the most frequently asked questions about youth sports. In R. H. Cox (Ed.), *Educating youth sport coaches: Solutions to a national dilemma.* Reston, VA: AAHPERD Publications.

Martens, R. (Ed.). (1978). *Joy and sadness in children's sports.* Champaign, IL: Human Kinetics.

Scanlan, T. (1984). Competitive stress and the child athlete. In J. M. Silva & R. S. Weinberg (Eds.), *Psychological foundations of sport.* Champaign, IL: Human Kinetics.

Smoll, F., Magill, R., & Ash, M. (Eds.). (1988). *Children in sport* (3d ed.). Champaign, IL: Human Kinetics.

Motor Behavior in the Adult Years

KEY TERMS

peak motor performance
chronological age
physiological age

regression
psychomotor slowing
longevity

PHASES OF MOTOR BEHAVIOR

Phase	Approximate age	Stage
	conception	
Reflexive/ Spontaneous	(3 mo.)	Prenatal
Rudimentary	birth	
		Infancy
	6 mo.	
	2 years	
Fundamental Movement		Early Childhood
	6	
Sport Skill		Later Childhood
	12	
Growth & Refinement		Adolescence
	18	
Peak Performance		Adulthood
	30	
Regression		
		Older Adulthood
	70	

■ *Objective 11.1*

Much of the information to this point in the text has provided the basis for understanding lifelong motor development by describing the biological and information-processing characteristics associated with motor behavior across the life span. Part Four began with an introduction to the developmental approach to motor performance and identified each of the contributing phases of motor behavior. In keeping with this approach, chapter 8 presented the initial phase of motor behavior by describing actions exhibited prior to birth. Subsequent chapters have traced the evolution of movement abilities and the acquisition of fundamental and basic sport skills through childhood and adolescence.

This chapter describes the final phases and associated behaviors in the lifelong process of motor development. Motor behavior during the adult years is characterized by two distinguishing milestones—peak performance, which is associated with biological maturity during early adulthood, and the emergence of performance regression, which is linked with the latter stages of aging.

The information presented in this chapter constitutes a brief review of those biological distinctions and motor performance characteristics that distinguish adulthood from the other periods of the total life span. A very timely discussion of the possible effects of physical activity on longevity is also presented.

PEAK MOTOR PERFOR-MANCE

■ *Objective 11.2*

Although adolescence is generally accepted as the most distinctive period of biological growth and motor skill development, it is in early adulthood that most individuals reach their peak of physical performance, health, and sexual maturity. It is also during the early adult years that sex differences in motor performance are maximized. Wide variation among young adults is common, but it is during this period in the life span that individuals attain **peak motor performance,** partly because the different physiological systems have become capable of working together so efficiently.

Growth and Physiological Function

Skeletal maturity is one example of continued growth into the adult years. Although most epiphyseal growth is completed in late adolescence (16 to 18 years), the long bones may continue to grow until approximately age 25, and the vertebral column until about age 30. This continued growth in the long bones may add up to one-fourth inch to the height of an individual (Tanner, 1978). In addition, certain areas of the braincase do not reach maturity until well into adulthood (Garn, 1980). This continued growth is evident when men (and women) who wear hats notice that as they age, they have to purchase larger sizes.

According to most sources, peak physiological function generally occurs between the approximate ages of 25 to 30 years of age (Brooks & Fahey, 1985; Eckert, 1987; McArdle, Katch, & Katch, 1991; Smith & Serfass, 1981). As a general rule, females tend to mature at the lower end (22 to 25 years) and males at the upper end (28 to 30 years) of the range. This is especially evident in three of the most influencing factors in motor performance: muscular strength, cardiorespiratory efficiency, and processing speed (reaction/movement time).

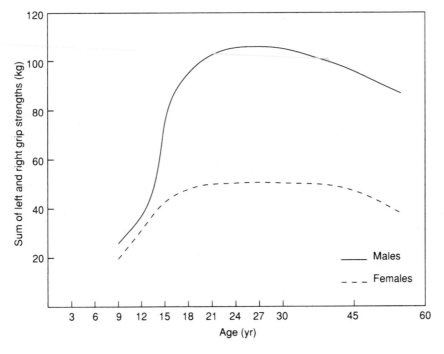

Figure 11.1
Changes in grip strength with age.

This figure is reprinted with permission from the *Research Quarterly for Exercise and Sport*, vol. 48, no. 109, 1977. The *Research Quarterly for Exercise and Sport* is a publication of the American Alliance for Health, Physical Education, Recreation and Dance, 1900 Association Drive, Reston, VA 22091.

Maximum strength for both sexes is generally achieved at a time when the muscular cross-sectional area is the largest, which is between the approximate ages of 20 and 30 years. Maximal strength for most muscle groups takes place between the ages of 25 and 29 years. Support for this assertion is shown in figure 11.1, which illustrates that grip strength values peak between the ages of 25 to 29 years. It is also around the age of 25 years that peak performance in grip strength-endurance is reached (Burke et al., 1953).

Peak physiological function in maximum oxygen consumption and cardiorespiratory work for both sexes typically occurs during the third decade. Several reviews and studies report peak values within the range of 25 to 29 years (Astrand & Rodahl, 1986; deVries & Housh, 1994; Quirion et al., 1987). Figure 11.2 represents a composite of data from 17 published studies (normalized for body weight) and shows the approximate peaking of maximum oxygen consumption around the age of 30 and the steady decline thereafter. Figure 11.3 is a frequently cited illustration depicting maximum work at various ages based on a two-step climbing test. Based on this information, peak work performance for females is said to occur around the age of 25 years, and for males, approximately three years later. In the early 1960s, Shock and colleagues conducted what is now referred to as one of the classic aging and physical performance studies (Shock, 1962). Included in that series of experiments was a test to determine the amount of "work" that males could do using a crank ergometer. Among the physiological variables measured were oxygen consumption and cardiac output. The researchers found that work rate peaked around the age of 28 years.

Figure 11.2
Changes in maximum
oxygen consumption
with age.

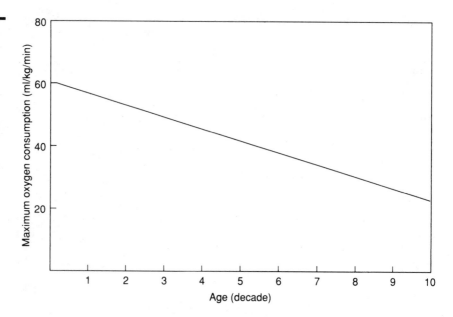

Figure 11.3
Relationship of age to
exercise tolerance (in
foot-pounds of work per
minute).

Source: Data from Master,
American Heart Journal,
10:495–510, 1935. C. V. Mosby,
St. Louis, MO.

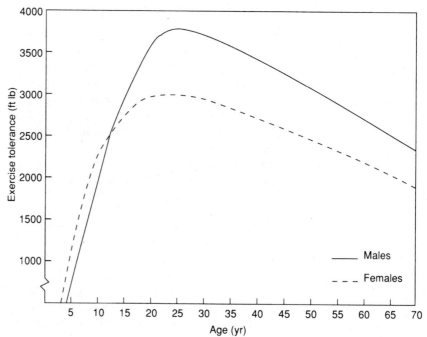

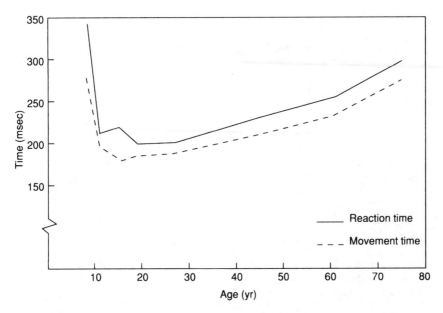

Figure 11.4
Reaction time and
movement time as a
function of age.

The importance of neurophysiological function and processing speed in the performance of a motor act is obvious. Assessments of successful motor performance are frequently based on the combination of perceptual recognition (attention), speed of memory, and neuromuscular response time (reaction/movement time). Processing speed, as determined by reaction time, improves with age up to the early adult years, with peak performance occurring in the twenties (Hodgkins, 1962; Wilkinson & Allison, 1989) (fig. 11.4).

Motor Performance Characteristics

Peak physiological function, which occurs in the approximate range of 25 to 30 years, closely parallels maximal motor performance. A general observation of the sporting world quickly leads to the speculation that college athletes are better than those of high school age and that few professional or Olympic-caliber athletes participate at these levels beyond their early thirties.

■ *Objective 11.3*

Since the majority of physical growth and development occurs during adolescence, the increases in motor performance during the adult years are due primarily to training, practice, experience, and motivation. Peak proficiency may be studied from the perspective of a population that is highly trained (i.e., athletes) and seeks maximum motor performance, or from the perspective of a population of average or "normal" ability. Although considerable diversity and variability exists within these groups, and "normal" is difficult to define, certain trends can be identified.

With the intent of determining age at peak athletic performance, Schultz and Curnow (1988) analyzed records and national ranking from track and field, swimming, baseball, tennis, and golf. In track and field, for example, they examined Olympic results from 1896 through 1980. Overall, the mean age of winners in seven events was about 25 years of age. Comparing running events, the average

Table 11.1 Age at Peak Performance for Selected Events

(Age)	Male	Female
17		Gymnastics
18		Swimming
19		
20	Swimming	
21		Diving
22	Wrestling, Diving	Running short distance
23	Running short distance, Cycling	
24	Jumping, Boxing	Running medium distance
	Running medium distance	Tennis, Rowing
	Tennis	
25	Gymnastics, Rowing	
26		Shot put
27	Running long distance	Running long distance
28	Baseball, Weight lifting	
29	Shot put	
30		Golf
31	Golf	
32		

Source: Data from Schultz and Curnow, 1988, and Hirata, 1979.

age at peak performance increased with the length of the race: from 22 years in the sprints, age 24 for middle distances, and 27 for distances of 5,000 meters and greater. For professional baseball players, the mean age of peak performance was consistently around 27 years, based on pitching performances, batting averages, home runs, and runs batted in for nonpitchers (also see Schultz, Musa, Staszewski, & Siegler, 1994). Swimming performance appeared to peak between 18 to 20 years, and record performances in golf were achieved between the ages of 30 to 31 years. An examination of world professional tennis rankings suggested that 24 to 25 years of age was best for optimal performance. Females generally achieved peak performance about one to two years before males.

Similar findings were reported by Hirata (1979) in the researcher's examination of past records of Olympic medalists (1964–1976). This study included a larger variety of events (than Schultz et al.), including boxing, canoeing, cycling, gymnastics, wrestling, and weight lifting. Although the range of winners' ages was expectedly wide, the average male was approximately 26 years, and the average female was about 23 years old. Table 11.1 presents a composite of selected events for age at peak performance based on the findings of Schultz et al. and Hirata.

Further support for the notion that peak performance in this aspect of maximum physiological function occurs between the approximate ages of 25 to 29 years is provided by McArdle and colleagues (1991). The researchers plotted a graph of world record age-group times for males and females for 1986 (fig. 11.5). As noted, peak performance for males appears to be between the ages of 25 and 31 years, whereas maximum performance for females is around 25 to 29 years.

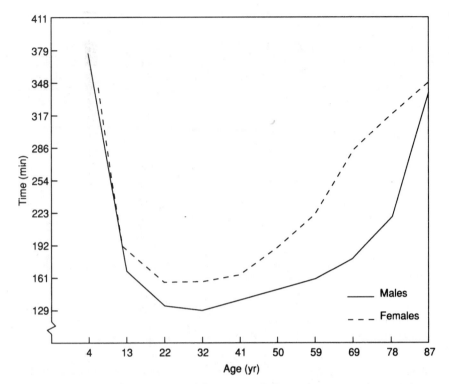

Figure 11.5
World record marathon times for males and females of different ages.

The conclusion drawn from reviews of the data relating to this topic has been that performance records requiring greater levels of strength, speed, and endurance are generally challenged by younger athletes. Further, peak proficiency in activities requiring a high skill level and experience but less vigorous physical demands is frequently attained by older athletes (Eckert, 1987; Schultz & Curnow, 1988). Although research in this area is limited, it suggests that maximum proficiency in relatively less physically demanding motor skills such as shooting, bowling, golf, and billiards generally occurs after the age of 30 years.

Another perspective of maximum physical and motor proficiency during adulthood may be formed by observing a more normalized (i.e., not highly trained) segment of the population. Unfortunately, there is a dearth of information concerning the physical and motor proficiency of nonathletic adults. However, available information does allow a comparison of fitness levels between high school students (17-year-olds) and college students (21-year-olds). As shown in table 11.2, the average health-related (i.e., based more on physiological efficiency than skill) fitness performance scores of the 21-year-olds are greater than or equal to those of the younger age group. Superiority is especially evident in the two test items associated with aerobic performance.

Eckert (1987) reported similar findings after making a comparison of national norms (50th percentile) of 1961 data between 17-year-olds and college students. Included in the comparison were seven health- and skill-related fitness

Table 11.2 A Comparison of Average Fitness Performance Scores

Test item	Age		
	17 years		21 years
One-mile run (min:sec)	8:20	male	7:05
	10:34	female	9:23
9-minute run (yd)	2,169	male	2,225
	1,729	female	1,761
Sit-ups	44	male	44
	35	female	37
Sit and reach (cm)	31	male	33
	36	female	36

Sources: Data from The National Children and Youth Fitness Study, *Journal of Physical Education, Recreation and Dance, 56,* (1), 47, 1985; and Norms for College Students Health-Related Physical Fitness Test, American Alliance for Health, Physical Education, Recreation and Dance, 1985.

items: pull-ups, sit-ups, shuttle run, standing broad jump, softball throw, 50-yard dash, and the 600-yard run-walk. For the males, college student values were superior in five of the seven items and equal in the remaining two (pull-ups and 50-yard dash). A comparison of female norms also disclosed higher scores for the college group in five of the seven test items. However, values for pull-ups and the softball throw, both of which involve a higher degree of strength in comparison to the other items, were higher for the 17-year-olds.

Although the information presented is quite limited and represents a specific segment of the population (one that is relatively active but not highly trained) it does suggest that peak motor performance, in general, is reached after adolescence (17 to 18 years). Unfortunately, data are unavailable on these fitness and motor skill characteristics among relatively inactive segments of the population to use in comparison. One may speculate that due to the influencing effects of a physically active lifestyle and practice on motor performance after adolescence, performance values for nonactive young adults are not likely to be as high as those for moderately active individuals.

Considerations of Race

Race implies a biologically distinct group of humans. In considering race as a factor in development and motor performance, caution in drawing any final conclusions is warranted. One reason for caution is that race is pure in very few instances. For example, an African American is likely to have ancestors (thus some genetic endowment) of, for example, African, West-Indian, Caucasian, and Spanish descent. Therefore, any practical comparisons made between general categories of race are usually quite "generalized." Second, there is considerable variation within racial groups and much overlap among groups being compared. Any valid observations should view race in a broader framework that is biocultural (Malina & Bouchard, 1991). As emphasized throughout this text, development

Table 11.3 Selected Characteristics of the African American Athlete

Anatomical

Greater: lower leg length/thigh length
    biacromial/biiliac breadth ratios (narrower hips relative to shoulders)
    arm length; relative to sitting height and stature

Physiological

Greater: lean body mass
    density of bone
    proportion of type II (fast-twitch) muscle fibers
    use of VO_2max (uses higher percentage)

Sources: Data from Himes, 1988, Ama, 1986, Schutte, 1984, and Bosch, 1990.

and motor performance depend on several factors, of which heredity and the environment are primary agents. Chapter 13 will provide additional insight in regard to race and various environmental contexts. The intent of this section is to briefly describe some biological factors associated with race as reflected in peak motor performance of running and jumping events. The vast majority of research in this area has been conducted on the comparison between Caucasians and African Americans; such will be the focus of this discussion. Given the increasing proportion in the United States of other racial groups, such as Hispanics and Asian Americans, future reports including these groups should be forthcoming.

Observations and Biological Considerations   In a feature, titled "White Men Can't Run," in a leading running magazine, the author notes that on the all-time list for 100 meters, 44 of the top 50 performers are sprinters of African origin (Burfoot, 1992). A strong presence in other running and jumping events also exists, including the showing of Kenyans in distance running. The question is, Aside from the cultural bias for such events, are specific athletes of African descent genetically gifted to perform these activities, compared to Caucasians? As noted in the general comparison of the sexes (chapter 10), certain anatomical and physiological characteristics are conducive to specific running and tasks. Over the years, several research studies have compared Africans and African Americans with Caucasians, producing some rather consistent findings (table 11.3).

In general, runners and jumpers of African descent would appear to have a physiological and biomechanical advantage in such events by possessing less body fat, narrower hips, thicker thighs, longer legs, lighter calves, a greater proportion of type II (fast-twitch) muscle fibers (for sprints), and greater VO_2max efficiency (for distance runs), in comparison to Caucasians. Though most attention seems to be on sprinters, one indication of excellence in distance running was noted by Bosch et al. (1990) when a group of South Africans were able to use a higher percentage of their VO_2max than elite Caucasian marathoners while running a marathon on a treadmill.

In summary, these data provide some biological (genetic) insight into an explanation for excellence in "specific" running and jumping events by athletes of African origin. Keep in mind that there is considerable variation within racial groups and much overlap among groups being compared. In addition, numerous sporting events have not been mentioned, many of which depend on a uniquely different set of skills and physical attributes.

REGRESSION

■ *Objective 11.4*

The differences between **chronological age** and **physiological age** become increasingly evident with advancing age. In general physiological function, active individuals appear to be physiologically superior to their counterparts of the same chronological age (fig. 11.6). From a general observation of human performance and attitude, this phenomenon is also apparent. Whereas some people consider themselves "old" by the age of 30, others continue to pursue a highly active lifestyle through old age. Several major championships and records in such sports as track and field, tennis, baseball pitching, and golf have been held by individuals in their forties (McFarlan, 1989). One of the truly remarkable athletic feats by an older individual in recent years was accomplished by John Kelley. In the 1992 Boston Marathon at the age of 84 years he completed the course (26 miles, 365 yards) in close to 5 hours. It was the 61st time he had run this race, which he won in 1935 and 1945.

However, though there are numerous benefits from a healthy and active lifestyle during adulthood, it appears that no matter how well people take care of themselves, the effects of advancing age will result in **regression** of physiological processes and motor performance.

Biological Regression and Motor Performance

■ *Objective 11.5*

After an individual reaches peak maturity at approximately 30 years of age, most physiological factors begin to show decrements at a rate of about 0.75% to 1% a year. This is a general trend of decline, however; differences between individuals and the function of specific organs can vary considerably. There is some suggestion from the research community that up to 50% of the loss usually attributed to physiological aging may be due to inactivity and other poor health habits. The decline in biological capacity and associated motor performance is generally characterized by decrements in cardiorespiratory function, muscular strength, neural function, balance, and flexibility. Additional age-related factors that may affect one's motor performance are changes in the skeletal system (bone tissue) and increased body fat.

For a summary of selected biological changes that occur with aging after peak maturity, refer back to table 4.4. The following is a summary of those changes and a description of associated decrements in motor performance.

Cardiorespiratory Function Cardiovascular capacity as measured by maximal oxygen consumption (maximal aerobic power) declines approximately 30% between the ages of 30 to 70 years (fig. 11.2). Closely associated with this general physiological decrement is a decrease in cardiac output (stroke volume

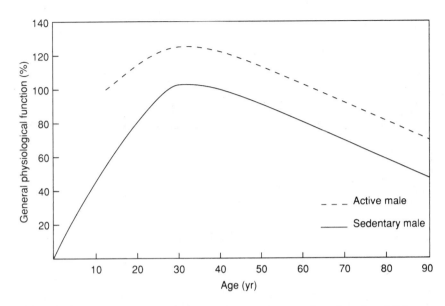

Figure 11.6
General physiological
function across the life
span for active and
sedentary males.

and maximal heart rate) and respiratory (pulmonary) function. In addition to
diminished cardiac capacity, respiratory function decreases with advancing age
as evidenced by an estimated 40% loss in vital capacity between the ages of 30
and 70.

Parallel to the loss in aerobic capacity after peak maturity is a slightly
greater decline in anaerobic power (about 40% by age 70). This loss is linked, in
part, to the aging body's diminishing capacity to "turn over" lactic acid accumulation during short bouts of intense physical activity. Other significant cardiovascular changes include increased blood pressure and blood flow resistance. Thus
aging produces significant decreases in the body's oxygen transport and extrac-
tion capacities.

Motor performance decrements in physical work capacity tasks and high-
level aerobic activities clearly follow the general aging trend. Examples of this
trend are shown in figures 11.3, 11.5, and 11.7. Figure 11.3 depicts changes in
work capacity (exercise tolerance) in a laboratory setting, and figures 11.5 and
11.7 illustrate age-related changes in world record marathon times and masters-
level freestyle swimming events. It may also be suggested from the data shown
on marathon and swim times that regardless of training experience, performance (aerobic capacity) diminishes with advancing age. However, highly
trained endurance athletes normally have a maximal aerobic and anaerobic capacity that is about twice as high as that of an average person (Astrand & Rodahl, 1986). Evidence also suggests that athletes who continue to train experience a decline in maximum oxygen consumption that is approximately half of
that seen in sedentary persons (Bruce, 1984; Fleg, 1986). Training and lack of
regular aerobic activity can significantly modify this aspect of the physiological function.

Figure 11.7
Mean performances for
short and long distance
masters-level freestyle
swimming events.
Source: Data from Hartley and
Hartley, *Experimental Aging
Research*, 10:35–42, 1984,
Beech Hill Enterprises, Inc.,
Southwest Harbor, ME.

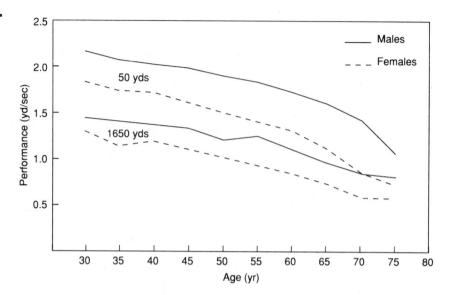

Muscular Strength The primary factor in the loss of muscular strength after peak performance is the estimated 25% to 30% decrease in muscle mass that occurs between the ages of 30 and 70 years. This general loss of body tissue is characterized by specific decreases in the size and number of muscle fibers, loss of biochemical capacity, increases in connective tissue and fat, and a general dehydration in body cells. In addition to affecting isometric and dynamic strength, these changes also hamper mobility and speed of movement, especially among the elderly.

The general decline with advancing age in maximal muscular strength seems to parallel the loss in muscle mass. That is, there is a 25% to 30% decline in strength between the ages of 30 and 70 years. It has also been suggested that the rate of decline in leg and trunk muscles in both sexes is greater than in arm and hand-grip muscles. As evidenced in figure 11.1, grip strength decreases very slowly after peak performance and does not show a marked decline until after the approximate age of 40 years. The total loss for this aspect of muscular strength from 30 to 60 years does not usually exceed 10% to 20% of maximum, with the degree of decrement being somewhat greater in females than in males. However, an analysis of the literature does suggest greater losses among older adults. In a study of adult dominant hand-grip strength values that included subjects as old as 90 years, Shock (1962) reported a decline of slightly less than 50% from age 35 to 90. The decrement in grip-strength endurance, usually measured by the amount of force that can be held for one minute, is not as great as the decline in grip strength; estimates suggest a decrease of about 30% from age 20 to 75 years.

Interestingly, a study of maximal grip strength and grip-strength endurance among males, all of whom performed similar work in a machine shop, revealed no change in strength or endurance from age 22 to 62 (Petrofsky & Lind, 1975). These data suggest that the changes found in a more typical population may be due largely to disuse rather than to aging. Although activity may slow down the effects of aging, however, there is little question that sizable decrements occur in old age, regardless of activity level.

Neural Function   Age-related decrements in neuromuscular performance may be partially explained by a significant reduction in brain cells, increased synaptic delay, a 10% to 15% decrease in nerve conduction velocity, and an estimated loss of 37% in the number of spinal cord axons. These factors, along with those associated with the decrement in muscular strength are responsible for what is described as **psychomotor slowing,** one of the most general and significant changes in older persons. Welford (1984) suggests that although muscular and neuromuscular factors contribute to slowing with age in speeded motor tasks, most of the decrement can be attributed to central mechanisms.

Evidence of age-related psychomotor slowing has been reported with several laboratory motor tasks based on reaction time and speed of movement (e.g., Haaland, Harrington, & Grice, 1993; Light & Spirduso, 1990; Lupinacci et al., 1993; Williams, 1990). Eckert (1987) notes that, in general, greater reductions in both reaction time and movement speed tend to occur in the lower parts of the body in contrast to the areas of most frequent use, such as the fingers. The rate of psychomotor slowing is also affected by the complexity of the motor task, with tasks of greater complexity being associated with increased rates of decline. That is, as movement complexity increases, the effects of adult aging on reaction time increase (Light & Spirduso, 1990; Lupinacci et al., 1993).

Data reported by Williams (1990) clearly support this contention. The research showed that between the ages of 50 and 90 years there was a 32% decrement in speed of simple discrete and repetitive arm/hand movements. In contrast, a 65% decrease in speed was found in the performance of more complex sequential arm/hand/finger movements. The researcher also found substantial decrements of about 30% in timed bilateral and unilateral object manipulation tasks. Evidence also suggests that fine motor actions that require steadiness, such as writing digits and letters, undergo even greater decrements (68% to 78%) during the adult years. For a more comprehensive review of aging and eye-hand coordination refer to Williams (1990).

In general, after performance peaks during the twenties, a gradual slowing of the body occurs in relation to reaction time and speed of movement (i.e., movement time), with more marked increases occurring after the age of 60 (fig. 11.4). Thus, the general trend is one of increasing proficiency, then peak performance, and finally the psychomotor slowing in finger, hand, and arm speed tasks with age.

The suggestion that a healthy, physically active lifestyle may reduce the decrement in psychomotor slowing (i.e., CNS deterioration) does have some support in research findings (Rikli & Edwards, 1991). In studies that have compared groups of physically active and nonactive young and old persons on reaction and movement time tasks, similar results have been reported (Rikli & Busch, 1986; Spirduso, 1975). That is, reaction and movement time scores for the active

Figure 11.8
Movement time
characteristics in young
(Y), old (O), active (A),
and nonactive (NA)
subjects.

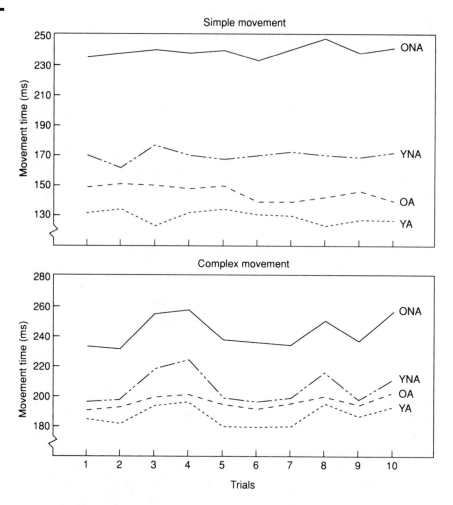

groups (be they young or old) were considerably faster than the corresponding age group that was less active. As one would expect, when scores for the age groups were averaged across activity levels, the young groups were faster than their older counterparts. Both studies also found that scores for the nonactive older group were substantially slower than those of the other groups. Figure 11.8 shows the movement time characteristics for simple and complex motor tasks.

What about the effects of training (practice) on reaction time and speed of movement among the elderly? Evidence has been documented that practice with videogame playing among elderly individuals (who averaged 70 years of age) improved the speed of response (i.e., reaction time) and, in essence, reversed the age-related decline in processing (Clark, Lanphear, & Riddick, 1987). This research suggested that central mechanisms (e.g., information-processing strategies), rather than neuromuscular factors, were most influenced by the practice. This conclusion is similar to Welford's remarks that most of the decrement with aging can be attributed to central mechanisms. Unfortunately, studies that have investigated the influence of specific exercise training (i.e., aerobic, strength) on reaction time and speed of movement among the elderly have been inconclusive (Panton, Graves, & Pollock, 1990).

Balance Associated with the age-related decline in neuromuscular capacity is the loss of balance (stability). Any movement requiring balance involves the successful integration of several anatomical, muscular, and neurological functions. The deterioration of any of these functions may impede the ability to maintain equilibrium. For example, loss of brain cells (primarily in the cerebellum and brain stem) with advancing age may hinder the capacity to use proprioceptive information. That is, the kinesthetic information received concerning the position of the body's various parts is less accurate, thus inhibiting precise postural control. From approximately 40 years of age there is a gradual deterioration of vestibular nerve cells. Sensory cell loss may begin as early as the fifth decade in the semicircular canals and be quite marked in the saccule and utricle, as well, after the age of 70.

Visual perception also plays a significant role in maintaining stability, especially when moving. The decrement in visual acuity for most people begins around the age of 40 and by 60 years the ability to focus on objects at close distance is greatly affected. A loss in the ability to judge depth is also particularly evident after age 60. Although performance (balance) and symptomatic indications of vestibular malfunction (vertigo, dizziness) in older adults are evident, the specific underlying mechanisms responsible for these responses have not been clearly defined. It is likely, however, that a number of causes may be involved with balance loss in older adults; they include weak muscles, limited range of motion, abnormal reflexes, visual/vestibular deficits to central sensory integration, deficiencies in motor programming, and motor control difficulties (Manchester et al., 1989).

A significant loss in the ability to establish and maintain balance can affect virtually all gross motor activities. However, the research technique most frequently used to study changes in balance is the measurement of postural sway while performing various balancing tasks in the upright position. Investigations into this aspect of postural control have spanned 3 to 96 years of age. In general, research findings suggest that the magnitude of the sway tends to be greater (i.e., less stability) in the very young and the very old (Hasselkus & Shambes, 1975). More specifically, after the approximate age of 31, postural sway appears to increase with advancing age (Era & Heikkinen, 1985; Overstall et al., 1977).

A common characteristic noted among the young child and older person is the tendency to utilize a wider base of support to enhance efforts to gain and maintain stability. The literature regarding sex differences in postural sway is mixed; that is, there are indications that sway is greater in females after the age of 60, whereas other reports suggest greater sway in males and no sex difference for older persons.

Falling is one of the major health concerns of older adults that has been linked in part to the decrement in balance. Research has shown that falls are the cause of more than two-thirds of the accidental deaths in persons over 75 years of age (Azar & Lawton, 1964). With advancing age, the ability to recover from instability is slower and less efficient than in younger years (Woolacott, Shumway-Cook, & Nasher, 1982). Overstall and colleagues found in their study of individuals aged 60 to 96 years, that the incidence of falls increases with advancing age. It was also noted that females incur about twice as many falls as males in late adulthood. Factors other than lack of balance that have been mentioned as contributors to falling include decreased levels of physical activity, stroke, arthritis, low blood pressure, muscle weakness, poor eyesight, ineffective walking pattern (especially, reduced foot raise), and adverse effects of medication (Campbell, Borrie, & Spears, 1989). Compared to males, females tend to inherit a much greater possibility of hip fracture as a result of falling due to structural and bone characteristics. The angle of insertion of the head of the femur into the hip socket creates a structural weakness that is more conducive to hip fracture. Females also lose considerably more bone mineral mass and have a much greater incidence of osteoporosis in late adulthood. Kreutzfeldt, Haim, and Bach (1984) report that the annual incidence of hip fracture per 1,000 individuals of similar age (over 80 years) is 8.4 for males and 19.8 for females.

Somewhat of a paradox exists in regard to the difference between males and females in the frequency of falls. Information presented earlier in this text suggested that females have the structural advantage (lower center of gravity and wider base relative to height) in maintaining stability. One example of this advantage is evidenced by superior scores in balance beam walking. One could speculate rather safely that, at worst, females should be no more prone to falling than males. According to Eckert (1987), the fact that females have potentially greater stability but less balance raises the possibility of a sex difference in balance-related sensory input and/or environmental influences. The research literature provides little support for a significant sex difference in kinesthetic mechanisms (including vestibular functions) or visual processes. In fact, what information is available appears to favor females for balancing mechanisms. Although speculative, it seems plausible that though the female physique is more suited to balancing tasks, practice and environmental influences are also significant factors in optimal balance development. To illustrate, females generally perform better than males on balancing tasks during childhood—a time when the anatomical structure of females (a higher center of gravity and longer legs) should not provide an advantage. Based on this hypothesis, a possible explanation for the sex difference in frequency of falls in old age could be that higher participation levels by males in physical activities are conducive to maintaining efficient stability.

Included in the Rikli and Busch (1986) study of active and inactive young and old females was a very difficult balancing task, the one-foot stand for time (60-second limit). The active older group performed significantly better than inactive older persons, indicating that physical activity level may influence balancing ability. It was also noted that both young groups (active and inactive) performed the task equally well and did considerably better than the active older group, suggesting that a healthy, active lifestyle does not prevent the age-related decrement in balancing performance but may change the rate of decline.

Flexibility   One of the most obvious changes associated with advancing age is the loss of joint mobility (range of motion). Closely linked to this general decline are changes in the joints and muscles, and the presence of degenerative joint diseases. A smooth functioning of the joints is made possible by the strength and elasticity of the tendons and ligaments (forms of connective tissue encompassing the joint) and the synovial fluid. After the early adult years or later (for some individuals), the joints become less stable and mobility diminishes. In addition, collagen fibers and synovial membranes degrade, and the viscosity of the synovial fluid decreases. As a result of these decrements, the joints become stiff, which may decrease the range of motion (flexibility). However, there is no conclusive evidence that biological aging inherently causes decreased flexibility. A considerable amount of research also links the decline in range of motion with degenerative joint diseases such as osteoarthritis and osteoarthrosis. There are indications that more than two-thirds of individuals 55 to 64 years of age have signs of osteoarthrosis in some joint.

The inherent difficulty with using a general performance curve to describe changes in flexibility across the life span is that activity level is a better indicator of flexibility than age. Though exceptions have been reported, most research suggests that general range of motion increases steadily through childhood and that flexibility starts to decline around 10 years of age in males and 12 years of age in females.

Again, a strong influencing factor is individual activity level, which may vary considerably within any age group. The available information suggests, however, that after the onset of puberty (early teens), flexibility begins to decline, more so in males than in females. This trend is supported in part by evidence indicating that as body surface area increases during adolescence, flexibility decreases, and the possibility that the older population generally participates less in activities demanding a wide range of motion.

The three factors that appear to have the greatest influence on flexibility during adulthood are physical activity level, the inherent effects of aging, and degenerative joint diseases. The effects of aging on joint and muscle tissue have been verified, as well as the presence of joint diseases in the majority of elderly persons. Results of the Rikli and Busch (1986) investigation of active and inactive young and old females demonstrated clearly that physical activity level was a strong factor in maintaining flexibility in old age. In this study, the old active females performed significantly better on a sit-and-reach (trunk flexibility) test than their inactive counterparts of the same age group. In fact, the older active group demonstrated slightly greater flexibility than the young inactive group.

Changes in Skeletal Tissue and Body Fat   It is typical for aging humans to increase in body weight and fat and to show a decrease in height and bone mass.

The average female loses a striking 30% of bone mass between the approximate ages of 35 and 70. Bone mass loss among males generally does not begin until about age 55 and reaches a smaller total loss of between 10% and 15% by age 70. With advancing age also comes a slight decrease in height, due primarily to bone mass loss, increased kyphosis (rounding of the back), compression of intravertebral disks, and deterioration of vertebrae.

For the average person, gains in body weight and fat begin gradually in the late twenties and continue to increase until approximately the age of 55 or 60, when they show signs of decreasing. The typical increase in body weight and fat from age 20 to 60 years for both sexes is approximately 15%. This value is highly variable and may be considerably less among highly active, diet-conscious individuals. Lean body mass also tends to decrease with age. This overall effect is due primarily to bone mass loss and the reduction of muscle mass. As noted earlier, in most individuals there is a marked deterioration in muscle mass with aging.

The effects of skeletal and body composition changes on motor performance are rather generic; that is, rather than affecting the performance of a specific type (group) of motor activities, these changes influence the overall ability to perform a wide scope of motor skills. For example, bone mass loss results in bone with less density and tensile strength. Therefore, it is generally weaker and more prone to injury and fracture. Due to the increased likelihood of injury, many older persons limit their physical exertion levels and modify movement patterns for fear of physical harm. The presence of kyphosis (rounding of the back) may also significantly alter basic and advanced movement patterns (e.g., swimming, running, throwing).

An increase in adult body fat of 10% to 15% may have implications for physiological functions and motor performance. In general, a significant increase in body fat impairs cardiac function by increasing the mechanical work that the heart must perform, thus affecting aerobic and muscular endurance activity. Numerous research studies involving young and older adults have found percent body fat to be inversely (negatively) related to the ability to move the total body weight. This is especially true for motor performance tasks requiring horizontal acceleration (e.g., running) or vertical lifting of the total body (e.g., jumping). Theoretically this relationship is based on the fact that body fat adds to the total mass of the body without contributing to the force-producing capability. Therefore, the additional body fat gained during adulthood becomes excess weight to be moved during locomotor activity.

Changing Movement Patterns

■ *Objective 11.6*

Walking   Although the information on age-related changes in movement patterns during the adult years is quite limited, a number of studies have focused on walking—the most utilized of any movement pattern. In general, a deterioration in walking gait occurs with advancing age beginning with late adulthood (over 60 years). The stereotypic gait of an elderly person is normally characterized by slow, short, shuffling steps in which there is little range of motion, and a slumping of the head, shoulders, and trunk. Differences in performance among individuals of similar age can be considerable, however.

Murray and associates have conducted several investigations on gait patterns in individuals 20 to 65 years of age (e.g., Murray, Kory, & Sepic, 1970). In summary, when older (ages 60 to 65) males were compared to younger persons (ages 20 to 25), the older males displayed *shorter steps* (i.e., linear distance traveled of alternate feet), *shorter stride lengths* (i.e., linear distance traveled of same foot), *increased out-toeing,* and *less pelvic rotation and ankle extension.* Out-toeing is a technique used to improve lateral stability. Similar results were also found among females. When only "able-bodied" older persons were used in the comparison with younger individuals fewer differences were noted—namely, shorter steps, a shorter stride length, and increased out-toeing.

Other significant changes that have been noted with advancing age are *step height* (foot raise) and *walking speed* (Aniansson, 1980). Compared to the adult norm, the ability to raise each foot during walking (step height) decreases in individuals older than 70. As one might expect after reviewing the information on falling frequency, older females in particular have difficulty with step height.

By what standard should normal walking speed be judged? One suggestion of particular relevance to older adults is that the minimum velocity needed to cross a traffic intersection be used as one criterion for measuring "normal" speed. Based on this criterion, Aniansson determined that the normal walking speed of the general adult population is about 1.4 meters per second. However, the average normal, functional walking speed of adults 70 and older was found to be much slower than the criterion speed; the average speed was 1.2 meters per second for males and 1.1 meters per second for females.

Another consideration among older adults in relation to gait pattern is the psychological factor. It may be speculated that many elderly persons shuffle cautiously using short steps in fear of falling. Evidence for this speculation is supported to some degree by the findings of a study reported by Willmott (1986). The researcher found that among elderly hospital patients, gait speed and step length were significantly greater on a carpeted surface compared to a vinyl surface. Much more confidence and very little fear was expressed about walking on the carpeted surface in comparison to the vinyl.

In an excellent review of research on walking among the elderly, Craik (1989) notes that several issues remain unclear and need to be addressed before an accurate description of the effect of aging on walking behavior can be provided. Although the finding is rather consistent that velocity slows with advancing age, more research is needed to address the specific causes of the decline.

Running and Jumping   There are some indications that basic running and jumping patterns also change with aging. Some older persons participate in competitive sporting events (e.g., Senior Olympics, Senior Games, marathons, triathlons) and engage in a regular routine of fitness-oriented activities, but most older individuals do not include running and jumping or other basic sport skills (e.g., throwing, catching, kicking) in their lives. In an unpublished study by Adrian (cited in Adrian & Cooper, 1989), the kinematics (movement pattern characteristics) of healthy, active females aged 60 to 80 were identical to university track team females during a paced run. However, when the sprint patterns were compared, there was little similarity. More specifically, the younger females exhibited greater leg flexion and extension. Other information suggests that older runners do not tuck (flex) their "recovery" leg as completely, display a shorter stride, and take a greater number of strides to complete the distance (Nelson, 1981).

Associated with the general age-related psychomotor slowing and decrements in anaerobic function and strength is a decline in running speed. Verification of this trend was reported by Nelson (1981) after comparing young (20-year-olds) and older (ages 58 to 80) females on jogging and running speed. The older females were significantly slower at both the jogging pace (1.85 meters per second versus 3.93 meters per second) and at maximum running speed (2.60 meters per second versus 6.69 meters per second). Other evidence of an age-related decline (among trained individuals) can be found in the results of the U.S. National Senior Olympics (1989). Without exception, there was a decrease in speed with advancing age in the 100-meter dash in 12 age-group categories (six male and six female) for participants aged 55 to over 80 years.

In regard to jumping very little has been studied among the adult population and, in particular, the elderly. The kinematics of the vertical jump of college-aged and elderly females (over age 60) have been compared (Klinger et al., 1980). Results suggested that, in general, the patterns of both groups were quite similar. However, the older persons displayed less knee flexion and could not extend their legs (at the knee joint) as quickly as the younger individuals.

Throwing and Jumping   Also included in the Klinger et al. study was a comparison of several throwing and striking patterns (tennis backhand, batting). In each case, the extension velocities achieved by the older individuals were slower than those of the younger group. It should be noted, however, that in many instances the older individuals had no athletic background or had gone years without practicing any of the skills. Those who were identified as more active displayed a more sophisticated movement pattern (e.g., better coordination, moved faster, greater range of motion) than those in their age group who were less active.

In a more recent series of studies by Williams, Haywood, and VanSant (1990, 1991), similar findings were reported for the overarm throw in the elderly (ages 63 to 78 years). That is, ball velocities approximated speeds usually generated by 8- to 9-year-olds, which was accompanied by a relatively shorter backswing (i.e., less range of motion). In addition, those individuals who participated in throwing-oriented sports in their past exhibited better movements. As one might expect, men generally had better form and threw faster than the females.

Some generalizations may be made concerning the changing of movement patterns in adulthood. In general, the average elderly person shows a definite slowing of movements and less flexion and extension compared to his or her younger counterpart. However, among active older persons the decrement in speed is not as great and the characteristics of the movement pattern are generally well maintained from younger years.

Figure 11.9
Jim Law, "America's
fastest man age 65+"
Courtesy of Helen Harris
Associates.

Numerous assertions have been presented in this text in regard to the potential benefits of physical activity to growth, development, and motor performance. Discussions have affirmed that physical activity and exercise can have positive effects on bone and muscle tissue, body composition, and cardiorespiratory development. Performance characteristics in the form of strength, aerobic and anaerobic function, flexibility, coordination, balance, speed, and numerous motor skills have all been shown to progress through the psychomotor mode. Several assertions were also made in this chapter concerning the contribution of a healthy, physically active lifestyle toward delaying and modifying the effects of aging on various physiological functions and motor performance. In essence, physical activity was suggested as an important factor in the association between physiological age and chronological age.

PHYSICAL ACTIVITY AND LONGEVITY

■ *Objective 11.7*

Although the effects of physical activity on motor development are generally positive and in some instances dramatic, what about its influence on **longevity** (length of life)? Over the last 25 years or so, a vast amount of research has been conducted and reported to the public concerning factors associated with human health and life expectancy (e.g., smoking, high cholesterol, coronary heart disease, high blood pressure, nutrition, exercise). More recently, what appears to be some very convincing evidence has been reported by several studies asserting that regular (habitual) physical activity and physical fitness of a *moderate* level is associated with increased longevity in comparison to sedentary and less active lifestyles. One of the most significant and common findings from this research is that physical activity is inversely associated with the rate of certain diseases and early mortality.

One of the most comprehensive investigations collected data on 16,936 Harvard alumni, aged 35 to 74 years, with 12 to 16 years of follow-up (Paffenbarger et al., 1986). The physical activity level for each subject was measured as the total amount of energy expended (in kilocalories) at physical activity (e.g., sports, yard work, walking) per week. Results indicated that physical activity level is inversely related to total mortality. That is, death rates declined steadily as physical activity (energy expenditure) increased. This trend continued with the increase in expenditure from less than 500 kilocalories to 3,500 kilocalories per week; beyond that amount the rates increased slightly. In males, mortality rates were one-fourth to one-third lower among those individuals expending at least 2,000 kilocalories per week, compared to less active males.

Perhaps most important, the researchers further concluded that regardless of whether the individual smoked, had high blood pressure, experienced extreme gains or losses in body weight, or their parents had died early, mortality rates were considerably lower among the more physically active. This general finding was confirmed by Powell (1987) after a critique of 43 studies. The researcher concluded that lack of regular physical activity contributes to heart disease in a cause-and-effect relationship, with sedentary individuals almost twice as likely to develop heart disease as the most active persons. Other studies using general adult populations have also reported an inverse relationship between coronary heart disease, death, and regular physical activity (e.g., Kannel, 1967; Leon et al., 1987; Morris et al., 1980). In addition to the inverse trends (of heart disease and mortality) with leisure-time habitual physical activity, similar results have also been associated with occupational activity.

All of the aforementioned studies used physical activity as determined by various units of measurement (e.g., kilocalories expended, hours participated) as the factor with which to compare rates of disease and death. An alternative approach that is perhaps more specific and objective is the use of physical fitness level. Most studies identify the level and type of physical activity, but few investigations provide the additional insight of comparing the physiological (physical fitness) state of the body to rate of disease and death.

One of the landmark investigations using this approach was conducted by Blair and associates (1989) and reported in the *Journal of the American Medical*

Association. The study consisted of more than 13,000 males and females, ranging from 20 to over 60 years of age, who were tested for aerobic fitness, then assigned to one of five physical fitness groups (low-fit to high-fit) based on age, sex, and fitness test results. Results indicated a strong, graded, and consistent inverse relationship between physical fitness and mortality due to all causes (i.e., smoking, cholesterol level, parent history, cardiovascular disease, and cancer). When comparing least-fit to most-fit groups, the chances of a low-fit individual experiencing an early death were 3.44 times greater for males and 4.65 times more likely among females. However, one of the most important conclusions of the study was that the fitness standard associated with increased longevity may be attained by individuals who engage in regular *moderate* exercise that approximates a brisk walk of about 30 minutes several times a week.

Two disturbing statistical figures associated with these results estimate that approximately 30% of adults are quite sedentary (Caspersen, Christenson, & Pollard, 1986), and with increasing age there is a progressive and predictable decline in exercise participation (Brooks, 1987).

In summary, it appears from the data that regular physical activity of a moderate level of intensity can "prolong" life. Longevity, however, is more associated with the prevention of early mortality then extending one's potential life span. McArdle et al. (1991) conclude that "while the maximum life span may not be extended greatly, more active people tend to survive to that "ripe old age" (p. 71). Refer to Blair (1993) in the Suggested Readings for an excellent review of the healthful effects of physical activity.

SUMMARY

Motor behavior during the adult years is characterized by two distinguishing milestones: peak performance (associated with biological maturity) and regression. Peak physiological function, for the most part, occurs between the approximate ages of 25 to 30 years of age, with females tending to mature at the lower end (22 to 25 years) and males at the upper end (28 to 30 years) of the range. This trend is evident in three of the most influencing factors in motor performance: muscular strength, cardiorespiratory efficiency, and processing speed. Closely associated with peak physiological function is maximal motor performance in several activities.

A basic fact of development during the adult years is the difference between chronological and physiological age. However, it appears that no matter how well people take care of themselves, some degree of regression in physiological processes and associated motor performance is inevitable. After peak maturity at approximately 30 years of age, most physiological functions begin to show decrements at a general rate of 0.75% to 1% per year. The decline is characterized by decrements in cardiorespiratory function, muscular strength, neural function, balance, and flexibility. Other major forms of biological regression include bone tissue loss (especially for females), an increase in body fat, and a decrease in muscle mass. There is some suggestion that much of the loss in physiological aging may be attributed to inactivity and poor health habits.

During late adulthood a deterioration in specific movement patterns may also occur. Although wide differences in walking, running, jumping, throwing, and striking performance exist, in general, elderly persons exhibit a definite slowing of movements with less flexion and extension compared to younger persons.

Physical activity, especially habitual participation, has been shown to have positive effects on several aspects of growth, development, and motor performance. In recent years, several research studies have verified that habitual health-related physical activity of moderate intensity is positively associated with a prolonged life span in comparison to sedentary and less active lifestyles.

SUGGESTED READINGS

Blair, S. N. (1993). 1993 C. H. McClay Research Lecture: Physical activity, physical fitness, and health. *Research Quarterly for Exercise and Sport, 64* (4), 365–376.

Brooks, G. A., & Fahey, T. D. (1985). *Exercise physiology.* New York: Macmillan.

Craik, R. (1989). Changes in locomotion in the aging adult. In M. H. Woollacott & A. Shumway-Cook (Eds.), *Development of posture and gait across the life span.* Columbia, SC: University of South Carolina Press.

deVries, H. A., & Housh, T. J. (1994). *Physiology of exercise* (5th ed.). Madison, WI: Brown & Benchmark.

McArdle, W. D., Katch, F. I., & Katch, V. L. (1991). *Exercise physiology* (3rd ed.). Philadelphia: Lea & Febiger.

Ostrow, A. C. (Ed.). (1989). *Aging and motor behavior.* Indianapolis, IN: Benchmark Press.

Smith, E. L., & Serfass, R. C. (Eds.). (1981). *Exercise and aging: The scientific basis.* Hillside, NJ: Enslow.

Spirduso, W. W., & Eckert, H. M. (Eds.). (1989). *The academy papers: Physical activity and aging, 22.* Champaign, IL: Human Kinetics.

Whitbourne, S. K. (1985). *The aging body.* New York: Springer-Verlag.

PART ◦ *five* ------------------------------

--------➤ *chapter*

12 Assessment

Part Five consists of a single chapter designed to provide the reader with insight into the concepts and techniques associated with motor assessment. Along with a discussion of the

Assessment

purposes and considerations for selecting and implementing a wide variety of instruments, this chapter features examples of the most widely used instruments available.

12

Assessment

KEY TERMS

assessment
measurement
evaluation
norm-referenced standards

criterion-referenced standards
product-oriented assessment
process-oriented assessment

Table 12.1 Comprehensive Physical-Motor Assessment

Biological Growth	Development (level of functioning)	Motor Behavior (performance)
Body mass	Cardiorespiratory	Reflex behavior
Height	Muscular strength/endurance	Spontaneous movement behavior
Body weight	Flexibility	Rudimentary behavior
Physique (body build)	Coordination	Fundamental movement behavior
Posture	Speed	Sport skill behavior
Anthropometric measures	Agility	
Circumference	Power	
Breadth/Length	Visual	
Skeletal maturity (age)	Balance	
Body composition	Reaction time	
Body fat	Kinesthetic	
Lean body mass	Temporal (Rhythm) timing	
Secondary sex characteristics	Menarche	

The operational definition of motor development used in this text suggests that it is the study of the process of change in the growth, development, and motor behavior (performance) across the life span. Assessment, in general, provides the opportunity to observe, document, and interpret this change, as well as to determine the growth and developmental status of the individual at that particular time in the life span. Motor assessment can be as diverse as the field of motor development itself. Table 12.1 presents an example of items that may be used in a comprehensive physical-motor assessment, grouped according to each of the different perspectives of growth, development, and performance from which motor development may be observed and assessed. An aspect that could be added to the assessment of performance is the evaluation of either the process (qualitative) or product (quantitative) characteristics of the movement action. These perspectives were described in chapter 9 under the discussion of fundamental movement skills.

■ *Objective 12.1*

An example of the dynamic nature and potential for assessment through motor performance may be illustrated using the skill of running. From a development, or level of functioning, perspective, running may be used as an indicator of cardiorespiratory function, speed, agility, general coordination, and reaction time. From the viewpoint of motor performance, running may be judged either by its process characteristics, or movement pattern, or by the product that is produced (e.g., performance times in the 50-yard dash or the marathon).

Numerous component-specific and task-specific tests are available for assessing the physical and motor behavior characteristics of individuals at virtually all ages, but unfortunately no assessment batteries include norms reflecting the lifelong perspective. Most batteries were developed to assess a diversity of age-span characteristics (e.g., motor, mental, social) displayed during the primary growth and development years of infancy through childhood.

This chapter provides the reader with an understanding of the basic terms, purposes, and considerations for motor assessment. Also presented is a description and review of selected popular assessment instruments. For additional information refer to the Suggested Readings at the end of this chapter.

BASIC TERMINOLOGY OF ASSESSMENT

■ *Objective 12.2*

Before the student of motor development can grasp a functional understanding of motor assessment in regard to the purpose, selection, and administration of assessment instruments, a review of basic terminology is warranted.

Assessment is a process that involves both measurement and evaluation. **Measurement** refers to the collection of information on which a decision is based. In measurement, various instruments and techniques are used to collect data. A *test* is a specific type of measurement. These instruments may take the form of requested motor performance, observation of behavior, or questions asked on paper or in an interview setting. Height, body weight, and leg length are examples of measurements that are not tests.

Evaluation is the process of decision making regarding the value or worth of collected information. Support for such judgments is more commonly based on either norm-referenced or criterion-referenced standards. For cases in which such standards are not available or an overall judgment is needed for several measurements representing a diverse spectrum of traits, evaluation is frequently thought of as the process that qualitatively (subjectively) appraises the quantitative data. For example, the results of three groups of tests (e.g., physical growth, fundamental motor skill "process" characteristics, motor performance "product" results) may be combined to produce a single assessment score.

The two most widely used types of standards are criterion-referenced and norm-referenced standards. **Norm-referenced standards** involve the hierarchical ordering of individuals. Assessment instruments that use this form of evaluation are basically quantitative evaluations designed to compare an individual's characteristics with those of other persons of similar sex, age, and socioeconomic group. Quality norm-referenced tests are based on statistical samplings of hundreds or, ideally, thousands of individuals.

The most common descriptors used to communicate norm-referenced standards are *percentile norms*. This type of norm describes the percentage of a given group that can be expected to score above or below a given value. For example, 40 modified sit-ups in a 1-minute period for a boy 10 years of age is at the 75th percentile. This means that only 25% of the boys in this age group did more sit-ups, and about 75% did less. For purposes of practical interpretation, the 50th percentile mark (and above) is often used as the level of acceptability. Norm-referenced standards are particularly useful for making comparisons among individuals when the situation requires a degree of selectivity, such as in identifying excellence or extremely low status values. Most assessment batteries are based on norm-referenced standards.

Criterion-referenced standards are concerned with the degree to which an individual achieves a specified level of development, motor performance, or physical status. This type of standard has received considerable attention in recent years, as evidenced by its use with most national physical fitness assessment batteries. In this case, the value of the criterion standard represents the acceptable level of performance, which is usually based on normative data and expert judgment. For example, the criterion standard for sit-ups for a 10-year-old boy is 34, and the acceptable level of flexibility on the sit-and-reach test is 25 centimeters, or 2 centimeters beyond the toes. In contrast to norm-referenced standards, individuals are compared to the criterion value and not to others of the same sex and age. Criterion-referenced standards are useful for setting performance standards for all students. However, a limitation of its use is that the tester knows only when individuals have met an acceptable performance level, not to what degree the person actually performed.

Some assessments are specific to motor behavior, such as fundamental movement skill performance; these are referred to as either product-oriented or process-oriented instruments. With the **product-oriented assessment,** the examiner is primarily interested in the performance outcome, or the product of the behavior. Product values are described as quantitative data such as the yards that a ball is thrown, running velocity in seconds, and feet and inches jumped.

In contrast, **process-oriented assessment** involves the measurement and evaluation of the characteristics of the process, or form, such as those used in fundamental movement patterns. These types of data are most often qualitative rather than quantitative. Although product-oriented data can be used with either norm- or criterion-referenced standards, they are more commonly associated with population norms. On the other hand, process-oriented assessment instruments are most often criterion-referenced; that is, individuals are compared to themselves in relation to established criteria.

PURPOSES OF ASSESSMENT

■ *Objective 12.3*

Assessment instruments are employed for many reasons and the applications of the information obtained are varied. Whether the setting be in schools, child care centers, university laboratories, or in the home, assessment provides vital services in the areas of teaching, learning, and science. Although most people associate measurement and evaluation with grading, such as is practiced in the schools, this is perhaps one of the least useful reasons to assess. Regardless of the setting and type of program, the most important factor is that assessment be conducted with a specific purpose. Some of the more widely accepted purposes for assessment are explained in the following discussion.

Diagnosis/screening. Some of the more frequently utilized assessment instruments are known as diagnostic/screening tools (e.g., Apgar scale, Denver Developmental Screening Test). Assessments such as these are designed to provide a means of differentiating normal-functioning individuals (usually newborns to middle childhood) from those who may not be developing normally. If weaknesses are detected, this information is then used to determine whether the individual should be referred for additional testing or remedial work. Basic screening procedures at the beginning of any physical education or motor development program are an excellent means of identifying individuals who may have special needs.

Determine status. Perhaps the most commonly stated purpose of assessment is to determine the status, progress, or achievement of individuals. This information may be used for a multitude of reasons related to more specific purposes. One of the indispensable practices of any instructional program is to determine whether objectives have been reached. Assessment is also essential for providing feedback on the individual's physical status or level of performance.

Placement. The results of assessment instruments are frequently used to place individuals in classes or groups according to their motor abilities. A rather common practice within instructional programs is to arrange individuals into homogeneous groups. In several cases this practice enhances the teaching or learning process (e.g., swimming, gymnastics, fitness classes).

Program content. After student behaviors and characteristics have been assessed, results can be used to plan program content and guide individual progress. Assessment information is a considerable aid in writing student and program objectives that are reasonable and challenging.

Program evaluation. Program directors and administrators frequently use assessment results to evaluate the program and thereby determine the possible need for change. One of the more common changes that occurs is a modification of program standards.

Construction of norms and performance standards. Measurement, especially among large populations, provides data that can be used in establishing norms and performance standards. Group norms enable comparisons to be made with other groups, such as those based on a nationwide sample. When criterion standards for specific performance tasks are not available, local norms (or a combination of local and national) can be used to establish them. Norms and performance standards are instrumental in providing motivation and interest within a program.

Research. Measurement and evaluation that are conducted scientifically can add significantly to the body of motor development research and provide information upon which educational programs can improve. Research is an important responsibility of any outstanding program.

Prediction. Theoretically some aspects of assessment may be used to a limited degree to predict future performance. That is, specific information may be useful to directors and individuals in selecting activities they are most likely to master. For example, an individual with a high maximal oxygen consumption may decide to participate in triathlons or other endurance events.

Motivation of individuals. Assessment instruments can be used as motivational tools. Individuals who know how well they are performing and how they can improve are likely to be better motivated than those who receive no feedback.

After the purposes for assessment have been established, additional considerations are warranted, especially in regard to instrument selection. Along with deciding what physical and motor characteristics are to be measured (which usually accommodate program objectives), a review of available instruments should be conducted. In addition to including the desired assessment items, only instruments that have an acceptable level of validity, reliability, and objectivity should be selected.

Validity refers to the degree to which an assessment item measures that which it is intended to measure. For example, if the purpose of the assessment is to measure aerobic capacity, experts have determined that a 1-mile run or 12-minute run is a valid indicator of this characteristic, while a half-mile run is not.

Reliability is a measure of the repeatability of assessment results. An individual's scores should not differ markedly on repeated administrations of the same assessment. Factors related to repeatability are consistency (dependability) of the instrument that is used to measure the individual's status (e.g., skinfold calipers, grip dynamometers), the examiner's instructions, and the subject's effort. One should also be aware that a test can be reliable without being valid, whereas a valid test must also be reliable.

The *objectivity* of a test refers to the degree of agreement among examiners. That is, a test that is completely objective is one that will be scored identically by different examiners. The level of acceptability of these basic test characteristics can vary widely, even among the most popular assessment instruments. The statistical term related to the comparison of test characteristics and repeated occurrences is known as a correlation; a correlation coefficient of .80 or above is generally deemed quite acceptable.

Additional considerations for the selection of an assessment instrument include the following:

1. Is the instrument norm-referenced or criterion-referenced? Care should be taken to determine if the sampled population is representative of the population of individuals in the program.
2. Is the instrument feasible to administer? This consideration may involve several things, including time, cost, equipment, space requirements, and personnel needed. Since time is often a perpetual problem facing the administrators of programs that have both large and small numbers of participants, instruments should be evaluated in terms of time economy.

Another basic consideration is available funds. Some assessment instruments cost hundreds of dollars and require expensive equipment, whereas others may accomplish the basic objectives with only minor compromises at a much lower price. For example, the best measurement of aerobic capacity is maximal oxygen consumption. However, thousands of dollars worth of equipment is needed to carry it out and each test requires about 45 minutes to administer. Less accurate but acceptable and more feasible measures (field tests) are the 1-mile

CONSIDER-ATIONS FOR PROPER ASSESSMENT

■ *Objective 12.4*

walk/run and 12-minute walk/run (using norm- or criterion-referenced standards). In general, however, the cost factor should consider accuracy, durability, and long-term use.

When reviewing assessment instruments it will also become evident that varying levels of training and expertise are required to administer the items and interpret the data. Thus, additional training and practice may be needed. In the interest of time economy, more than one examiner is usually imperative. This may call for considerable planning in order to provide reliable and objective assessment.

For a more thorough treatment of test development as a science, refer to *Psychological Testing* (Kaplan & Saccuzzo, 1993) and *Standards for Educational and Psychological Tests* published by the American Psychological Association (APA, 1985). Both are excellent sources of information and guidance.

ASSESSMENT INSTRUMENTS

The possibilities of assessing the various aspects of physical growth, development, and motor behavior are numerous (table 12.1). In the areas of physical growth and physiological assessment alone, there are hundreds of instruments. Many of these have been mentioned or described throughout the text (e.g., skinfolds, anthropometric measures, various physiological tests). The focus of this discussion will be on some of the more widely used product-oriented motor assessment batteries and process-oriented instruments. An *assessment battery* refers to a group of several tests standardized on the same population.

Motor behavior assessment items have been included in general child development batteries, along with mental, language, and social entries, since the 1920s (Gesell, 1925). Since that time numerous instruments have been developed. However, while the quantity of instruments seems sufficient, it is the quality that has been suspect. Most motor behavior items have minimal if any rationale for development and are based on generalized product (quantitative) characteristics (i.e., how far, how fast, how many, etc). Other global assessment items merely ask if the child can throw a ball, kick a ball forward, stack cubes, or balance on one foot. Although some of these types of assessments may have an acceptable level of validity and reliability, they are quite limited from a motor development perspective, especially in describing the process of change. In recent years greater focus has centered on the qualitative characteristics of motor behavior as evidenced by the development of process-oriented assessment instruments.

Only a selected number of assessment instruments are described on the following pages. The intent of this discussion is to provide the student of motor development with a glimpse into the diversity of assessment items and formats available. For additional information on specific instruments, a list of recommended sources is provided in the latter part of this chapter. Refer to Suggested Readings for in-depth descriptions of these and other physical and motor assessment tests.

Table 12.2 The APGAR Scale

	0	1	2
Appearance (skin color)	White or blue	Limbs blue, body pink	Pink
Pulse (rate)	No pulse	100 beats/min	More than 100 beats/min
Grimace (reflexive grimace initiated by stimulating the plantar surface of the foot)	No response	Facial grimaces, slight body movement	Facial grimaces, extensive body movement
Activity (muscle tone)	No movement, muscles flaccid	Limbs partially flexed, little movement, poor muscle tone	Active movement, good muscle tone
Respiratory effort (amount of respiratory activity)	No respiration	Slow, irregular respiration	Good, regular respiration, strong cry

From V. A. Apgar, "A Proposal for a New Method of Evaluation of a Newborn Infant" in *Anesthesia and Analgesia: Current Research,* 32:260–267. Copyright © 1953 Williams & Wilkins, Baltimore, MD. Reprinted by permission.

The following examples represent standardized instruments that have received considerable recognition for the assessment of newborns, infants, and school-age populations.

The APGAR Scale   *The APGAR scale* (Apgar, 1953) is used in delivery rooms all over the United States and in other countries. The scale is typically administered to newborns one minute after delivery and again after five minutes. APGAR is an acronym that stands for appearance (skin color), pulse, grimace (reflex irritability), activity (muscle tone), and respiratory effort. Each of these characteristics is rated on a scale of 0 to 2 in which 2 denotes normal function, 1 denotes reduced function, and 0 denotes seriously impaired function (table 12.2). Thus, a total score of 10 is perfect.

Newborn Assessments

■ *Objective 12.5*

Approximately 90 percent of all newborns score 7 or higher. A score of 4 or less is generally an indication that immediate medical assistance is needed. The most controversial assessment item on the scale is skin color. Few newborns receive a 2 at one minute, and of course, the measure is invalid with dark-skinned babies. Despite these limitations the Apgar scale is noted as a reliable and fairly accurate indicator of later neuromuscular difficulties.

Standardization of the instrument showed that newborns with the lowest scores had the highest mortality rates and that the device was useful in predicting infant death rate (Apgar & James, 1962). A high percentage of infants found to have neuromuscular or developmental problems by 1 year of age had low Apgar scores at birth (Drage et al., 1966).

The Gestational Age Assessment   The *Gestational Age Assessment (GAA)* (Ballard, Novak, & Driver, 1979) provides a simplified scoring system (from a medical perspective) for clinically determining maturation of newly born babies.

NEUROMUSCULAR MATURITY

NEUROMUSCULAR MATURITY SIGN	SCORE						RECORD SCORE HERE
	0	1	2	3	4	5	
POSTURE							
SQUARE WINDOW (WRIST)	90°	60°	45°	30°	0°		
ARM RECOIL	180°		100°-180°	90°-100°	<90°		
POPLITEAL ANGLE	180°	160°	130°	110°	90°	<90°	
SCARF SIGN							
HEEL TO EAR							
						TOTAL NEUROMUSCULAR MATURITY SCORE	

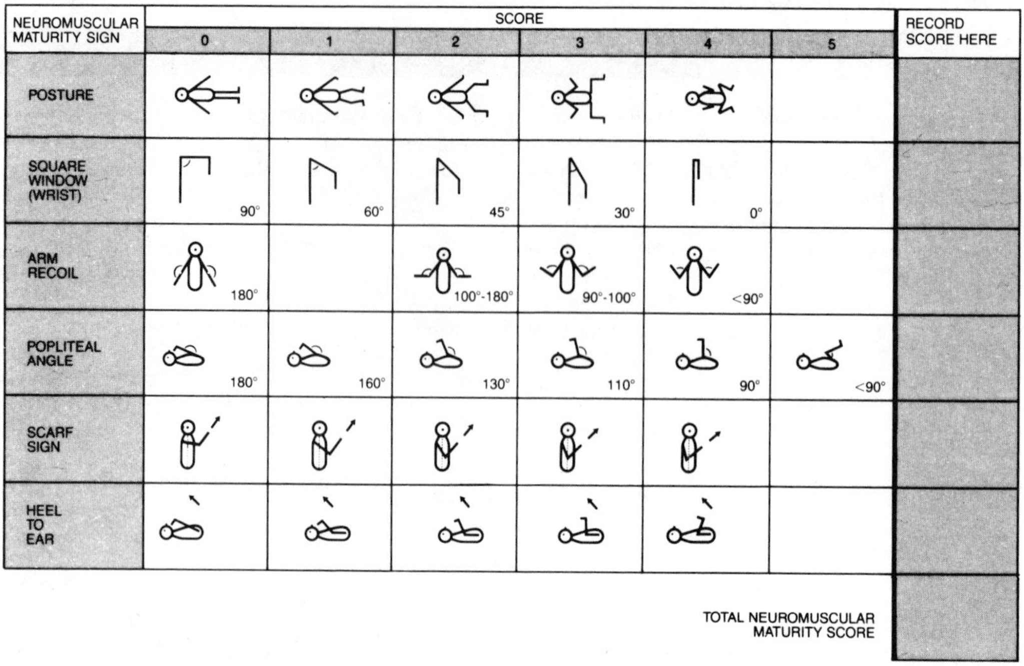

Figure 12.1
The neuromuscular maturity assessment portion of the GAA.

An overall newborn maturity rating is provided based on scores derived from the assessment of physical and neurological maturity. The physical maturity assessment consists of seven items rated according to physical appearance (maturity level) with a score of 0 (lowest) to 4 or 5. Assessment areas include skin, plantar creases, breast, ear, head, and genitals. The neurological assessment is somewhat unique in that it measures passive muscle action maneuvers rather than primitive newborn reflexes (fig. 12.1). Two rather consistent principles of motor development are evident in the criteria. One is that neurological change consists of a replacement of extensor muscle tone by flexor tone; the other is that change generally proceeds in a cephalocaudal direction.

Reliability data based on a comparison of the GAA with gestational age (as determined from dates and the results of a more extensive and established instrument) have proved to be quite acceptable (Ballard, Novak, & Driver, 1979). It is recommended that the assessment be administered between 30 and 42 hours after birth; this allows the newborn time to stabilize after delivery and is the time during which reliability is highest. A significant amount of supervised training is required to accurately administer the GAA.

Table 12.3   Reflex and Behavioral Items for the Brazelton Neonatal Behavioral Assessment Scale

	Behavioral Items	
Reflex Behaviors	*Specific*	*General*
1. Plantar grasp	1. Response decrement to light	1. Alertness
2. Hand grasp	2. Response decrement to rattle	2. General tonus
3. Anal colonus	3. Response decrement to bell	3. Motor maturity
4. Babinski	4. Response decrement to pinprick	4. Cuddliness
5. Standing	5. Inanimate visual orientation response	5. Consolability with
6. Stepping	(focusing and following an object)	intervention
7. Placing	6. Inanimate auditory orientation	6. Peak to excitement
8. Incurvation	response (reaction to an	7. Rapidity of buildup
9. Crawling	auditory stimulus)	8. Irritability
10. Glabulla	7. Animate visual orientation (reaction)	9. Activity (alert states)
11. Tonic deviation of head	8. Animate auditory orientation	10. Tremulousness
and eyes	(reaction to a voice)	11. Startle
12. Nystagmus	9. Animate visual and auditory	12. Lability of skin color
13. Tonic neck	orientation (reaction to a person's	13. Lability of states
14. Moro	face and voice)	14. Self-quieting activity
15. Rooting (intensity)	10. Pull-to-sit	15. Hand-to-mouth facility
16. Sucking (intensity)	11. Defensive movements	
17–20. Passive movements		
(right arm, left arm,		
right leg, left leg)		

Source: Data from Als, et al., 1979.

The Brazelton Behavioral Assessment Scale   Among the unique characteristics of the *Brazelton Behavioral Assessment Scale* (Brazelton, 1984) is that it provides an assessment two or three days after the baby is born, then another assessment on day 9 or 10, after the baby is home from the hospital and has had the opportunity to relax from the stress of delivery. The Brazelton was designed to measure three primary items: subtle central nervous system functions, temperamental differences, and the capacity for interaction. Typically the scale is administered to at-risk babies or those for whom there is a concern about development.

The instrument generates results in three areas. One consists of 20 reflex items designed to screen for neurological soundness; another is the neonate's "interactive behavioral repertoire," which is assessed with 26 items (table 12.3). The third area of assessment, which is not as closely linked to motor behavior, is a rating of "attractiveness" behavior (i.e., the baby's contribution to and readiness for interaction) and "need for stimulation" (i.e., the need for and use of stimulation as it relates to organizing responses).

In the reflex assessment, each reflex is scored 0 (no response), 1 (low), 2 (medium), or 3 (high level of response). Babies with three or more low scores are typically referred for additional neurological assessment.

The 26 behavioral items are considered the most important assessment area. These items may be grouped into four dimensions:

1. Interaction capacities: the capacity to attend to and process environmental events.

2. Motoric capacities: the ability to maintain muscle tone, control motor behavior, and accomplish integrated acts, such as placing the thumb in the mouth and sucking.

3. Organizational capacities in respect to state control: the ability to maintain a calm, alert state when facing increased stimulation from events such as a light, bell ring, or pinprick.

4. Organizational capacities in the physiologic response to stress: the capacity to master the physiologic demands of immaturity, such as the ability to inhibit startles and tremors while attending to stimuli.

Each item is scored on a continuum of 1 to 9. A score of 1 represents the lowest state, while the ideal state varies from item to item and lies somewhere between 1 and 9 (table 12.3). The baby's performance may also be rated on each item with a more simplistic "good," "adequate," or "deficient."

Although evidence of the Brazelton's validity has been verified (Als et al., 1979), its test-retest reliability and standardization have been questioned. The reliability problem is due primarily to the fact that the instrument assesses the neonate's current status, which can change quite rapidly. One of the problems plaguing standardization is that newborns from different cultures tend to behave differently (Brazelton et al., 1979). Two primary reasons that the Brazelton scale is not used extensively is that only highly trained examiners normally found at large medical centers are deemed competent to administer the instrument, and it is recommended that the instrument not be used with premature babies (thus excluding a large risk group). However, there is evidence that the instrument has utility as a research device and in helping identify high-risk infants (Wodrich, 1990).

Infant Assessment

The Gesell Schedules   Historically, the *Gesell Schedules* (Gesell, 1925) were the first attempts at measuring an infant's developmental status. The first version of the schedules consisted of four categories of assessment items: (a) motor behavior, including postural control, locomotion, prehension, drawing, and hand control; (b) language behavior; (c) adaptive behavior, which contains several tasks requiring motor abilities; and (d) personal and social behavior. Recent revision of the schedules left the original categories unchanged with the exception that within the motor behavior category, fine motor is distinguished from gross motor behavior. Although the instrument is designed to assess individuals from 1 to 36 months of age, it is known primarily as an infant assessment scale. Table 12.4 shows a sample of the adaptive and motor behavior items for 36 to 44 weeks.

Although standardization and interpretation problems limit the use of the Gesell (Wodrich, 1990), it has found support among pediatricians and in some medical settings.

Table 12.4   Revised Gesell Developmental Schedules

Name:		Age: *Key Age*	Date:	Case #:
36 Weeks		*40 Weeks*		*44 Weeks*

HO		HO	Adaptive	HO	
	Cup and cubes: hits cube against cup (*44w) Pellet: index finger approach Pellet and bottle: holds bottle, grasps pellet (*52w)		Cubes: matches 2 cubes (*56w) Pellet and bottle: points at pellet through glass (*15m) Bell: grasps by handle Bell: pokes at clapper Bell: waves Ring-string: sees connection, pulls string		Cup and cubes: cube into cup, no release (*48w) Cube and paper: uncovers cube (toy) Pellet and bottle: approaches pellet first Formboard: removes round block easily Formboard: bangs or releases round block near hole (*48w)

			Gross Motor		
	Sit: indefinitely steady Sit: to prone with ease Prone: creeps on hands and knees (*56w) Prone: to sitting Stand: pulls to feet at rail (*52w) Stand: lifts, replaces foot at rail (*52w)		Rail: cruises, using two hands (*44w) Rail: retrieves toy from floor (*52w) Stand: lets self down with control		Sit: pivots in sitting Walks: two hands held (*48w) Rail: cruises, using one hand (*52w)

			Fine Motor		
	Cube: radial digital grasp Pellet: scissors grasp (*40w)		Pellet: grasps promptly Pellet: inferior pincer grasp (*48w) String: scissors grasp (*48w)		Cube: crude release (*56w) String: plucks promptly Pegboard: removes large round eventually (*52w) Pegboard: removes small round eventually (*52w) Pegboard: removes large square eventually (*52w)

*Pattern replaced by more mature one at later age. H = History; O = Observation.
From H. Knoblock, F. Stevens, and A. Malone, *Manual of Developmental Diagnosis*, pp. 64–65, Developmental Evaluation Materials, 1980, Houston, TX. Reprinted by permission of Hilda Knoblock.

The Bayley Scales of Infant Development   Originally published in 1933, the *Bayley Scales of Infant Development* (Bayley, 1969, 1993) are regarded as having outstanding psychometric properties for infant assessment (e.g., Kaplan & Saccuzzo, 1993). The Bayley-II (1993) is a norm-referenced, standardized instrument designed for use with children from 1 to 42 months and consists of three sections: the Mental Scale, Motor Scale, and Behavior Rating Scale.

The Behavioral Rating Scale contains 30 ratings that are completed after the evaluation of the mental and motor items. The record is a systematic way of

Table 12.5   Bayley-II Motor Scale Items

Group	Motor Area (number of items)
Gross motor	Head control (10) Hands posture (2) Turning (3) Imitative behavior (2) Preambulatory skills (4) Body positioning (10) Sitting (8) Stairs (6) Balance (9) Walking (16) Jumping (5) Throwing (1)
Fine motor	Grasping (9) Drawing (4) Pencil grasp/Writing (6)
Ungrouped (e.g., retains ring, buttons one button, tactile discrimination)	(16)
Motor quality (behavioral rating scale)	(8)

summarizing such behavioral observations as attention/arousal, orientation/engagement, emotional regulation, and motor quality.

The Motor Scale (psychomotor development index) includes items related to the progression of postural control, gross motor abilities such as sitting, standing, walking, and stair climbing, as well as abilities involving fine motor control. In Bayley's words the Motor Scale was "designed to provide a measure of the degree of control of the body, coordination of the large muscles and finer manipulatory skills of the hands and fingers" (Bayley, 1969, p. 3). A list of items of the Motor Scale is shown in table 12.5.

As noted earlier, the Bayley scales are known as one of the best infant assessments available. They can make accurate, reliable, and valid measures of an infant's current ability status. Their use as a research instrument has also received wide support. The scales are relatively expensive and because they are somewhat difficult to administer, a significant amount of training is required. Thus the scales are confined mostly to specialists who work extensively with infants.

Assessment of School-Age Populations

The Denver Developmental Screening Test   The DDST (1990) is one of the most extensively used screening instruments for children from birth to 6 years of age. As a screening device, the instrument's purpose is to quickly identify those

Table 12.6 Sample Motor Items from the DDST

Area	Items
Gross motor	Rudimentary posture control: of head, neck, trunk; rolling over, sitting, standing with support, pulling to stand, cruising, standing alone, walking
	Fundamental manipulative behaviors: kicking, throwing, tricycle riding, catching
	Fundamental locomotor behaviors: jumping in place, long jumping, hopping
	Fundamental stability abilities: balancing on one foot, heel-toe walk
Fine motor/adaptive	Rudimentary manipulative behaviors: visual tracking, reaching, grasping and releasing, cube stacking, scribbling, copying, draw-a-person

individuals who are likely to have significant developmental delays. The instrument is not an IQ test, nor is it an in-depth assessment device. The latest version (Denver II) consists of 125 items that are divided into four assessment areas:

1. Personal/social. Assesses the child's ability to get along with people and to provide personal care. Sample items: smiles spontaneously at 3 months, drinks from cup at 11 ½ months, and dresses without supervision by 4 years.
2. Language. Assesses the ability to hear, understand, and use language. Sample items: laughs at 2 months, combines two different words at 20 months, and gives first and last name by 2 ½ years.
3. Fine motor/adaptive. Assesses the child's ability to use vision and the hands to pick up objects and draw.
4. Gross motor. Assesses the ability to perform rudimentary and fundamental gross motor skills.

Table 12.6 shows a sample of fine motor/adaptive and gross motor items from the DDST.

A rather unique and interesting characteristic of the DDST form is the bar graph for each item that represents the age at which the standardization population (norm group) performed the specific task (fig. 12.2). As shown, the bar indicates the ages at which 25%, 50%, 75%, and 90% of the norm group performed the task. Consistent with its stated purpose, the DDST looks for delays rather than producing a numerical score. A developmental delay is described as any item failed that is completely to the left of the age line, indicating that the child failed an item passed by 90% of the norm group.

Research concerning the DDST's reliability and validity has been generally supportive, but the fact that the norm group was limited to children from Denver, Colorado, has been suggested as a limitation. The concern is that the DDST norms may not reflect the racial and socioeconomic levels of the country. However, the DDST is generally recognized as a valuable screening instrument capable of providing a great deal more insight than clinical observation or global descriptions from parents in the search for children who may have developmental problems.

Figure 12.2
Percentage of normal
children passing the
item.

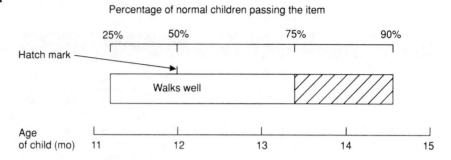

A short form of the DDST, the *Revised-Prescreening Developmental Questionnaire* (R-PDQ, 1986), consists of 24 to 25 questions designed for four age groups (0 to 9 months, 9 to 24 months, 2 to 4 years, 4 to 6 years). The instrument can be administered by the parent or caregiver and is designed primarily to make guardians more aware of the development of their children and to screen out those children likely to have problems with the longer DDST.

The Basic Motor Ability Tests-Revised   The BMAT-R was designed to assess motor behavior in children 4 to 12 years of age (Arnheim & Sinclair, 1979). The assessment battery consists of 11 tests for evaluating the general areas of fine and gross motor ability, coordination, balance, and selected physical fitness components (table 12.7).

Advantages of this instrument are that it is relatively easy to administer and requires little training. The authors of the test point out that a single examiner can test one child in about 15 to 20 minutes and a group of up to five children in approximately 30 minutes. Norms were established on children from diverse backgrounds and the retest reliability for the entire battery is quite high (.93). However, because inadequate information is available concerning validity and the instrument consists of a relatively few general items (as is the case with most motor assessment instruments) its use is limited, except perhaps as a screening device to discriminate between "normal" children and those with physical and motor deficiencies.

The Peabody Developmental Motor Scales   The PDMS (1983) were designed to assess gross and fine motor development in children ages birth through 7 years. Their use includes: the identification of children with delayed or abnormal motor development, comparison of skills to a general population sample, and assessment of abilities across time, or in response to intervention.

The Gross Motor Scale consists of 170 items representing five skill categories: reflexes, balance, nonlocomotion, locomotion, and receipt and propulsion of objects. The Fine Motor Scale, consisting of 112 items, is divided into four skill categories: grasping, hand use, eye-hand coordination, and manual dexterity. Both scales provide 16 to 17 age levels with 6 to 10 items at each level. Items are scored on a 3-point scale (0, 1, and 2), with a score of 1 (the middle)

Table 12.7 BMAT-R Test Items

Item	Stated Purpose of Test
Bead stringing	Bilateral eye-hand coordination and dexterity
Target throwing	Eye-hand coordination as associated with throwing
Marble transfer	Finger dexterity and speed of hand movement
Back and hamstring stretch	Flexibility of back and hamstring muscles
Standing long jump	Strength and power in the thigh and lower leg muscles
Face down to standing	Speed and agility
Static balance	Static balance with eyes open and closed
Basketball throw	Arm and shoulder girdle explosive strength
Ball striking	Coordination associated with striking
Target kicking	Eye-foot coordination
Agility run	Ability to rapidly move the body and alter direction

representing emerging behavior, but not proficiency. Unfortunately, specific criteria for a score of 1 are not provided for each item. Norms are provided with developmental motor quotients and age equivalents.

The PDMS were normed on children in 20 states representing a diversity of social conditions. Validity and reliability findings have shown the PDMS to be equal to, or better than the Bayley Scales (1969) and Gesell Schedules. No special qualifications are required to administer the PDMS, which can be completed on a single child in about 30 minutes (both scales). Although they have been used widely among child developmentalists, the PDMS have also received considerable attention among those who work with preschool children with special needs (Zittel, 1994).

The Bruininks-Oseretsky Test of Motor Proficiency Of the available product-oriented motor assessment instruments, the BOT is heralded as one of the best. Since its introduction (Bruininks, 1978), the BOT has undergone substantial review from the psychological and motor development community, resulting in the general conclusion that it is a well-normed, technically acceptable measure of gross and fine motor ability for children $4\frac{1}{2}$ to $14\frac{1}{2}$ years of age (e.g., Haubenstricker & Sapp, 1985; Sabatino, 1985).

The BOT consists of 46 items that yield both a comprehensive index of motor proficiency and separate composite scores for fine and gross motor abilities. These items are clustered into eight subtests that are designed to measure running speed and agility, balance, bilateral coordination, strength, upper limb coordination, response speed, visual-motor control, and upper limb speed and dexterity (table 12.8). An acceptable 14-item BOT short form that provides a brief survey of the child's motor proficiency is also available.

Table 12.8 Test Items from the BOT

General Category	Subtest Area	Items
Gross motor skills	1. Running—Designed to measure running speed.	Shuttle run
	2. Balance (8 times)—Designed to measure static and dynamic balance abilities.	Static balance (3 items) Dynamic balance (5 items)
	3. Bilateral coordination (8 items)—Designed to measure simultaneous coordination of upper and lower limbs.	Foot and finger tapping (3 items) Jumping in place (4 items) Drawing lines and crosses simultaneously (1 item)
	4. Strength (3 items)—Designed to measure upper arm and shoulder girdle strength, abdominal strength, and leg strength.	Standing broad jump Sit-ups Knee push-ups (boys under 8 and all girls) Full push-ups (boys 8 and older)
Gross and fine motor skills	5. Upper limb coordination (9 items)—Designed to measure gross and fine eye-hand coordination.	Ball bounce and catch (2 items) Catching tossed ball (2 items) Throwing a ball (1 item) Touching swinging ball (1 item) Precise upper limb movements (3 items)
Fine motor skills	6. Response speed (1 item)—Designed to measure response to a moving object.	Yardstick drop (1 item)
	7. Visual-motor control (8 items)—Designed to measure coordination of eye-hand movements.	Cutting (1 item) Drawing and copying (7 items)
	8. Upper limb speed and dexterity (8 items)—Designed to measure hand and finger dexterity, hand speed, and arm speed.	Penny placing (2 items) Card sorting (1 item) Bead stringing (1 item) Peg displacing (1 item) Lines and dots (3 items)

The BOT is an individually administered test that takes 45 to 60 minutes to complete; the short form takes about 15 to 20 minutes. Among its several merits as an assessment instrument is that its norms are provided in the form of standard scores, percentile ranks, stanines, and age equivalents for the complete battery. Interpretation tables representing age intervals of six months permit the conversion of subtest scores into subtest standard scores. Practitioners have noted that one disadvantage of the BOT is that results are not too helpful for programming.

In summary, though the BOT has limitations in overall motor assessment because it is not capable of detecting developmental change as with process-oriented instruments, it is a well-developed, well-standardized test battery of value as a research tool and diagnostic device.

A variety of assessment instruments are designed to examine the process (i.e., qualitative) characteristics of movement. Inherent in these instruments is the intent to identify an individual's current movement qualities and compare the actions to an established developmental sequence. (A discussion of the research and applied information concerning developmental sequence in regard to specific fundamental movement skills was presented in chapter 9 and it may be helpful to review that information. Particularly relevant are the discussions related to stage theory and Roberton's intratask component theory of identifying change in motor behavior.) Generally accompanying process-oriented instruments are criterion-referenced standards that, by design, compare individuals to themselves as well as to the criterion standard rather than to others as with norm-referenced standards.

Keep in mind that the instruments described in the following pages represent practical attempts to observe, measure, and assess the process characteristics of movement. Research endeavors in this field are relatively recent and not a great deal has been established concerning the validity of several developmental movement pattern sequences. To more accurately assess the developmental change of an individual's movement pattern characteristics (though it is generally not practical), film analysis over time is desirable. For additional information regarding movement analysis using film, various scientific instruments, and computers, refer to Adrian and Cooper (1989).

Since the use of sophisticated instrumentation and film analysis are needed to detect precise changes, most process-oriented instruments use a more global analysis. That is, the developmental sequence of a specific skill is identified using three to five "total body" stages (e.g., immature, elementary, mature), rather than making a developmental level analysis of specific body segments. There are also instruments that use identifiers that are considered to be relatively subjective; these are useful primarily as global screening and diagnostic devices.

Test of Gross Motor Development   The TGMD (Ulrich, 1985) represents one of the best examples, to date, of a practical, easily administered instrument developed to assess the sequence and qualitative aspects of motor skill behavior. Designed for use with children 3 to 10 years of age, the TGMD provides assessment of 12 of the most commonly practiced locomotor and manipulative motor skills. The locomotor skill assessment includes running, leaping, horizontal jumping, hopping, galloping, skipping, and sliding. The manipulative skill section includes two-handed striking, stationary ball bouncing, catching, kicking, and overhand throwing. A sample of the qualitative assessment format is presented in table 12.9.

The TGMD is considered quite comprehensive in the number and diversity of the motor skills that it assesses; it can also be administered with a minimum amount of special training and takes only about 15 minutes to complete per child. Another attractive characteristic is that it provides both norm-referenced and criterion-referenced interpretations. The instrument has well-established validity and reliability characteristics and, for an instrument of this type, is standardized on a diverse and relatively large normative sample. Langendorfer (1986) provides

Process-Oriented Assessment Instruments

■ *Objective 12.6*

Table 12.9 The TGMD Assessment Format

Skill	Equipment	Directions	Performance Criteria	1st	2nd
Locomotor Skills					
Run	50 feet of clear space and masking tape, chalk, or other marking device	Mark off two lines 50 feet apart Instruct student to "run fast" from one line to the other	1. Brief period where both feet are off the ground	1	
			2. Arms in opposition to legs, elbows bent	0	
			3. Foot placement near or on a line (not flat-footed)	1	
			4. Nonsupport leg bent approximately 90 degrees (close to buttocks)	0	
Horizontal jump	10 feet of clear space and tape or other marking device	Mark off a starting line on the floor, mat, or carpet Have the student start behind the line Tell the student to "jump far"	1. Preparatory movement includes flexion of both knees with arms extended behind the body	1	
			2. Arms extend forcefully forward and upward, reaching full extension above head	0	
			3. Take off and land on both feet simultaneously	1	
			4. Arms are brought downward during landing	0	
Object Control Skills					
Kick	8–10 inch plastic or slightly deflated playground ball, 30 feet of clear space, tape or other marking device	Mark off one line 30 feet away from a wall and one that is 20 feet from the wall Place the ball on the line nearest the wall and tell the student to stand on the other line. Tell the student to kick the ball "hard" at the wall	1. Rapid continuous approach to the ball	1	
			2. The trunk is inclined backward during ball contact	0	
			3. Forward swing of the arm opposite kicking leg	0	
			4. Follow-through by hopping on the nonkicking foot	0	
Overhand throw	3 tennis balls, a wall, 25 feet of clear space	Tell student to throw the ball "hard" at the wall	1. A downward arc of the throwing arm initiates the windup	0	
			2. Rotation of hip and shoulder to a point where the nondominant side faces an imaginary target	0	
			3. Weight is transferred by stepping with the foot opposite the throwing hand	0	
			4. Follow-through beyond ball release diagonally across body toward side opposite throwing arm	0	

1 = performs task correctly 2 of 3 trials
0 = does not perform task correctly 2 of 3 trials
From Ulrich, 1975, the Test of Gross Motor Development. Reprinted by permission of Pro-Ed, Inc., Austin, TX.

on a diverse and relatively large normative sample. Langendorfer (1986) provides an extensive review of the TGMD.

Fundamental Movement Pattern Assessment Instrument Of the more established process-oriented instruments, the FMPAI (McClenaghan & Gallahue, 1978) is one of the easiest to use. It was designed to measure the present developmental status of children and to assess movement pattern change over time. The instrument includes assessment of five fundamental movement skills: running, horizontal jumping, throwing, catching, and kicking. After the individual's movement pattern is observed, it is analyzed by "body segment action" and classified according to one of three stages (for that body action) to depict the developmental continuum. These stages are the initial stage, the elementary stage, and the mature stage (table 12.10). In essence, this is a form of the component approach to movement pattern assessment.

The instrument includes good visual illustrations of each stage and easy-to-use scoring aids. In contrast to the Test of Gross Motor Development, the developmental sequences of the FMPAI were not developed from or standardized on a population sample but are based on a review of the biomechanical literature. Nonetheless, the authors report acceptable validity and reliability. In summary, the FMPAI is a carefully developed observation instrument that allows for the documentation of change over time and by individual comparison.

Developmental Sequence of Fundamental Motor Skills Inventory The DSFMSI also provides process analysis of the commonly practiced fundamental locomotor and manipulative motor skills of children (Seefeldt & Haubenstricker, 1976). Included in the assessment are locomotor-walking, running, jumping, hopping, and skipping, and manipulative-throwing, catching, striking, kicking, and punting. Each motor skill is described within a developmental sequence that includes four or five stages ranging from (stage 1) immature to mature (stage 4 or 5). During the evaluation process children are matched to both visual and verbal descriptions of each stage.

Though the psychometric properties of the test have not been reported, the developmental sequences are based on longitudinal and cross-sectional data that have very useful application to teaching.

Hughes Basic Gross Motor Assessment Known for its practical application in the field, the Hughes test is a process/product-oriented assessment designed for use with children 5 ½ to 12 ½ years of age (Hughes, 1979; Hughes & Riley, 1981). The test consists of items intended to evaluate a selection of locomotor, balance, throwing, dribbling, and other ball-handling skills. Although the norm population did not include children from various ethnic groups, overall reliability has been reported as quite good and the validity acceptable. Scoring is with a four-point rating scale based on quality of movement using stated deviations from "normal" performance. Interrater reliability is also reported to be quite good (.97). The test is administered individually and uses commonly available

Other Commonly Used Instruments

Table 12.10 Sample of Body Action Assessment of the FMPAI

	Running		
	Initial	*Elementary*	*Mature*
Leg action (side view)	—Leg swing is short, limited —Stiff, uneven stride —No observable flight phase —Incomplete extension of support leg	—Length, swing, and speed increase —Limited but observable flight phase —Support leg extends more completely at takeoff	—Length of stride is at its maximum; speed of stride is fast —Definite flight phase —Support leg extends completely —Recovery thigh is parallel to ground
Arm action	—Stiff, short swing; varying degrees of elbow flexion —Tend to swing outward horizontally	—Arm swing increases —Horizontal swing is reduced on backswing	—Swing vertically in opposition to the legs —Arms are bent in approximate right angles
Leg action (rear view)	—Swinging leg rotates outward from the hip —Swinging foot, toes outward —Wide base of support	—Swinging foot crosses midline at height of recovery to rear	—Little rotary action of recovery leg and foot

From *Fundamental Movement* by Bruce A. McClenaghan and David L. Gallahue. Copyright © 1978 by Saunders College Publishing, a division of Holt, Rinehart and Winston, Inc. Reprinted by permission of David L. Gallahue.

equipment for several tasks. In summary, the Hughes test is useful in the field for evaluating minor motor problems and identifying those children who require further assessment and possible remediation.

Ohio State University Scale of Intra-Gross Motor Assessment   The SIGMA is a criterion-referenced instrument designed to assess fundamental motor skills of children of preschool to 14 years of age (Loovis & Ersing, 1979). The motor skills included are walking, running, jumping, hopping, skipping, throwing, catching, striking, and kicking. Also included are ladder and stair climbing. Each skill is described on a separate sheet in four levels of developmental sequence. Table 12.11 shows a sample of this format for the skill of stair climbing.

Aside from face validity with the literature, the instrument's other psychometric properties have been questioned. However, the authors report that the SIGMA can be administered with acceptable accuracy in formal as well as in an informal free play setting. An additional strength of the SIGMA is the accompanying Performance Base Curriculum (PBC). The PBC provides the much-needed link between assessment and program intervention. The curriculum is an objective-based instructional program that provides activities to accommodate each developmental level within each skill.

Purdue Perceptual Motor Survey   The PPMS is not a test; rather, as the name implies, it is a behavior assessment survey. The developers, Roach and Kephart

Table 12.11   Sample Process Analysis Format of SIGMA

Skill of Stair Climbing
Test Equipment: Series of Stairs

Level I	*Level II*	*Level III*	*Level IV*
When positioned at the bottom of a series of steps, the child demonstrates one of the following behaviors:	When standing at the bottom of a series of steps, the child, with or without the aid of the railing or wall, walks up and down five steps and demonstrates the following behaviors in two out of three trials:	When standing at the bottom of a series of steps, the child, with the aid of the railing or wall, walks up and down five steps and demonstrates the following behaviors in two out of three trials:	When standing at the bottom of a series of steps, the child independently walks up and down five steps and demonstrates the following behaviors in two out of three trials:
1. creeps up five steps and slides down from step to step on buttocks 2. creeps up and down five steps 3. walks up and back down while in a hands-feet position like animal walking	1. walks up using two-foot landing (mark time pattern) 2. walks down either in the same manner or by sliding from step to step on the buttocks	1. walks up using an alternate-stepping pattern 2. walks down in the same manner or by using a two-foot landing (mark time pattern)	1. walks up using an alternate-stepping pattern 2. walks down using an alternate-stepping pattern

Reprinted by permission of Mike Loovis.
Note: Examiner cannot serve as an aid in place of railing or wall.

(1966), state that "at its present level of development, the survey should be regarded merely as an instrument which allows the examiner to observe a series of perceptual-motor behaviors and to isolate areas which may need further study" (p. 10). The instrument was designed for children 6 to 10 years of age (grades 1 through 4) and is administered individually in about 45 minutes. The array of motor components that the survey measures is outlined in table 12.12.

Item validity and the relatively small norm group sample for the PPMS have been criticized, but its reliability scores have been acceptable. The survey has been recognized for several years as an effective observational instrument for discriminating between children with motor functioning problems and those in the normal range of behavior.

McCarthy Scales of Children's Abilities   Of the numerous instruments that are known primarily for their assessment of cognitive abilities of children, the *McCarthy Scales of Children's Abilities* warrant mention in regard to its perceptual and motor components. The McCarthy is designed for children ages 2 ½ to 8 ½ years of age but is known primarily as a preschool assessment device of considerable value with good psychometric properties (Wodrich, 1990).

Table 12.12 Test Items from PPMS

Motor Area	Item
Balance and posture	Walking forward, backward, sideways on board; performing a series of tasks evaluating ability to jump, hop, and skip while maintaining balance.
Body image and differentiation	Identification of body parts, imitation of movement, obstacle course activities, Kraus-Weber Test, angels-in-the-snow.
Perceptual-motor match	Making circle (on chalkboard), double circle, lateral line, and vertical line; rhythmic writing tasks.
Ocular control	Ocular pursuits of both eyes, right eye, left eye, and convergence.
Form perception/organization	Drawing several geometric forms on a single sheet of paper.

The instrument offers three general indexes of behavior: a general cognitive index, which consists of verbal, perceptual/performance, and quantitative subareas; a memory index; and a motor index. The memory and motor indexes are interpreted separately from the general cognitive index. The perceptual/performance and motor index components have specific relevance to motor behavior, in general, and include the following task behaviors:

1. Perceptual/Performance: block building, puzzle solving, tapping sequence, right-left orientation, draw-a-design, draw-a-child, and conceptual grouping
2. Motor: leg coordination, arm coordination, imitative action, draw-a-design, draw-a-child

Compared with similar measurement instruments, the objective scoring for the motor tasks on the McCarthy scales has been well received. The tasks approximate the motor screening given by a specialist but offer the advantage of objective scoring and derived scores. It is perhaps due at least in part to this advantage that researchers in recent years have used parts of the McCarthy motor index as variables in several scientific studies. The entire battery takes approximately 45 to 60 minutes per child, and it is suggested that individuals with specific training and background administer the instrument.

Frostig Developmental Test of Visual Perception   Designed for children 4 to 8 years of age, the Frostig Developmental Test of Visual Perception (Frostig, 1963, 1966) is known primarily as a visual perception screening tool. The instrument measures five operationally defined perceptual skills: eye-hand coordination (drawing various lines); figure-ground (identifying intersecting and hidden forms); constancy of shape (recognition of geometric figures presented in a variety of conditions); position in space (discrimination of reversals and rotations of

Table 12.13 Summary Information on Selected Instruments

Instrument	Age Range (yr)	Type of Items	Available
Denver Developmental Screening Test	Birth–6	Product (screening: fine motor; adaptive: gross motor)	DDM, Inc. P.O. Box 20037 Denver, CO 80220-0037
Basic Motor Ability Tests-Revised	4–12	Product (fine and gross motor, coordination, balance, physical fitness)	In Arnheim, D., & Sinclair, A., *The Clumsy Child,* St. Louis: C. V. Mosby, 1979
Bruininks-Oseretsky Test of Motor Proficiency	4.5–14.5	Product (gross motor, coordination, motor fitness, balance, visual-motor)	American Guidance Service Circle Pines, MN 55014
Peabody Developmental Motor Scales	Birth–7	Product (fine and gross motor)	DLM Teaching Resource Corp. One DLM Park Allen, TX 75002
Test of Gross Motor Development	3–10	Process (basic locomotor and manipulative skills)	PRO-ED 5341 Industrial Oaks Blvd. Austin, TX 78735
Hughes Basic Gross Motor Assessment	5.5–12.5	Combination (balance, gross motor, basic locomotor and manipulative skills)	G. E. Miller, Inc. 484 South Broadway Yonkers, NY 10705

Table 12.14 Fitness Assessment Batteries

Name	Level	Components	Address
AAHPERD Physical Best	Ages 5–18	Cardiovascular Body composition Strength/endurance Flexibility	AAHPERD Publications 1900 Association Dr. Reston, VA 22091
AAHPERD Functional Fitness Test For Older Adults	60+ years	Cardiovascular Strength/endurance Flexibility Agility/dynamic balance Coordination	AAHPERD Publications
Fitnessgram	Grades K–12	Cardiovascular Strength/endurance Flexibility Body composition Agility	Fitnessgram Institute for Aerobics Research 12330 Preston Rd. Dallas, TX 75230
Presidential Physical Fitness Award Program	Ages 6–17	Cardiovascular Flexibility Strength/endurance Agility	Department of Health & Human Services The President's Council on Physical Fitness and Sports Washington, D.C. 20001

figures); and spatial relationships (analysis of simple forms and patterns). The instrument can be administered to a small group in less than one hour. One noted limitation of the test is that the normative sample was selected only from Caucasian schoolchildren from middle-class homes.

Table 12.13 summarizes some information about a few of the more recommended and available motor assessment instruments. Table 12.14 lists selected "field-oriented" tests in physical fitness test batteries. Refer to Baumgartner and Jackson (1995) and Safrit (1990) in the Suggested Readings for single test and "laboratory-oriented" recommendations.

SUMMARY

Assessment provides the opportunity to observe, document, and interpret change across the life span, as well as to determine the growth and developmental status of an individual at a particular time. The possibilities for physical and motor assessment as associated with growth, development, and motor performance are vast. Although hundreds of component-specific and task-specific tests are available to assess the various characteristics of physical and motor development, there are no assessment batteries available that reflect the lifelong perspective.

Assessment involves both measurement (collection of information) and evaluation (process of judgment). The two most widely used types of standards by which performance is judged are norm-referenced standards (which order individuals on a quantitative basis) and criterion-referenced standards (which specify a level by which individuals are judged). The assessment of movement skill performance is usually conducted with either process-oriented (form characteristics) or product-oriented (performance outcome) assessment instruments.

Some of the more widely accepted purposes for conducting assessments are: for diagnostic/screening, to determine status, for placement, to determine program content, for program evaluation, to construct norms and performance standards, for research, to predict performance, and to motivate individuals. After the purpose has been established, additional considerations are warranted in regard to instrument selection. Considerations should include a review of the instrument in regard to its psychometric properties (e.g., validity, reliability, norm sample), feasibility (e.g., cost, equipment, space, personnel needed, time), and level of training and expertise needed.

Although motor assessment items have been included in general child development batteries for more than 60 years, their general quality has been suspect. Most instruments are based on generalized product (quantitative) characteristics and provide little insight into the process of developmental change. These shortcomings have been addressed to some degree in recent years as evidenced by the introduction of several process-oriented instruments and motor-specific, product-oriented assessment batteries.

SUGGESTED READINGS

General Assessment

American Psychological Association, American Educational Research Association, National Council on Measurement in Education. (1985). *Standards for educational and psychological tests.* Washington, DC: Author.

Buros, O. K. (Ed.). (1987). *The tenth mental measurement yearbook.* Highland Park, NJ: Gryphon Press.

Kaplan, R. M., & Saccuzzo, D. P. (1993). *Psychological testing* (3rd ed.). Pacific Grove, CA: Brooks/Cole.

Osofsky, J. D. (Ed.). (1987). *Handbook of infant development.* New York: John Wiley & Sons.

Salvia, J., & Ysseldyke, J. E. (1991). *Assessment* (5th ed.). Boston: Houghton Mifflin.

Wodrich, D. L. (1990). *Children's psychological testing: A guide for nonpsychologists* (2nd ed.). Baltimore, MD: Brookes.

Physical and Motor Assessment

Adrian, M. J., & Cooper, J. M. (1989). *Biomechanics of human movement.* Indianapolis: Benchmark Press. (See chapter 8.)

Baumgartner, T. A., & Jackson, A. S. (1995). *Measurement for education: In physical education and exercise science* (5th ed.). Madison, WI: Brown & Benchmark.

Safrit, W. J. (1990). *Introduction to measurement in physical education and exercise science.* St. Louis: Mosby.

PART ∘ *six*

The final chapter of this text is devoted to presenting a description of the role of sociocultural factors in motor development. This information addresses the primary sociocultural

Sociocultural Influences on Motor Development

factors from infancy to old age that may influence the individual's participation in sport and physical activity.

Sociocultural Influences on Motor Development

KEY TERMS

socialization
culture
status
social class
social role
norm
socializing agent
Bronfenbrenner's
   ecological systems model

Parten's play model
peers
sex roles
sex-role stereotyping
self-esteem
age-related stereotyping
disengagement theory
activity theory
social breakdown-reconstruction theory

At birth, newborns are very much alike in that they have similar needs for nurturing, progress through the same general sequence in developing locomotion and speech, and must learn appropriate social behavior. However, children become different due to sociocultural influences, experience, and their biological uniqueness. Chapter 5 addressed some of the types of factors that affect growth and development, namely, maternal and external influences on prenatal development, nutrition, hormones, and the effects of physical activity (and exercise) on lifelong human development. This chapter focuses more specifically on the influences and importance of sociocultural factors on lifelong motor development.

BASIC TERMINOLOGY OF SOCIALIZATION

■ *Objective 13.1*

Although socialization is commonly linked with child and adolescent development, it is a lifelong process. As individuals vary their social environment, they must learn to adapt to new situations. **Socialization** is a term that refers to a set of events and processes by which individuals acquire the beliefs and behaviors of the particular society and subgroup in which they live and, in most cases, are born into. Many goals of socialization are common to all societies, but each society and subgroup develops its own unique practices and goals to maintain itself in its particular ecological context.

Culture, which is generally thought of as a subset of society, is the collection of specific attitudes, behaviors, and products that characterize an identifiable group of people. Since societies and cultures frequently differ, it is reasonable to assume that expectations would vary from culture to culture. Though it is obvious that differences exist between our society and those of other countries, it is also true that differences exist *within* our society between cultural groups. Some cultures place a higher priority on leisure, recreation, and sports than do others; this will have an effect on motor development of its members.

Four general concepts frequently used to describe the structure of socialization are status, class, role, and norm. **Status** refers to an individual's position in society. One individual can hold many positions—father, brother, coach, mentor, friend, athlete, and so on. Some positions are conferred on individuals due to their experience or expertise, whereas others are simply a function of sex (e.g., mother, wife) or age (e.g., family leader, grandparent). Individuals learn to assume the role associated with the position of status that has been established.

A complementary term to one's status in society is social class. **Social class,** also called socioeconomic status, refers to a grouping of people with similar economic, educational, and occupational characteristics. Cross-cultural studies often use terms like lower-, middle-, and upper socioeconomic class as a factor for making comparisons.

Accordingly, a **social role** is the particular behavior that an individual uses to fulfill a position of status. One may describe a role as the set of dynamics and expectations that complements the status; in essence, it is the "job description" that goes with a position. For example, a quarterback holds the status of team leader on the field. With that status, the individual is expected to do such things as organize the group and carry out the game plan. As role expectations for a specific position of status are developed and a relatively predictable set of characteristics are formed, norms are established.

A **norm** refers to a standard of behavior that would be expected from members of a similar group of society. For instance, common and expected behaviors are associated with specific groups of teachers and coaches in addition to their individual personalities.

Gallahue (1989) notes that "cultural socialization should be viewed as an interactive process between society and the individual with one intricately influencing the other" (p. 445). Socialization is a lifelong dynamic process that allows individuals to use their individual differences to make unique contributions.

SOCIALIZING PROCESSES AND AGENTS

■ *Objective 13.2*

Socialization involves acquiring knowledge and various types of skills, including those related to the motor domain. According to one of the dominant cognitive-social learning perspectives associated with child development (Ladd & Mize, 1983), three principal modes of socialization can be described: direct instruction, shaping, and modeling. All three are used to teach individuals about social roles they and others can play in the culture. Each is also mediated by the individual's *social cognition,* or understanding of the social world. *Direct instruction* deals primarily with concepts and ideas conveyed through language. This mode of socialization includes either specific or general information about how to act or what to say, and when to act or speak. *Shaping* refers to the social learning processes that allow an individual to benefit from experience. The concept that individuals can learn social behaviors by observing others who serve as models is known as *modeling.* Frequently associated with cognitive-social learning theory and the works of Bandura (1986) and Mischel (1984), it is suggested that individuals cognitively represent the behavior of others through observational learning and then possibly adopt, or model, this behavior themselves. Modeling can have a profound effect on the development of motor skills, especially during childhood and adolescence when individuals frequently attempt to model the behavior of famous athletes (Weiss, 1983).

■ *Objective 13.3*

Socialization depends on the contexts and settings in which it occurs. Individuals may be influenced by the overlapping factors of society at large, one or more subcultures, socioeconomic class, institutions, and individuals. Each can play a significant role in the socialization process. Complementary to the processes of socialization are the various agents that are influential in affecting social behavior and motor development.

A **socializing agent** may be defined as an individual, group, or institution that interprets culture to the individual. Although there are numerous subagents, the major socializing influences are the family, school, peer group, church, and community. Of this group, it is generally acknowledged that the family is the primary socializing agent responsible for the transmission of cultural content during the early developmental years. The family not only passes on its attitudes, prejudices, and health habits but also determines social class, ethnic origin, race, and religious beliefs. As individuals grow older, the school and peer group character-istically become stronger influences in socialization. More specific to motor de-velopment, teachers and coaches are in positions of powerful influence in the school and community. Another potentially strong agent (medium) through which individuals may be socialized is play and sports involvement. Subsequent discussions of the primary factors influencing development during childhood, adolescence, and adulthood will incorporate these primary agents.

One of the few comprehensive frameworks for understanding sociocultural influ-ences on socialization was developed by Bronfenbrenner (refer to chapter 1, Contextualist Approach). **Bronfenbrenner's ecological systems model** places the individual at a theoretical center surrounded by five environmental systems: microsystem, mesosystem, exosystem, macrosystem, and chronosystem. The most direct interactions are with the microsystem, the environment in which the person lives. Influential agents include the family, school, peers, church, and neighborhood play area. Within these contexts the individual is viewed as one who helps construct the environment.

■ *Objective 13.4*

Influences that occur as a result of relationships between different mi-crosystems are part of the mesosystem. For example, the relation of family expe-riences to school experiences, or church training to family experiences, falls within the mesosystem. Developmentalists suggest that it is important to observe the individual's behavior in multiple contexts to provide a more complete picture of social development.

The exosystem involves indirect influences from another social context that affect what the person experiences in an immediate setting, for instance, the school board and city government, both of which have control over the quality of physical education and recreational facilities available to individuals.

The macrosystem incorporates the broad-based attitudes and ideologies of the society and specific culture in which the individual lives. It is suggested that these influences may permeate virtually all of the other systems within the indi-vidual's ecological world. As mentioned earlier, different societies and cultures place varying levels of importance on such influential factors as education, leisure time, sports, and health promotion, all of which may have a powerful im-pact on motor development across the life span. The final influence is the chronosystem, which involves the patterning of environmental events and transi-tions over the life span; that is, the sociohistorical contexts. For example, females today are much more likely to participate in athletic activities than they were in the 1950s.

Socialization affects behaviors that include values, knowledge, social skills, and traits and is vital for optimal physical and motor development. Societies, cul-tures, or parents who do not expose their members or their children to physical activities, sports, and proper health practices are limiting the human potential for optimal development.

■ *Objective 13.5*

Figure 13.1
Parental beliefs and
attitudes toward a
healthy lifestyle are
important in socializing
the child.

The following discussion focuses on the primary sociocultural influences that can affect motor development through involvement in sports and other physical activity across the life span. Although it has been shown that sociocultural factors begin to show their influence on growth and development in the prenatal stages (when the health practices of the mother may be affected by socioeconomic status) and become more direct after birth, the focus of this discussion will be on the primary factors associated with early childhood through adulthood.

PRIMARY INFLUENCES DURING CHILDHOOD AND ADOLESCENCE
The Family

■ *Objective 13.6*

The primary sociocultural influences that have been shown to influence sports involvement and physical activity during childhood and adolescence are the family, peer relations in play and sport participation, and the relationship to coaches and teachers in the school and community. Additional factors to be discussed are sex-role expectations, self-esteem, and race.

The family is the socializing agent most directly responsible for the communication of cultural content to the growing child. This observation is relevant not only to socialization in general but also in regard to the child's movement endeavors (Snyder & Spreitzer, 1973). Significant changes in the family unit have taken place in recent decades, brought about by an increase in the number of working mothers and single-parent homes. In fact, the majority of mothers of children and youth now work outside of the home (Hoffman, 1989). In today's society, about one-fourth of American children live in a single-parent family, nine times out of ten with their mother (Glick & Lin, 1987). Despite all of these changes and their ramifications, the family is still the primary caregiver and socializing agent during infancy and childhood.

Of the many influences associated with the family, perhaps the most dominant are *parental beliefs and attitudes.* Parents who believe in the importance of an active, healthy lifestyle usually share it with their children. Many parents take their young offspring along to such settings as tennis courts, swimming pools, health clubs, and other recreational areas. With this positive attitude and the knowledge that physical activity is beneficial, it is usually the parent's wish that their children adopt these characteristics; consequently, these parents usually provide their children with early experiences.

The first contact that young children have with motor activities and sports normally takes place with the family. Along with taking their children to fitness and sporting activities, parents may give sport equipment and clothing as presents and introduce them to the basic skills and concepts of selected activities, all of which may begin at a very young age. As children grow older, they are encouraged to participate in more formal settings (e.g., youth sports) and to maximize their abilities. Sports involvement for the vast majority of eventual participants normally begins by the age of 8, and if parental interest is high, the likelihood of participation is much greater (Greendorfer, 1976; Snyder & Spreitzer, 1973).

As a result of early experiences and watching their parents participate in physical activity regularly over the years, children may also develop a positive attitude, thus becoming socialized into an active physical lifestyle. Participation in sport activities of individuals in later years (i.e., older childhood through adulthood) reflects, at least to some degree, their parent's beliefs, attitudes, and encouragement during earlier years. Although other influences are likely to be present, research findings indicate that the best predictor of participation in sports by adults is childhood and adolescent involvement (Greendorfer, 1979; Loy, McPherson, & Kenyon, 1978).

Research findings with regard to *individual family member influence,* though somewhat inconsistent, indicate that specific members of the unit very likely play differential roles in play and sport socialization. One body of literature suggests that the same-sex parent has the most influence on the extent of the child's involvement (Lewko & Ewing, 1980; McPherson, 1978; Snyder & Spreitzer, 1973). In the Lewko and Ewing study involving children 9 to 11 years of age, the investigators found that girls seemed to require a higher level of encouragement than boys to become involved. This perhaps reflects traditional sex-role expectations regarding sport involvement. There is also evidence that the opposite-sex parent influences involvement but that fathers are the primary influencing agent for both girls and boys (Greendorfer & Lewko, 1978). Evidence from this investigation also found that the role of the mother in influencing sport participation of either sex child was minimal, supporting the more traditional view that the father is responsible for initiating sport involvement of the offspring, especially for the male child.

The literature concerning sibling influence seems to be quite inconsistent. Some investigators and the empirical evidence would suggest that older brothers and sisters may be important agents in the sport socialization of younger siblings (Lewko & Ewing, 1980; Weiss & Knoppers, 1982). However, this finding has not been supported in other studies (Greendorfer & Lewko, 1978). Despite the inconsistencies, it is clear that the family, especially the father, is the significant agent in the socialization of their children to physical activity and sport.

Along with determining religious beliefs, race, and ethnic origin, the family also serves as the economic unit. Economic factors can affect nutritional practices and corresponding physical growth and development. There are also indications that socioeconomic status may be an influential factor in motor development. Although research on this topic is gravely lacking, it does appear as though children from low-income families (especially single-parent homes) may be limited in motor development experiences due at least in part to lack of financial resources, parental involvement time, and the parent's experience (Coakley, 1987). Living on a tightly budgeted low income may present a shortage of funds for such items as toys, sport equipment, and activity program enrollment fees.

There are also indications that children in low-income or single-parent homes may have limited exposure to different kinds of play activities and sports in comparison to children of higher income families. For instance, Greenspan (1983) found that age-group skiers, gymnasts, and swimmers come primarily from homes of upper-middle class socioeconomic status, whereas young boxers, wrestlers, and baseball players derive from less economically endowed families. Empirical evidence would also suggest that activities such as golf and tennis are associated with higher income families.

Another consideration is socioeconomic status as it relates to education and knowledge of the importance of motor activities in child and adolescent development. Low-income families, who generally lack higher education, may be less knowledgeable about the importance of physical activity to health and so may be less apt to consider motor activities as important to their children's development. On the other hand, some highly educated people may view the time spent on motor activities as detrimental to optimal cognitive development.

Peer Relations

The influence of peers and participation in play and sport activities become increasingly powerful socializing influences through the childhood years. As these factors increase their effects through later childhood and adolescence, family dominance tends to decrease on several socialization issues. From a motor development perspective, the influence of both of these agents are intertwined. Although research has not shown that participation in play and sport activities systematically produces positive social characteristics, it does suggest that when good consequences occur, they are tied to the social relationships found in these settings (Coakley, 1987).

Figure 13.2
Play is considered a primary mode of social development during childhood.

Play, as it relates to socialization and other aspects of human behavior, has been one of the most extensively studied areas of child development. In addition to stimulating peer relations, play has also been shown to advance cognitive development, increase exploration, release tension (Christie & Wardle, 1992; Santrock, 1992), enhance self-esteem (Weiss, 1987), influence attitude formation (Sage, 1986), and stimulate moral growth (Bredemeier & Shields, 1987). Play and sport activities take up a sizable amount of time during childhood and adolescence, providing many opportunities for enhancing physical and motor development. Although numerous research papers have been written on the topic, the term *play* remains an elusive concept. Descriptions of play range from an infant's simple display of a newly discovered perceptual-motor talent, to a parent and child catching a ball, to an older child's involvement in organized youth sports. Some describe play as pleasurable activity that is engaged in for its own sake.

Others suggest that there is no universally accepted definition because it encompasses such a diversity of activities.

■ *Objective 13.7*

Many years ago Parten (1932) identified categories of play (**Parten's play model**) that are still recognized by developmentalists today as relevant forms of social development; these are the categories of solitary, parallel, associative, and cooperative play. In *solitary play* the child plays alone or independently of those around him or her. Solitary play is characteristic of 2- and 3-year-olds. *Parallel play* is similar to solitary play in that the child still plays independently but with toys like those that other children are using. Thus parallel play mimics the behavior of children who are playing alongside. Although this type of play decreases in frequency with age, older preschool children can be observed in this behavior quite often.

Social interaction in the form of play with little or no organization is described as *associative play.* At this time in social development children seem to be more interested in associating with each other than in the tasks around them. Specific characteristic behaviors include follow-the-leader and invitation-type activities, along with borrowing and sharing toys. This type of play is often observed between the ages of 3½ to 5 years.

Around the age of 5 marks the beginning of purposeful *cooperative play,* representing social interaction in a group with a sense of group identity and organization. This category is characterized by formal games, competitive group activities, and activities in which children must work together to achieve a common goal. Cooperative play at this level is the precursor to forms of movement activities and sport observed during older childhood and adolescence. Observation also suggests that as children grow older, they tend to form more complex and larger social structures when at play (Crum & Eckert, 1985).

There are indications that play behavior, as described by these categories, has changed over the last half century. Some of this change has implications for motor development. Using Parten's categories as a model, Barnes (1971) found that children in the 1970s did not engage in as much associative and cooperative play as they did about 40 years earlier. It has been suggested that children may be expressing more passive solitary and parallel play behavior as a reflection of the extended periods of time that they spend watching television and playing video games. Related to this is the notion that play items are more attractive today and parents seem to encourage this type of play behavior. Children spend about 25 hours per week viewing television, which is more than any other activity except sleep (Huston et al., 1990). By the time the average adolescent completes high school, the individual will spend more time watching television (approximately 20,000 hours) than in the classroom (Santrock, 1992). Obviously, this type of social behavior can have strong implications for motor development.

Beginning at approximately age 5, individuals engage in physical activities that are more cooperative and complex (i.e., group oriented). The need for this type of play appears to increase with age (peaking during adolescence) due in part to the individual's strong social desire to be a part of a peer group. Cooperative play may take the form of informal or formal games (and sports). In informal activities, children usually get together on their own to play with equipment at hand, using makeshift or flexible rules.

Organized activities such as youth sport and school athletic programs often demand significant time commitments by the participant and family. It is not uncommon for parents to adjust meal and work schedules and weekend plans to get their children to practices and games. Along with probable financial requirements, these considerations, when compared to other family commitments and priorities, can be important factors in the degree of organized sport socialization. Estimates suggest that more than 30 million children, ages 6 to 16 years, participate in at least one organized youth sport program (Martens, 1978). In addition to being quite influential in developing social behaviors and stimulating peer relations, sport settings also provide excellent opportunities for advancing motor development and physical fitness.

One of the most frequently cited reasons for involvement in play and sport activities is the desire to interact with peers. **Peers** are individuals of about the same age or maturity level, and in some instances, social class. A rather common characteristic of the peer group during childhood and adolescence is that its members are usually of the same sex, which provides an extremely powerful socialization environment (Maccoby, 1990). Most peer relations arise from associations made in the school, particularly if the schools are attended by individuals from the same neighborhood. The desire to "fit in" and the need to belong and be identified as part of a group, team, or club can be compelling socialization forces. As individuals enter later childhood and adolescence, influence from the peer group becomes increasingly important, while that of the family and other social agents may diminish.

Coakley (1987) notes that children's involvement in either informal or organized sports is influenced by: (a) the availability of opportunities, (b) support from the family, peers, role models, and the general community, and (c) the child's self-perception as a participant. Of these influences, there are strong indications that the peer group is a dominant factor with respect to socialization into physical activity and sport from childhood through early adulthood (Greendorfer, 1977). Peers have the potential to reinforce family influence concerning socialization into physical activity and sport and to provide new considerations not stressed at home. There may also be situations when peer influence challenges those values set by other social agents (e.g., family, school, church).

Through childhood and adolescence this influence seems to be stronger in regard to involvement in team (group) activities, rather than individual sport endeavors (Kenyon & McPherson, 1973). However, regardless of the type of activity that the individual is drawn by peer influence to participate in, peer relations can be a powerful social agent in enhancing motor development.

Usually associated with the school and community social structures, teachers and coaches have the potential for being very influential agents in regard to sport involvement. In fact, in some situations where few other influential adults are

Coaches and Teachers

*Art
Linkletter*

available, these agents may become "significant others" for the child. Although some suspicion has been cast regarding the primary motives of coaches (e.g., too much emphasis on winning), research does support that in most cases these agents are genuinely concerned with helping members of the team to experience positive socialization outcomes (Weiss & Sisley, 1984). Unfortunately, no evidence has been presented on the extent to which coaches affect behaviors that transcend sport situations and influence individuals in later life.

In general, coaches and teachers appear to influence males more than females in regard to involvement in sport and physical activity (Greendorfer & Lewko, 1978). In 1976 female athletes reported that coaches and teachers influenced their involvement during adolescence but not in the earlier years or as young adults. In another early investigation, male athletes reported benefiting from these influences before, during, and after adolescence (Kenyon & McPherson, 1973). Empirical evidence would suggest that these characteristics have not changed significantly over the years for males, but those for females may be outdated considering more recent views and opportunities (e.g., university scholarships).

Sex-Role Expectations and Stereotyping

■ *Objective 13.8*

Sex roles are social expectations of how individuals should act and think as males and females. Along with being influenced by biological factors (primarily sex hormones), socializing agents such as family members, schools, peers, and the media can have a significant effect on this aspect of human behavior. In regard to family influence, fathers seem to play an especially important part in sex-role development and are more likely to act differently toward opposite-sex children than mothers (Fagot & Leinbach, 1987; Lamb, 1986).

Related to this aspect of sociocultural influence is **sex-role stereotyping,** the use of different methods to introduce boys and girls to physical activities and sports. Even as infants, girls tend to be handled more gently and protectively than boys. Infant boys, on the other hand, are perceived as tougher, stronger, and more athletic and are more often given toys requiring active play and the use of motor skills.

Toy selection and marketing are traditionally based on sex-role stereotyping and tied to socialization into physical activity and sport. For example, when a parent gives a basketball to a young boy, the gesture usually promotes physical activity in the form of shooting and dribbling, as well as introducing the child to the sport. Although basketball is also considered a female sport, there seem to be few promotion products that use females as sex-role models. By 3 years of age, most children have a firmly established preference for "sex-appropriate" play items (Conner & Serbin, 1977; Kuhn, Nash, & Brucken, 1978). Although biological disposition cannot be completely ruled out, early preferences for play items are probably strongly influenced by parents and significant others.

As children enter school they also tend to be treated differently. For instance, males are more likely to receive strong support from family, relatives, peers, teachers, and coaches to begin and continue in sports. For boys, sport involvement is viewed as part of the young man's path to manhood, whereas such endeavors are traditionally not linked to becoming a woman (Coakley, 1987). Because many girls are not encouraged to participate in skilled, vigorous motor activities, many never develop their skills to full potential. These attitudes also have an effect on the motivation to get involved seriously in sports and to train to achieve maximum performance.

During adolescence, sex-role expectations can be of primary influence in the individual's decision to get involved in sport activities. *Sex-role conflict* may be created when adolescents decide to participate in certain sport activities. Anthrop and Allison (1983) found in a survey of female high school athletes that 17% indicated that they had a significant problem with role conflict in that they were concerned that their participation in games may be viewed as a masculine endeavor. Once again, this reflects the traditional view of females in sport.

For males, however, involvement in sport seems to cause little sex-role conflict and is viewed as stereotypic of ideal male characteristics. Related research findings also suggest that males are more easily socialized into sports than females (Ostrow, Jones, & Spiker, 1981). Two reasons were given for this finding: First, there are relatively few female role models; second, the majority of the sports surveyed were considered to be "masculine." Participation in these sports can cause more sex-role conflict in females, therefore decreasing the likelihood of their involvement.

However, society's view that females are the "weaker" sex appears to be changing. In more recent years, female participation in sports has increased not only at the youth sport level but also in school and university programs (due in part to Title IX legislation). As a result, participation levels have increased, decreasing the performance differences between the sexes.

Also referred to as self-image and self-concept, **self-esteem** is the evaluative and personal judgment of one's self, a judgment not necessitating accuracy so much as one's belief that it is correct. Self-esteem is included in this section as a sociocultural factor because it is considered, in large part, a reflection of social support in the form of parents, peers, teachers, and coaches, etc. For example, a child highly praised for her physical ability by a teacher and preferred by peers as a teammate for games generally has a good sense of worth for this particular endeavor. Self-esteem is considered by some to be a critical index of mental health, with a high sense of worth during childhood being linked to satisfaction in later life. However, global self-esteem judgments should be viewed with caution, for an individual's sense of worth may vary according to different skill domains and areas of competence. For example, an individual may perceive that he is quite intelligent but inadequate in athletic and social abilities. Variation within skill domains is not uncommon. Some individuals may have a strong sense of worth in, for example, basketball but not in baseball or track. Indications are that feelings about self-esteem are relatively well established by middle childhood (Harter, 1982, 1987).

Within the context of youth sports involvement, researchers have found that such participation is generally associated with higher self-esteem (e.g.,

Self-Esteem

■ *Objective 13.9*

Anthrop & Allison, 1983; Hall, Durborow, & Progen, 1986; Kirshnit, Richards, & Ham, 1988). It would seem reasonable to expect a person with a good self-esteem regarding physical ability to be more likely to engage in complementary activities such as youth sports. On the other hand, research confirms that a person with a low sense of physical competence is more likely to avoid these opportunities (Weiss, 1993). Along with the internal factors associated with personality type and emotions, social interactions play a vital role in the development of self-esteem.

Whereas children under the age of about 10 years depend primarily on parents and the outcomes of contests for appraisals, those who are older tend to rely more on comparison to and appraisals provided by their peers (Horn & Hasbrook, 1987; Horn & Weiss, 1991). In addition, feedback from teachers and coaches may also be influencing factors. For example, research with male athletes 10 to 15 years of age suggests that they exhibit higher self-esteem while playing for coaches who give frequent encouragement and corrective feedback, especially if the individual begins with a relatively low sense of worth (Smith, Smoll, & Curtis, 1979). As one would expect, the type of feedback given to young athletes is also a factor in their state of self-esteem. With some participants, "general" comments and appraisals may not be as effective as feedback that is corrective, even if delivered in the form of criticism (Horn, 1985). Obviously, mental maturity, emotional state, manner of delivery, and rapport between athlete and coach are critical factors in effectiveness of communication and development of self-esteem.

Individuals with a high self-esteem in their physical abilities may be more apt to join in sport programs, but a large portion of children and youth, especially females, drop out (Brown, 1985; Kirshnit, Ham, & Richards, 1989). There are two prominent theories addressing this issue. One suggests that during adolescence, interest in sports conflicts with other opportunities more than it did in childhood (Gould et al., 1982). That is, adolescents have a greater number of desirable options during leisure time and more demands on how they use their time. A second theory emphasizes the negative aspects of youth sports and disenchantment with participation (Gould & Horn, 1984). Such experiences as lack of skill improvement, lack of playing time, dislike for the coach, pressure to perform (or win), and lack of success have been suggested as factors in dropping out.

Although different age groups may state different reasons for participating and continuing in youth sport programs (Weiss, 1993), the most commonly stated reasons include: motivation to demonstrate competency (Harter, 1981); desire to compete, liking the coaches, and pleasing family and friends (Brodkin & Weiss, 1990); and desire to affiliate with or make new friends, be part of a team, have fun, increase fitness, and improve skills (Weiss, 1993).

Possible Influences of Race

Although the research on this topic is quite sparse and the effects of socioeconomic factors are difficult to separate, there are indications that black and white children are socialized into sports differently (Braddock, 1980; Greendorfer & Ewing, 1981; Oliver, 1980). A generalization is that possible success in sports appears to take on more importance for black youths compared to their white counterparts, especially for males from lower socioeconomic homes.

Encouragement from family members, same-sex peers, coaches, teachers, and the community may be more influential for blacks than whites. It is suggested that some of this encouragement originates from the somewhat misleading perception that sports offer blacks more opportunities than other careers in society, therefore providing a more readily conceivable means for increasing social status. Harris and Hunt (1984) found that encouragement for involvement in sports among black youth gets stronger as they move through adolescence and begin to perceive major obstacles to success in careers outside of sports and entertainment. Adding to this view is the fact that although professional black athletes are strong role models for young blacks and whites, blacks have fewer role models from other occupations to emulate. Research also indicates that encouragement in sport involvement among lower socioeconomic black families is stimulated by the wish for their child's fame and fortune (Snyder & Spreitzer, 1983). Although little research on this topic is available and much needs to be learned concerning the effects of such influences on several aspects of human development, one could speculate that positive motor development benefits could arise from such encouragement and parental involvement.

■ *Objective 13.10*

Just as society and various cultures have expectations for individual behavior from infancy through adolescence, numerous sociocultural influences and expectations are also present through adulthood. Whereas home and school settings were dominant socializing factors prior to adulthood, these agents become less of a primary influence in establishing or changing behavior characteristics as individuals attain independence from the family and complete formal schooling. For the remainder of life, the primary agents are usually the media, peers, spouses, and other individuals in the community (e.g., doctors, class instructors). In addition to these influences, seasonal, personal, and situational factors that are continuous and dynamic over the life cycle (e.g., marriage, children, and employment situation) play an important role in adult sport socialization (Mihalik et al., 1989; Wankel, 1985). Just as there are varying sociocultural views and expectations for behavior concerning health practices and physical activity in earlier years, a variety of views and trends are evident during adulthood. It is rather obvious that society in general, as expressed through various media, places a premium on youth and good looks. The idea of living a zestful, active lifestyle while at the same time eating and drinking things that allow one to remain slim has been one of the central themes in advertising. Advertising, in today's world, is a powerful socializing agent. During early adulthood individuals can still be quite impressionable and the need for self-esteem and peer acceptance has led many to either take up or continue physical activity in some form. This is also the period in the life span when most individuals are at their healthiest and reach peak physical performance (around age 30). For many individuals, the notion that they are at their best and not slowing down physically needs to be tested on a regular

PRIMARY INFLUENCES DURING ADULTHOOD

■ *Objective 13.11*

basis. Obviously, however, many young adults have not been influenced or do not have the desire to engage in physical activities as part of their lifestyle. In fact, empirical evidence suggests that due at least in part to their relatively healthy status during this period in their lives, many individuals do not stop to think about how their lifestyles will affect their health in later years. College students specifically seem to have unrealistic and overly optimistic views about their future health and longevity. Unfortunately, many of those who are not healthy (e.g., overweight, unfit) at this time have established attitudes concerning physical activity that are difficult to overcome.

As individuals approach middle age (ages 35 to 40), concern for health and physical appearance becomes more evident. This is a time when the regression of physical skills and signs of physical aging begin for most individuals, causing more attention to be drawn to these characteristics. Part of the stimulus for personal concern is likely related to two sociocultural influences. As mentioned earlier, our society places a premium on a healthy, youthful physical appearance. This factor alone is responsible for millions of dollars spent each year nationally on health and beauty items and classes to help "preserve" youthfulness. Many individuals begin dying gray hair, undergo cosmetic surgery, and enroll in weight reduction and conditioning classes, all in effort to maintain a healthy, youthful appearance.

Another major influence that has drawn attention and concern to health status and promoted physical activity is the flood of educational information that comes by way of the media. This information is also conveyed by physicians, peers, spouses, and various community agents (e.g., class instructors, pharmacists). For several years there has been a national effort to make the general public aware of such health concerns as obesity, cardiovascular disease, high blood pressure, too much salt, high cholesterol, and lack of exercise.

Although recent trends seem to suggest that attitudes have changed in regard to involvement in physical activity and active sports among the middle-aged and elderly, many adults still decrease their level of involvement as they age (Brooks, 1987; Ostrow et al., 1981). Ostrow's study asked questions concerning 12 common sports in which respondents indicated that several activities (e.g., swimming, jogging, tennis, basketball) were decreasingly appropriate as one aged from 20 to 80 years. These views may be described as a form of **age-related stereotyping.** Although such activities as swimming, jogging, and tennis are currently encouraged as lifelong pursuits, there are sociocultural forces that expect aging adults (especially the elderly) to become more sedentary with age.

■ *Objective 13.12*   Presently there are three dominant social views of aging that have implications for involvement in physical activity. **Disengagement theory** suggests that as older adults slow down they gradually withdraw from society (Cumming & Henry, 1961). This is viewed as a mutual process in which the individual disengages from society and society detaches from the older person. Disengagement can be a consequence of several situational (e.g., loss of position in society, widowed), psychological, and biological factors. There is a reduction of social

Figure 13.3
An active lifestyle during later adulthood increases the likelihood of life satisfaction. Senior athletes pictured here are (left to right) Jean Benear, Jim Law, Jim Larsen, Dottie Gray, and Ben Knaub, sharing a congratulatory embrace at the biennial 1993 U.S. National Seniors Sports Classic IV.
Courtesy of: Helen Harris Associates.

interaction with aging and an increased self-preoccupation. These actions are thought to increase life satisfaction among older adults.

However, most research studies have found that when individuals continue to live active and productive lives as older adults, their satisfaction with life, in general, does not decline and may even go up (Schaie & Willis, 1991). According to **activity theory,** an active lifestyle during later adulthood increases the likelihood for life satisfaction (Neugarten, Havighurst, & Tobin, 1968). That is, the more active older people are, the less likely they are to age and show discontent with life. This theory suggests that middle-aged adults continue their roles and practices through late adulthood, and that if activities cease for some reason (e.g., retirement), it is important to find comparable substitutes.

A third view of aging is **social breakdown-reconstruction theory** (Kuypers & Bengston, 1973). This theory suggests that aging is promoted through negative psychological functioning, which involves a poor self-concept, a general lack of skills to deal with the world, and negative feedback from others and society. Negative feedback may take the form of *ageism,* or prejudice against older adults. Discrimination is often related to the perception that older persons are physically unfit, too rigid or feebleminded, useless, senile, and boring. For these reasons and more, they may be shunned socially, moved out of their jobs, or not hired for new ones. Unfortunately, in our society elderly individuals are generally not treated with the level of social respect shown such individuals in other countries such as Japan and China. Social breakdown-reconstruction theory suggests that we reorganize the social system so that elderly people are treated with more respect.

Although all three social theories would seem to describe adequately how different segments of society view and provide influences concerning a physically active lifestyle during older adulthood, Santrock (1989) probably sums it up best by stating that "All we know about older adults indicates that they are healthier and happier the more active they are" (p. 519).

In addition to these general social views of aging, several theories related to exercise motivation in older adults have been proposed. One of the more interesting perspectives proposes that Bandura's (1986) social cognitive theory may be an appropriate framework to study this behavior (Dzewaltowski, 1989). Using the triadic causation concept (the person, behavior, and context) as the basis for affecting change across the life span, three mechanisms are suggested as important mediators in influencing exercise motivation. Foremost is *self-efficacy,* defined as the individual's judgment of his or her capabilities to complete courses of action. Older adults who lack this characteristic usually drop out of or do not participate in exercise programs. As with younger populations, a strong influence in self-efficacy is interaction with various social agents. A second factor is the value older persons place on the outcome and the provision of incentives, which is associated with *outcome expectations.* The third mechanism, *self-evaluated dissatisfaction,* is illustrated by individuals who are motivated to exercise because they are not satisfied with their current level of performance (e.g., muscular strength, flexibility). Many older persons will continue exercising to avoid becoming dissatisfied if they quit. This perspective in general recognizes that in addition to psychological factors, social contexts and biological constraints are strong influences in exercise motivation in the elderly.

SUMMARY

Sociocultural influences have been shown to have a lifelong effect on involvement in physical activity and motor development. Socialization is a lifelong process that refers to a set of events and processes by which individuals acquire the beliefs and behaviors of a particular society and subgroup (culture). The general concepts of status, class, role, and norm are commonly used to describe components of the social structure.

Socialization involves acquiring knowledge and skills, including those related to the motor domain. Individuals may be taught these skills and information by direct instruction, shaping (experience), and modeling (observing others). Socialization is strongly influenced by various sociocultural agents that interpret culture to the individual. The major socializing agents are the family, school, peers, church, and the community. During the early developmental years, the family is the primary socializing agent responsible for the transmission of cultural content.

Bronfenbrenner has developed an ecological model designed to describe the various sociocultural influences. The individual is visualized in the center of five ecological environments: the microsystem (the most direct influences of family, school, etc.); the mesosystem (relationships between microsystem contexts); the exosystem (indirect influences); the macrosystem (broad-based attitudes and ideologies of society and culture); and the chronosystem (sociohistorical contexts).

The primary sociocultural influences during childhood and adolescence are the family, play and sport participation and peer relations, and coaches and teachers in the school and community. Although the family has undergone several changes in recent years, it still remains the dominant agent to the growing child. Of the many influences associated with the family, the most important are parental beliefs and attitudes toward involvement in physical activity. Parents who believe in a healthy, physically active lifestyle usually share it with their offspring and encourage their participation. Within the family unit, there are indications that specific members very likely play differential roles in play and sport socialization. All family members (parents and siblings) may play a significant role in influencing behavior, but the father appears to be the dominant agent in most families. Economic factors, as related to family income status, may also affect motor development. There are indications that children from low-income families (especially single-parent homes) may be limited in motor development experiences due at least in part to living on a tightly budgeted income and lack of parental involvement time, education, and experience.

Play and sport participation, both highly involved social practices, offer excellent opportunities for motor development. Play, recognized widely as a powerful form of social development, can be categorized as solitary, parallel, associative, and cooperative. As children grow older they normally progress from playing alone (solitary) to participation in organized group involvement (cooperative play), such as offered in most school and youth sport programs. There are indications that children and youth of today are involved in more passive forms of play and leisure (e.g., TV, video-type games) than their counterparts of about 40 years ago. Concern has been expressed that this type of behavior may have negative implications in regard to physical fitness (and obesity) and motor development.

Closely related to play and sport participation is the strong desire of individuals to interact with their peers. During childhood and adolescence, peer relations are usually of the same sex and are formed at the school and community through play and youth sport programs. As individuals enter later childhood and adolescence, influence from the peer group becomes increasingly important while that of the family and others may diminish. Peers have the potential to reinforce family influences regarding involvement in physical activity or to provide new considerations not introduced at home. Peer influence seems to be stronger in regard to involvement in team activities rather than individual sport endeavors.

Coaches and teachers are also strong socializing agents with a verified concern for helping individuals experience positive socialization outcomes. Past research studies indicate that whereas these agents are a strong influence on male involvement in physical activity and sport from childhood through the early adult years, encouragement with females has been limited mainly to the adolescent period. However, due to more recent changes in society's view of female involvement in sport and their increased opportunities (e.g., Title IX, university scholarships), this trend may be changing.

Society has expectations of how individuals should act and think as males and females beginning with infancy. Sex-role stereotyping, introducing boys and girls to physical activities and sport differently, is practiced by several social agents. Traditionally, male children tend to be given more active, sports-oriented play items and are encouraged very early to be involved in vigorous physical activity and sports. On the other hand, females are normally introduced to more passive play items and are not expected to strongly pursue athletic endeavors. As a result of these and other influences, young males and females develop a preference for "sex-appropriate" play items and physical activities. When individuals decide to get involved in physical activities that are not sex-appropriate in their view, a sex-role conflict can occur. However, society's view of sex-appropriate behavior for females does appear to be changing.

Self-esteem influences participation in play and sports involvement. Children who are praised by others generally have a good self-esteem and are more likely to participate in organized activities. Such participation is generally associated with higher self-esteem. Feedback from parents, peers, coaches, and teachers is critical to the development of self-esteem. However, a large portion of individuals are likely to drop out (especially females) of youth sports because of conflicts with other opportunities and undesirable features of the activity.

Although research on the issue of race and its influence on physical activity and sport involvement is sparse and difficult to separate from socioeconomic factors, there are indications that black and white children are treated differently. Generalizations contend that possible success in sports seems to take on more importance among black youths (compared to whites), especially males from low-income homes, and that encouragement from several social agents may be more influential. It is speculated that many black youths view sports as a more readily conceivable means for increasing social status, due in part to perceived barriers to other careers. There are also indications that blacks have fewer role models from other occupations to emulate. Although research is lacking, there may be some positive implications for motor development that arise from the additional encouragement to excel in sport activities.

The adulthood years are also a time of much social learning with the presence of several sociocultural influences and expectations. Beginning with early adulthood, the primary social agents are the media (education), peers, spouses, and individuals in the community. Other influences include personal, seasonal, and situational factors. Society in general, as expressed predominantly through the media, places a premium on youth and physical appearance. This influence, along with the positive habits formed in earlier years, causes many young adults to continue, improve on, or take up a physically active lifestyle. However, there is also a large segment of young adults who have not been influenced or who do not have the desire to engage in physical activities as part of their routine. This is due in part to lack of encouragement during their upbringing, present good general health, and optimistic views about future health and longevity.

As individuals approach middle age and enter later adulthood, concern for health status and physical appearance normally receives more attention. This is a time when physical skills are regressing and physical aging is becoming more evident. Society's emphasis on good health and youthfulness stimulates many to become involved in physical activity. Another strong agent is the extensive media promotion of information concerning the need for good health practices. Similar information may also be transmitted through peers, spouses, and members of the community.

Although recent trends seem to suggest that attitudes have changed in regard to involvement in physical activity among the middle-aged and elderly, many older adults still decrease their level of involvement as they age. There may be several causes for lack of involvement, but the most evident is society's traditional view of age-appropriate behavior (i.e., age-related stereotyping).

Related to this are the various social views of aging. Disengagement theory suggests that as older adults age they gradually withdraw from society and society detaches from them. Advocates of social breakdown-reconstruction theory contend that aging is promoted through negative psychological functioning. Negative feedback frequently takes the form of ageism, or prejudice against older persons. This theory suggests that society reorganize the system so that elderly people are treated with more respect. Activity theory seems to have the most support. According to this view, an active lifestyle during later adulthood increases the likelihood for life satisfaction. Implications include continuing or beginning a physically active lifestyle.

SUGGESTED READINGS

Dunning, E. G., Maguire, J. A., & Pearton, R. E. (1993). (Eds.). *The sports process.* Champaign, IL: Human Kinetics.

McPherson, B. D., Curtis, J. E., & Loy, J. W. (1989). *The social significance of sport.* Champaign, IL: Human Kinetics.

Rubin, K. H., Fein, G. G., & Vanderberg, B. (1983). Play. In P. H. Mussen (Ed.), *Handbook of child psychology* (Vol. 4, 4th ed.). New York: Wiley.

Sage, G. H. (1986). Social development. In V. Seefeldt (Ed.), *Physical activity and well-being.* Reston, VA: AAHPERD.

Santrock, J. W. (1995). *Life-span development* (5th ed.). Dubuque, IA: Wm. C. Brown.

Appendix A
Locating and Retrieving Information: Conducting a Literature Search

The purpose of this section is to assist you in locating information related to topics discussed in this text and to briefly survey selected sources of information. With the emergence of personal and large computer systems, much has changed (and continues to evolve rapidly) in regard to the efficiency and sophistication of information storage and retrieval systems. Most libraries today offer automated information retrieval services (AIRS) that use elaborate computer systems and databank services (providers of numerous groups of important databases) to locate and retrieve information much more rapidly than an individual could do manually through card catalogs and other reference materials. Such services enable users to search virtually the entire body of recent information (within the last 10 to 15 years) in hundreds of indexes, abstracting services, and directories. In some instances, a personal computer with modem or online connection can perform many of these same operations.

Search Guidelines

Before undertaking a literature search using an AIRS, you need to decide the following:

1. The subject, scope, and limitations of your search.
2. The key words that best represent the subject of your search.
3. The time span to be covered by your search (most databases go back about 10 to 15 years).
4. The retrieval system(s) (databank) to be used and databases, indexes, abstracts, or other sources you want to search. Many libraries provide special assistance for computer-based searches.

Frequently Used Sources of Information

(Consult with library staff/computer services for updates due to rapidly changing technology.)

Major Databanks (computer-based retrieval services)

BRS

CD-ROM (varied database sets)

CompuServe

DIALOG (over 400 databases online)

DATA-STAR

Abstracts and Indexes (databases)

Physical Education Index	*Index Medicus*
Dissertation Abstracts Online	*Medline*
PSYCLIT (Psychological Abstracts)	*Life Sciences Collection*
SIRC (Sport Information Resource Center)	*Science Citation Index*
Sport and Fitness Index	*Social Sciences Citation Index*
Child Development Abstracts	*Applied Social Science Index*
Biological Abstracts	*Index to Scientific Reviews*
ERIC	*Completed Research in HPER*
Education Index	

Electronic Bulletin Board (via Bitnet/Internet)

LISTSERV (electronic forum for specific topics related to motor development; to get a complete list of topics, send the following message: LISTSERV@HEARN.BIT-NET list

Journals (developmental focus)

British Journal of Developmental Psychology

Child Development

Developmental Psychobiology

Developmental Psychology

Developmental Review

Human Development

Infant Behavior and Development

Journal of Applied Developmental Psychology

Merrill-Palmer Quarterly

Journals (older adults)

Experimental Aging Research

International Journal of Aging and Human Development

Journal of Aging and Health

Journal of Applied Gerontology

Journal of Gerontology

Journal of Physical Activity and Aging

Psychology of Aging

Journals (other)

Annuals of Human Biology

Brain and Cognition

British Journal of Educational Psychology

British Journal of Psychology

Canadian Journal of Psychology

Cortex

Child Study Journal

Early Childhood RQ

Growth

International Journal of Neuroscience

Journal of Applied Physiology

Journal of Experimental Child Psychology

Journal of Human Movement Sciences

Journal of Human Movement Studies

Journal of Motor Behavior

Journal of Pediatrics

Neuropsychologia

Neuropsychological Review

Pediatrics

Psychological Bulletin

Psychological Review

Research Quarterly for Exercise and Sport

Science

General Subject Titles (key words)

Child development

Motor development

Kinesiology

Physical Fitness

Human Development

Growth (and development)

Perceptual motor development

Aging

Motor Skills

Sports

Appendix B
Projects

The supplemental activities in this section are included to provide the student with additional learning experiences designed to expand their understanding and appreciation for lifelong motor development. Most of the activities are "hands-on" experiences that make some application to the study of motor development and the developmental process. You may find it useful to combine projects where possible; for example, you may find that you can measure or assess a variety of different characteristics (e.g., growth, physiological, perceptual) using the same subject during one or two visits. Chapter references are provided to guide you to the place in the text where you can find relevant background information.

Chapter Reference (all)

1. Objective: to introduce literature search skills. After reading Appendix A, go to the library and search for two journal research articles that are not more than 5 years old on topics related to lifelong motor development. After reviewing both articles, write a summary for each one that includes the purpose of the study, methods used to carry it out, results and conclusions, and personal comments about the study in general.

Chapter Reference (1)

2. Objective: to provide "hands-on" experience using some of the methods of studying human behavior. Obtain permission to visit a day-care center, early childhood facility, or kindergarten and observe the behavior of a child 2 to 5 years of age. Using the technique of *naturalistic observation* in a play setting, record the behavior of the child over a period of 30 to 45 minutes (this may take more than one visit). Try to be unobtrusive; if noticed, pretend to be watching all of the children. After completing the observation, attempt to organize the information in the form of a summary of behavior characteristics. Additional possibilities include: (a) formulate a questionnaire and *interview* the parents about their child in general, and his or her play behavior, (b) conduct a more in-depth *case history* with the

parents' assistance, and (c) ask permission to interview the child, if he or she is willing and mature enough to respond sufficiently (between 4 and 5 years).

Chapter Reference (3)

3.   Objective: to develop skills in measuring and interpreting anthropometric data. Select a male and female adolescent (ages 12 to 17 years) who are similar in age. Using the measurement descriptions provided in chapter 3, collect anthropometric information on the physical characteristics in the following list. After data collection is complete, use the interpretative tables and figures provided in the chapter to assess each person. Then compare the two sets of data looking for indications of individual and sex-related differences.

Anthropometric Characteristics

Head circumference

Height

Body weight

Sitting height

Shoulder width

Hip width

Somatotype (use Sheldon's method or estimate subjectively)

Triceps skinfold (see the current fitness battery in chapter 12 for value)

Chapter Reference (4)

4.   Objective: to enhance the understanding of the physiological process by observing and participating in laboratory assessment. Class project: In cooperation with an exercise physiologist or qualified lab assistant, observe the techniques for determining maximal oxygen consumption (aerobic power), anaerobic power, and vital capacity. Ask the specialist to explain the physiological functions that are involved in each evaluation. Some class members may wish to volunteer as subjects.

Chapter Reference (4, 1)

5.   Objective: to provide practical experience in determining possible age- and sex-related physiological differences in a cross-sectional study. Select at least four different age groups (seek a wide age span) with a male and female representing each. Collect data on one or more of the following:

1-mile run (aerobic power); 50-yard dash (anaerobic power); pull-ups, flexed-arm hang, sit-ups, or grip dynamometer (muscular strength and endurance); and the sit-and-reach (flexibility). The Suggested Readings in chapter 12 list recommended sources of information on measurement procedures. In the final report, note the purpose of the study, methods, results (indicating age group averages, averages by sex, and comparisons by age and sex), and conclusions.

Chapter Reference (5)

6. Objective: to gain insight from practitioners regarding selected factors that may affect growth and development. Ask an obstetrician (childbearing specialist) or nurse practitioner to visit class and discuss prenatal care and factors that may affect prenatal development. Have a set of questions ready.
7. Objective: same as project #6. Ask an exercise physiologist to visit the class and discuss the effects of physical activity and exercise on the various physiological functions of the body (e.g., cardiorespiratory, muscular development and strength, flexibility, body weight).

Chapter Reference (6)

8. Objective: to develop skills in testing selected perceptual characteristics among young children. Arrange permission at a local early childhood facility (or with an individual parent) to conduct a series of simple perceptual tests on a child 4 to 6 years of age. Develop a checklist evaluation form that includes items from the following suggestions. After the data are collected, use the information to make judgments of the child's performance based on what you learned in chapter 6.
Evaluation item suggestions:
   a. Body awareness (identification of body parts). Ask the child to touch basic body parts without a demonstration (e.g., head, eyes, ears, chin, knees). Use the checklist to record yes or no.
   b. Directional awareness (identification of right/left and various dimensions of external space). Ask the child to distinguish or point to various dimensions (e.g., right/left, up/down, over/under, front/back, in/out, top/bottom). Use the checklist to record yes or no.
   c. Lateral preference (hand preference). Place a pencil (writing while seated) and ball (throwing while standing) at the midline of the body. Ask the child to pick up the object and perform the task. Allow two trials for each task; if the right hand is used on both trials, record right-handed; if the right is used once and the left once, record nonestablished (mixed); if the left hand is used on both trials, record left-handed.

Lateral preference (foot preference). Using the same basic procedures (midline and preference scoring), determine foot preference by kicking a stationary ball and drawing imaginary letters with toes (while standing).

d. Visual perception (spatial orientation/size constancy). Using an open area (at least a 20-by-20-foot space) with the child standing, place three balls of the same size (medium) about 12 to 15 feet away from the child and approximately 1 foot apart. Ask the child the following questions: Which ball is closest to you? Which ball is farthest away? Are the balls the same size? If balls of different colors are available, also ask about this characteristic (color perception).

Visual perception (whole-part perception). Using figure 6.3, point to the figure (happy face made of household items) and ask the child to describe what he or she sees. Interpret the child's description.

e. Vestibular awareness (static and dynamic balance). Ask the child to balance on one foot, walk on a balance beam (2 to 4 inches wide and 6 to 7 feet long), and hop (traveling). Record the number of seconds of sustained activity.

Chapter Reference (6, 1)

9. Objective: to provide practical experience in determining possible age-related differences in selected aspects of perceptual development.

a. An optional supplement to project #8. Conduct a *cross-sectional study* of children 4, 5, and 6 years of age (or 5, 6, and 7) on lateral preference and balancing characteristics. If possible, try to use at least 10 children per age group. It may be more conducive to work in pairs or small groups.

b. In reference to balancing performance, the project could also include a collection of data on a group of adolescents (13 to 15 years), adults (25 to 30 years), and elderly individuals (over 60 years) to be used in comparison with the children's scores.

Chapter Reference (7, 1)

10. Objective: to observe the attention characteristics of children in a practical setting. Arrange permission to visit children 3 and 7 years of age (or 4 and 7) while they are being given some sort of instruction (classroom or outside). To maximize naturalistic observation, try to place yourself in an unobtrusive position. During each instructional session (at least two different times for about 30 minutes each), note the general attention characteristics of the group. For example, are they selectively attending to the instruction?

Are they working on the task provided? Are the older children more attentive? Were there conditions that affected their attention?

11. Objective: to provide experience in determining possible age-related differences in short-term memory. Using a life-span cross-sectional design (e.g., subjects aged 4, 9, 14, 25, 60+), administer a memory span test and compare the averages of each group (5 to 10 per group). Use a random table of digits, 1 through 9, and begin by presenting three to five digits verbally (three for the 4-year-olds). Present one per second, then immediately ask the subject to repeat the list in order. Record the highest number recalled after two trials. If the entire list is recalled successfully, increase the number of digits presented by two until the individual is unsuccessful. Arrange for two to four students per group to collect data on a specific age group, and share the results with the other student groups or have them present the data in class in visual form.

12. Objective: to gain practical insight into the life-span perspective of change in reaction time. Using a life-span design similar to project #11, administer simple reaction time tests and compare group averages. Because only a few reaction time instruments are likely to be available, this project may best be conducted in small groups or as an option among several projects.

Chapter Reference (8)

13. Objective: to provide practical experience in observing and evaluating characteristics associated with early movement behavior. After locating and receiving permission to study an infant 3 to 9 months of age (preferably 4 to 5 months), compile a list of reflexive, stereotypic, and rudimentary behaviors characteristic of a child of that age (refer to chapter 8). During your visits (of approximately three or four sessions of 45 to 60 minutes), record observed behavior and try to elicit the characteristic reflexes and rudimentary movements (with the aid of a parent or guardian).

Chapter Reference (9, 10, 12)

14. Objective: to increase insight and develop skills related to process- and product-oriented assessment of selected motor skills. Select a child who is between 5 and 6 years of age. Using the mature movement pattern (process) characteristics as a point of reference (chapter 9) and product (performance) values (chapter 10), assess the child's movement behavior. If possible, use videotape or film to record movement pattern form to allow repeated viewing. Following are some items that may be selected for assessment:

Process (characteristics)	Product (performance)
running (sprint 30–40 yd)	speed (yd/sec)
jumping (standing long jump)	distance (in)
hopping (10–12 ft)	consecutive hops/distance (ft)
skipping (25–30 ft)	
throwing (one-arm/overarm) tennis ball	distance (ft)
catching (two-handed) tennis ball	
striking (batting/two-handed) whiffle ball suspended by rope or tee at waist level	
kicking rubber #8 stationary ball	distance (ft)
bouncing/dribbling (rubber #8 ball)	distance (ft)
climbing (6–7 ft ladder)	

Chapter Reference (9)

15. Objective: to gain experience in observing behavior and administering tests associated with manual manipulation.
    a. Select three children approximately 3, 5, and 8 years of age. Administer the test of finger opposition. Let the subjects have one practice trial, then record three 30-second trials per person. Calculate the average and compare subjects. Optional tasks include picking up pennies or match sticks and placing them in a container, stringing beads, and moving pegs. The number completed in 30 seconds is the measure of performance.
    b. Obtain permission to study a 4- to 5-year-old child's fine motor manipulation behavior. Review the information and developmental milestones associated with this age range: Look at construction and self-help skills (e.g., buttons shirt, laces shoes), writing grips (e.g., power grip, tripod), and drawing or writing skills (e.g., ability to draw a circle, lines, square, and other figures). Develop a checklist of characteristic behaviors likely to be observed or elicited during visits. Most of the expected behaviors can be observed in either the school or home setting. However, an interview with the aid of a parent is recommended if more information is needed.

Chapter Reference (12)

16. Objective: to gain experience administering a formal motor assessment test or battery of tests. Depending on the assessment materials available, allow students in groups of two to four to administer one of the easier assessment instruments (e.g., DDST, PDMS, TGMD, FMPAI, or the BMAT-R).

Chapter Reference (13)

17. Objective: to observe and evaluate children engaged in various forms of play. Arrange permission to visit an early childhood or day-care facility where there are children 2 to 5 years of age. Observe and record play behavior using Parten's four categories at each of these age levels over a period of two or three days.

References

Acredolo, L. P. (1978). Development of spatial orientation in infancy. *Developmental Psychology, 14,* 224–234.

Acredolo, L. P., & Hake, J. L. (1982). Infant perception. In B. B. Wolman (Ed.), *Handbook of developmental psychology.* Englewood Cliffs, NJ: Prentice-Hall.

Adams, J. A. (1971). A closed theory of motor learning. *Journal of Motor Behavior, 3,* 111–150.

Adams, R. D. (1980). Morphological aspects of aging in the human nervous system. In J. E. Birren & R. B. Sloane (Eds.), *Handbook of mental health and aging* (pp. 149–160). Englewood Cliffs, NJ: Prentice-Hall.

Adrian, M. J., & Cooper, J. M. (1989). *Biomechanics of human movement.* Indianapolis: Benchmark.

Alexander, M. J., Ready, A. E., & Fougere-Mailey, G. (1985). The fitness level of females in various age groups. *Canadian Journal of Health, Physical Education, and Recreation, 51,* 8–12.

Allen, J. (1978). *Visual acuity development in human infants up to 6 months of age.* Unpublished doctoral dissertation, University of Washington, Seattle.

Allen, P. A., & Crozier, L. C. (1992). Age and ideal chunk size. *Journal of Gerontology, 47* (1), 47–51.

Als, H., Tronick, E., Lester, B. M., & Brazelton, T. B. (1979). Specific neonatal measures: The Brazelton Neonatal Assessment Scale. In J. D. Osofsky (Ed.), *Handbook of infant development.* New York: John Wiley & Sons.

Ama, P. F. M., Simoneau, J. A., Boulay, M. R., Seresse, O., Theriault, G., & Bouchard, C. (1986). Skeletal muscle characteristics in sedentary Black and Caucasian males. *Journal of Applied Physiology, 61,* 1758–1761.

American Academy of Pediatrics. (1983). Weight training and weight lifting: Information for the pediatrician. *The Physician and Sportsmedicine, 11,* 157–161.

American Academy of Pediatrics. (1987). Physical fitness and schools. *Pediatrics, 80* (3), 449–450.

American College of Sports Medicine. (1988). Physical fitness in children and youth. *Medicine and Science in Sports and Exercise, 20,* 422–423.

American Psychological Association. (1985). *Standards for educational and psychological tests.* Washington, DC: Author.

Andres, R., & Tobin, J. D. (1977). Endocrine system. In C. E. Finch & L. Heyflick (Eds.), *Handbook of the biology of aging* (pp. 357–378). New York: Van Nostrand Reinhold.

Aniansson, A. (1980). *Muscle function in old age with special reference to muscle morphology, effect of training and capacity in activities of daily living.* Goteborg, Sweden: Goteborg University.

Annett, M. (1970). The growth of manual performance and speed. *British Journal of Psychology, 61,* 545–558.

Annett, M. (1978). *A single gene explanation of right and left handedness and braindedness.* Coventry, England: Lanchester Polytechnic.

Annett, M. (1985). *Left, right, hand, and brain: The Rightshift theory.* London: Laurence Erlbaum Associates.

Anthrop, J., & Allison, M. T. (1983). Role conflict and the high school female athlete. *Research Quarterly for Exercise and Sport, 54,* 104–111.

Apgar, V. A. (1953). A proposal for a new method of evaluation of the newborn infant. *Current Research in Anesthesia and Analgesia, 32,* 260–267.

Apgar, V. A., & James, L. S. (1962). Further observations on the newborn scoring system. *American Journal of Diseases of Children, 104,* 419–428.

Apt, L., & Gaffney, W. L. (1982). The eyes. In A. M. Rudolph (Ed.), *Pediatrics* (17th ed., p. 1779). Norwalk, CT: Appleton-Century-Crofts.

Arlin, P. K. (1975). Cognitive development in adulthood: A fifth stage? *Developmental Psychology, 11,* 602–606.

Arlin, P. K. (1977). Piagetian operations in problem finding. *Developmental Psychology, 13,* 297–298.

Arnheim, D. D., & Sinclair, W. A. (1979). *The clumsy child: A program of motor therapy* (2nd ed.). St. Louis: Mosby.

Astrand, P-O., & Rodahl, K. (1986). *Textbook of work physiology: Physiological bases of exercise* (3rd ed.). New York: McGraw-Hill.

Atkinson, R. C., & Shiffrin, R. M. (1971). The control of short-term memory. *Scientific American, 225,* 82–90.

Ayres, A. J. (1966). *Southern California sensory-motor integration tests.* Los Angeles, CA: Western Psychological Services.

Ayres, A. J. (1978). *Southern California sensory-motor integration test manual.* Los Angeles, CA: Western Psychological Corporation.

Azar, G., & Lawton, A. (1964). Gait and stepping factors in the frequent falls of elderly women. *Gerontologist, 4,* 83–84, 103.

Baddeley, A. D. (1986). *Working memory.* New York: Oxford University Press.

Baddeley, A. D., & Hitch, G. (1974). Working memory. In G. H. Bower (Ed.), *The psychology of learning and motivation: Advances in research and theory* (Vol. 8, pp. 47–48). New York: Academic Press.

Bahrick, L. E., Walker, A. S., & Neisser, U. (1981). Selective looking by infants. *Cognitive Psychology, 13,* 377–390.

Bailey, D. A., Malina, R. M., & Rasmussen, R. L. (1978). The influence of exercise, physical activity, and athletic performance on the dynamics of human growth. In F. Falkner & J. M. Tanner (Eds.), *Human growth* (Vol. 2, pp. 475–505). New York: Plenum Press.

Ballard, J. L., Novak, K. K., & Driver, M. (1979). A simplified score for assessment of fetal maturation of newly born infants. *Journal of Pediatrics, 95* (5), 769–774.

Bandura, A. (1977). *Social learning theory.* Englewood Cliffs, NJ: Prentice-Hall.

Bandura, A. (1986). *Social foundations of thought and action: A social cognitive theory.* Englewood Cliffs, NJ: Prentice-Hall.

Bandura, A. (1989). Social cognitive theory. In R. Vasta (Ed.), *Six theories of child development.* Greenwich, CT: JAI Press.

Banks, M. S., & Salapatek, P. (1981). Infant pattern vision: A new approach based on the contract sensitivity function. *Journal of Experimental Child Psychology, 31,* 1–45.

Banks, M. S., & Salapatek, P. (1983). Infant visual perception. In P. H. Mussen (Ed.), *Handbook of child psychology* (Vol. 2). New York: Wiley.

Baranowski, T., Thompson, W. O., Durant, R. H., Baranowski, J., & Puhl, J. (1993). Observations on physical activity in physical locations: Age, gender, ethnicity, and month effects. *Research Quarterly for Exercise and Sport, 64* (2), 127–133.

Barnes, K. E. (1971). Preschool play norms: A replication. *Developmental Psychology, 4,* 99–103.

Bar-Or, O. (1983). *Pediatric sports medicine for the practitioner: From physiologic principles to clinical applications.* New York: Springer-Verlag.

Barr, H. M., Streissguth, A. P., Darby, B. L., & Sampson, P. D. (1990). Prenatal exposure to alcohol, caffeine, tobacco, and aspirin: Effects on fine and gross motor performance in 4-year-old children. *Developmental Psychology, 26* (3), 339–348.

Barrett, T. R., & Wright, M. (1981). Age-related facilitation in recall following semantic processing. *Journal of Gerontology, 36,* 194–199.

Barsch, R. H. (1965). *Achieving perceptual-motor efficiency.* Seattle, WA: Special Child Publications.

Bartlett, F. C. (1932). *Remembering: A study in experimental and social psychology.* Cambridge, MA: Cambridge University Press.

Bayley, N. (1969). *Manual for the Bayley Scales of infant development.* New York: Psychological Corporation.

Bayley Scales of Infant Development (Bayley II). (1993). San Antonio, TX: The Psychological Corporation.

Bell, A. D., & Variend, S. (1985). Failure to demonstrate sexual dimorphism of the corpus callosum in childhood. *Journal of Anatomy, 143,* 143–147.

Berg, K. M., & Smith, M. C. (1983). Behavioral thresholds for tones during infancy. *Journal of Experimental Child Psychology, 35,* 409–425.

Bernstein, N. (1967). *The coordination and regulation of movements.* London: Pergamon Press.

Berry, M. (1982). The development of the human nervous system. In J. W. T. Dickerson & H. McGurk (Eds.), *Brain and behavioral development.* London: Surrey University Press.

Beunen, G., Malina, R. M., Van't Hof, M. A., Simons, J., Ostyn, M., Renson, R., & VanGerven, D. (1988). *Adolescent growth and motor performance: A longitudinal analysis of Belgian boys.* Champaign, IL: Human Kinetics.

Bhatia, V. P., Katiyar, G. P., & Agarwal, K. N. (1979). Effect of intrauterine nutritional deprivation on neuromotor behavior of the newborn. *Acta Paediatrica Scandinavica, 68,* 561–566.

Birch, H. D., & Lefford, A. (1967). Visual differentiation, intersensory integration and voluntary motor control. *Monographs of the Society for Research in Child Development, 32,* 1–87.

Birnholz, J. C., & Benacerraf, B. R. (1983). The development of human fetal hearing. *Science, 222,* 516–518.

Blair, S. N., Kohl, H. W., Paffenbarger, R. S., Jr., Clark, D. G., Cooper, K. H., & Gibbons, L. W. (1989). Physical fitness and all-cause mortality: Prospective study of healthy men and women. *Journal of American Medical Association, 262,* 2395–2401.

Blank, M., & Bridger, W. (1974). Cross modality transfer in nursery school children. *Journal of Experimental Child Psychology, 58,* 277–282.

Blote, A. W., & Van Haastern, R. (1989). Developmental dimensions in the drawing behavior of pre-school children. *Journal of Human Movement Studies, 17,* 187–205.

Boone, P. C., & Azen, S. P. (1979). Normal range of motion of joints in male subjects. *Journal of Bone and Joint Surgery, 16A,* 756–759.

Bornstein, M. H. (1984). Perceptual development. In M. H. Bornstein & M. E. Lamb (Eds.), *Developmental psychology: An advanced textbook.* Hillsdale, NJ: Erlbaum.

Bosch, A., Goslin, B. R., Noakes, T. D., & Dennis, S. T. (1990). Physiological differences between black and white runners during a treadmill marathon. *European Journal of Applied Physiology, 61,* 68–72.

Botwinick, J., & Storandt, M. (1974). *Memory, related functions and age.* Springfield, IL: Charles C Thomas.

Bower, T. G. R. (1966). The visual world of infants. *Scientific American, 215,* 80–97.

Bower, T. G. R. (1976). Repetitive processes in child development. *Scientific American, 235* (5), 38–47.

Braddock, J. (1980). Race, sports, and social nobility: A critical review. *Sociological Symposium, 30,* 18–38.

Bradshaw, J., & Nettleton, N. (1983). *Human cerebral asymmetry.* Englewood Cliffs, NJ: Prentice-Hall.

Branta, C. F. (1993). Qualitative research symposium in motor behavior: An overview. *1993 NASPSPA Abstracts, Sport and Exercise Psychology, 15,* S7.

Branta, C. F., Haubenstricker, J., & Seefeldt, V. (1984). Age changes in motor skills during childhood and adolescence. *Exercise and Sport Sciences Reviews, 12,* 467–520.

Branta, C. F., Painter, M., & Kiger, J. E. (1987). Gender differences in play patterns and sport participation of North American youth. In D. Gould & M. R. Weiss (Eds.), *Advances in Pediatric Sport Sciences: Vol. 2. Behavioral issues* (pp. 25–42). Champaign, IL: Human Kinetics.

Brazelton, T. B. (1978). Introduction. In A. J. Sameroff (Ed.), Organization and stability of newborn behavior: A commentary on the Brazelton Neonatal Behavioral Assessment Scale. *Monographs of the Society for Research and Child Development, 43,* 1–13.

Brazelton, T. B. (1979). Specific neonatal measures: The Brazelton Neonatal Behavior Assessment Scale. In J. D. Osofsky (Ed.), *The handbook of infant development.* New York: Wiley.

Brazelton, T. B. (1984). *Neonatal behavioral assessment scales*

(2nd ed.). Philadelphia: J. B. Lippincott.

Bredemeier, B. J., & Shields, D. L. (1987). Moral growth through physical activity: A structural developmental approach. In D. Gould and M. R. Weiss (Eds.), *Advances in pediatric sport sciences* (Vol. 2). Champaign, IL: Human Kinetics.

Broadhead, G. D., & Church, G. E. (1985). Movement characteristics of preschool children. *Research Quarterly for Exercise and Sport, 56,* 208–214.

Broca, P. (1865). Sur la faculté du language articulé. *Bulletin of Social Anthropology, 6,* 493–494.

Brocklehurst, J. C., Robertson, D., & James-Groom, P. (1982). Clinical correlates of sway in old age: Sensory modalities. *Age and Aging, 11,* 1–10.

Brodkin, P., & Weiss, M. R. (1990). Developmental differences in motivation for participating in competitive swimming. *Journal of Sport and Exercise Psychology, 12,* 248–263.

Broekhoff, J. (1986). The effect of physical activity on physical growth and development. In G. A. Stull & H. M. Eckert (Eds.), *The academy papers: Effects of physical activity on children.* Champaign, IL: Human Kinetics.

Bronfenbrenner, U. (1986). Ecology of the family as a context for human development: Research Perspectives. *Developmental Psychology, 22,* 723–742.

Bronfenbrenner, U. (1989). Ecological systems theory. In R. Vasta (Ed.), *Annals of Child Development* (Vol. 6). Greenwich, CT: JAI Press.

Bronfenbrenner, U. (1993). Ecological systems theory. In R. H. Wozniak (Ed.), *Development in context.* Hillsdale, NJ: Erlbaum.

Bronson, G. W. (1982). Structure, status and characteristics of the nervous system at birth. In P. Stratton (Ed.), *Psychobiology of the human newborn.* New York: Wiley.

Brooks, C. M. (1987). Adult participation in physical activities requiring moderate to high levels of energy expenditure. *Physician and Sportsmedicine, 15,* 118.

Brooks, G. A., & Fahey, T. D. (1985). *Exercise physiology: Human bioenergetics and its applications.* New York: Macmillan.

Brooks-Gunn, J. (1991). Maturational timing variations in adolescent girls, antecedents of. In R. M. Lerner, A. C. Petersen, & J. Brooks-Gunn (Eds.), *Encyclopedia of Adolescence.* New York: Garland.

Brown, A. L., Bransford, J. D., Ferrara, R. A., & Campione, J. C. (1983). Learning, remembering, and understanding. In R. H. Mussen (Ed.), *Handbook of child psychology* (Vol. 3). New York: Wiley.

Brown, A. L., & Smiley, S. S. (1978). The development strategies for studying texts. *Child Development, 49,* 1076–1088.

Brown, B. A. (1985). Factors influencing the process of withdrawal by female adolescents from the role of competitive age group swimmer. *Sociology of Sport Journal, 2,* 111–129.

Brown-Sequard, C. E. (1877). Dual character of the brain: The Toner lectures (Lecture II). *Smithsonian miscellaneous collections* (pp. 1–21). Washington, D. C.: Smithsonian Institute.

Bruce, R. A. (1984). Exercise, functional aerobic capacity and aging-another viewpoint. *Medicine and Science in Sports and Exercise, 16,* 18.

Bruce, R. D. (1966). *The effects of variation in ball trajectory upon the catching performance of elementary school children.* Unpublished doctoral dissertation, University of Wisconsin, Madison.

Bruininks, R. H. (1978). *Bruininks-Oseretsky test for motor proficiency.* Circle Pines, MN: American Guidance Service.

Bryant, P. E., Jones, P., Claxton, V., & Perkins, G. M. (1972). Recognition of shapes across modalities by infants. *Nature, 240,* 303–304.

Bukatko, D., & Daehler, M. W. (1995). *Child Development* (2nd ed.). Boston: Houghton Mifflin.

Burfoot, A. (1992). White men can't run. *Runner's World,* August, 90–95.

Burg, A. (1968). Lateral visual field as related to age and sex. *Journal of Applied Psychology, 52,* 10–15.

Burke, W. E., Tuttle, W. W., Thompson, C. W., Janney, C. D., & Weber, W. J. (1953). The relation of grip strength and grip-strength endurance to age. *Journal of Applied Physiology, 5,* 628–630.

Buros, O. (1987). *The tenth mental measurement yearbook.* Highland Park, NJ: Gryphon Press.

Burton, H. W., & Davis W. E. (1992). Optimizing the involvement and performance of children with physical impairments in movement activities. *Pediatric Exercise Science, 4,* 236–248.

Bushnell, E. W. (1982). The ontogeny of intermodal relations: Visions and touch in infancy. In R. Walk & H. Pick (Eds.), *Intersensory perception and sensory integration* (pp. 5–36). New York: Plenum.

Bushnell, E. W. (1985). The decline of visually guided reaching during infancy. *Infant Behavior and Development, 8,* 139–155.

Butler, G. E., McKie, M., & Ratcliffe, S. G. (1990). The cyclical nature of prepubertal growth. *Annals of Human Biology, 17*(3), 177–198.

Cacoursiere-Paige, F. (1974). Development of left-right concept in children. *Perceptual Motor Skills, 38,* 111–117.

Campbell, A. J., Borrie, M. J., & Spears, G. F. (1989). Risk factors for falls in a community-based prospective study of people 70 years and older. *Journal of Gerontology, 44*(4), 112–117.

Carson, L., & Wiegand, R. L. (1979). Motor schema formation and retention in young children: A test of Schmidt's schema theory. *Journal of Motor Behavior, 11,* 247–251.

Carter, J. E. L. (1980). *The Health-Carter somatotype method* (rev. ed.). San Diego: San Diego State University.

Carter, J. H. (1982). The effects of aging on selected visual functions: Color vision, glare sensitivity, field of vision, and accommodation. In R. Sekuler, L. P. Hutman, & C. Owsley (Eds.), *Aging and human visual function.* New York: Alan R. Liss.

Case, R. (1985). *Intellectual development: Birth to adulthood.* Orlando, Florida: Academic Press.

Case, R., Kurland, D. M., & Goldberg, J. (1982). Operational efficiency and the growth of short-term memory span. *Journal of Experimental Child Psychology, 33,* 386–404.

Caspersen, C. J., Christenson, G. M., & Pollard, R. A. (1986). Status of the 1990 physical fitness and exercise objectives—evidence from NHIS 1985. *Public Health Report, 101,* 587–592.

Cavanaugh, J. C., & Borkowski, J. G. (1980). Searching for metamemory-memory connections: A developmental study. *Developmental Psychology, 16,* 441–453.

Centers for Disease Control. (1992). Vigorous physical activity among high school students—United States, 1990. *Morbidity and Mortality Weekly Report, 41*(3), 33–36.

Cerella, J. (1990). Aging and information-processing rate. In J. E. Birren & K. W. Schaie (Eds.), *Handbook of the psychology of aging* (3rd ed.). New York: Academic Press.

Chasnoff, I. J. (1991). *Cocaine versus tobacco: Impact on infant and child outcome.* Paper presented at the biennial meeting of the Society for Research in Child Development, Seattle.

Cherry, E. C. (1953). Some experiments on the recognition of speech with one and with two ears. *Journal of the Acoustical Society of America, 25,* 975–979.

Chi, M. (1976). Short-term memory limitations in children: Capacity or processing deficits? *Memory and Cognition, 4,* 559–572.

Chi, M. (1977a). Age differences in memory span. *Journal of Experimental Child Psychology, 23,* 266–281.

Chi, M. (1977b). Age differences in the speed of processing: A critique. *Developmental Psychology, 13,* 543–544.

Chi, M. (1978). Knowledge structures and memory development. In R. Siegler (Ed.), *Children's thinking: What develops?* Hillsdale, NJ: Erlbaum.

Christie, J. F., & Wardle, F. (1992). How much time is needed to play? *Young Children, 47,* 28–33.

Clark, J. E. (1988). Development of voluntary motor skill. In E. Meisami & P. S. Timitas (Eds.), *Handbook of human growth and developmental biology* (Vol. I). Boca Raton, FL: CRC Press.

Clark, J. E., Lanphear, A. K., & Riddick, C. C. (1987). The effects of videogame playing on the response selection processing of elderly adults. *Journal of Gerontology, 42,* 82–85.

Clark, J. E., & Phillips, S. J. (1993). A longitudinal study of intralimb coordination in the first year of independent walking: A dynamical systems analysis. *Child Development, 64*(4), 1143–1157.

Clark, J. E., Whitall, J., & Phillips, S. J. (1988). Human interlimb coordination: The first 6 months of independent walking. *Developmental Psychobiology, 21*(5), 445–456.

Clarke, H. H. (Ed.). (1975). Joint and body range of movement. *Physical Fitness Research Digest, 5,* 16–18.

Clarke-Stewart, A., Perlmutter, M., & Freedman, S. (1988). *Lifelong human development.* New York: John Wiley & Sons.

Clifton, R. K., Muri, D. W., Ashmead, D. H., & Clarkson, M. G. (1993). Is visually guided reaching in early infancy a myth? *Child Development, 64*(4), 1099–1110.

Coakley, J. J. (1987). Children and the sport socialization process. In *Advances in pediatric sport sciences.* Champaign, IL: Human Kinetics Publishers, Inc.

Cole, M., & Cole, S. R. (1993). *The development of children.* New York: W. H. Freeman.

Coley, I. L. (1978). *Pediatric assessment of self-care activities.* St. Louis: Mosby.

Colling-Saltin, A. S. (1980). Skeletal muscle development in the human fetus and during childhood. In K. Berg & B. O. Eriksson (Eds.), *Children and exercise IX* (pp. 193–207). Baltimore: University Park Press.

Collins, J. K. (1976). Distance perception as a function of age. *Australian Journal of Psychology, 28,* 109–113.

Conner, J. W., & Serbin, L. A. (1977). Behaviorally based masculine- and feminine-activity-preference scales for preschooler: Correlates with other classroom behaviors and cognitive tests. *Child Development, 48,* 1411–1416.

Corballis, M. C. (1983). *Human laterality.* New York: Academic Press.

Corballis, M. C., & Morgan, M. J. (1978). On the biological basis of human laterality: I. Evidence for a maturational right-left gradient. *Behavioral and Brain Sciences, 2,* 261–269.

Corbin, C. B. (1980). *A textbook of motor development* (2nd ed.). Dubuque, IA: Wm. C. Brown.

Coren, S. (1993). The lateral prefancy inventory for measurement of handedness, footedness, eyedness, and earedness: Norms for young adults. *Bulletin of the Psychonomic Society, 31*(1), 1–3.

Coren, S., & Porac, C. (1977). Fifty activities of right-handedness: The historical record. *Science, 198,* 631–632.

Coren, S., Porac, C., & Duncan, P. (1981). Lateral preference behaviors in preschool children and young adults. *Child Development, 52,* 443–450.

Cowan, W. (1978). Aspects of neural development. In R. Porter (Ed.), *Neurophysiology* (Vol. 111, No. 17, pp. 149–191). Baltimore, MD: University Park Press.

Cowan, W. (1979). The development of the brain. *Scientific American, 241,* 112–133.

Cragg, B. (1974). Plasticity of synapses. *British Medical Bulletin, 30,* 141–144.

Craik, F. I. M. (1977). Age differences in human memory. In J. E. Birren & K. W. Schaie (Eds.), *Handbook of the psychology of aging.* New York: Van Nostrand Reinhold.

Craik, F. I. M. (1979). Levels of processing: Overview and closing comments. In L. S. Cermak & F. I. M. Craik (Eds.), *Levels of processing in human memory.* Hillsdale, NJ: Erlbaum.

Craik, F. I. M., & Lockhart, R. (1972). Levels of processing: A framework for memory research. *Journal of Verbal Learning and Verbal Behavior, 11,* 671–676.

Craik, F. I. M., & Simon, E. (1980). Age differences in memory: The roles of attention and the depth of processing. In L. W. Poon, J. L. Fozard, L. S. Cermak, D. Arenberg, & L. W. Thompson (Eds.), *New directions in memory and aging.* Hillsdale, NJ: Erlbaum.

Craik, R. (1989). Changes in locomotion in the aging adult. In M. H. Woollacott & A. Shumway-Cook (Eds.), *Development of posture and gait across the life span.* Columbia, SC: University of South Carolina Press.

Cratty, B. J. (1986). *Perceptual and motor development in infants and children* (3rd ed.). Englewood Cliffs, NJ: Prentice-Hall.

Crum, J. F., & Eckert, H. M. (1985). Play patterns of primary school children. In J. E. Clark & J. H. Humphrey, *Motor development: Current selected research* (Vol. 1, p. 99). Princeton, NJ: Princeton Book Co.

Cummings, E., & Henry, W. (1961). *Growing old.* New York: Basic Books.

Cummings, M. F., Van Hof-van Duin, J., Mayer, D. L., Hansen, R. M., & Fulton, A. B. (1988).

Visual fields of young children. *Behavioral Brain Research, 29,* 7–16.

Cunningham, D. A., Paterson, D. H., & Blimkie, C. J. R. (1984). The development of the cardiorespiratory system with growth and physical activity. In R. A. Bileau (Ed.), *Advances in Pediatric Sport Sciences, 1,* 85–116.

Dargent-Paré, C., De Agnostini, M., Mesbah, M., & Dellatolas, G. (1992). Foot and eye preferences in adults: Relationship with handedness, sex, and age. *Cortex, 28,* 343–351.

Day, N. (1991). *Effects of alcohol and marijuana on growth and development.* Paper presented at the biennial meeting of the Society for Research in Child Development, Seattle.

Deach, D. (1950). *Genetic development of motor skills in children two through six years of age.* Unpublished doctoral dissertation, University of Michigan, Ann Arbor.

Delacato, C. H. (1959). *Treatment and prevention of reading problems.* Springfield, IL: Charles C Thomas.

Delacato, C. H. (1966). *Neurological organization and reading.* Springfield IL: Charles C Thomas.

de Lacoste, M. C., Halloway, R. L., & Woodward, D. J. (1986). Sex differences in the fetal human corpus callosum. *Human Neurobiology, 5,* 93–96.

Dempster, F. N. (1981). Memory span: Sources of individual and developmental differences. *Psychological Bulletin, 89,* 63–100.

Denckla, M. B. (1973). Development of speed in repetitive and successive finger movements in normal children. *Developmental Medicine and Child Neurology, 15,* 635–645.

Denckla, M. B. (1974). Development of motor coordination in normal children. *Developmental Medicine and Child Neurology, 16,* 729–741.

Denney, N. W., Jones, F. W., & Krigel, S. W. (1979). Modifying the questioning strategies of young children and elderly adults. *Human Development, 22,* 23–36.

Denver Developmental Screening (Denver II). (1990). W. K. Frankenburg & J. B. Dodds. Denver: DDM.

DeOreo, K. D. (1975). Dynamic balance in preschool children: Process and product. In D. M. Landers (Ed.), *Psychology of sport and motor behavior II: Proceedings of the North American Society for Psychology of Sport and Physical Activity* (pp. 575–584). University Park: Pennsylvania State University.

DeOreo, K. L. (1974). The performance and development of fundamental motor skills in preschool children. In D. M. Landers (Ed.), *Psychology of motor behavior and sport.* Champaign, IL: Human Kinetics Press.

DeOreo, K. L., & Keogh, J. (1980). Performance of fundamental motor tasks. In C. B. Corbin (Ed.), *A textbook of motor development* (2nd ed.). Dubuque, IA: Wm. C. Brown.

DeOreo, K. L., & Wade, M. G. (1971). Dynamic and static balancing ability of preschool children. *Journal of Motor Behavior, 3,* 326–335.

de Schonen, S. (1977). Functional asymmetries in the development of bimanual coordination in human infants. *Journal of Human Movement Studies, 3,* 144–156.

Deupree, R. H., & Simon, J. R. (1963). Reaction time and movement time as a function of age, stimulus duration, and task difficulty. *Ergonomics, 6* (4), 403–411.

deVries, H., & Housh, T. (1994). *Physiology of exercise* (5th ed.). Madison, WI: Brown & Benchmark.

Dietz, W. H., Jr., & Gortmaker, S. L. (1985). Do we fatten our children at the TV set? Obesity and television viewing in children and adolescents. *Pediatrics, 75,* 807–812.

Dixon, R. A., Kurzman, D., & Friesen, I. C. (1993). Handwriting performance in younger and older adults: Age, familiarity, and practice effects. *Psychology and Aging, 8*(3), 360–370.

Dobbing, J. (1976). Vulnerable periods of brain growth and somatic growth. In D. F. Roberts (Ed.), *The biology of human fetal growth.* New York: Halsted Press.

Dobbs, A. T., & Rule, B. G. (1989). Adult age differences in working memory. *Psychology and Aging, 4*(4), 500–503.

Dohrmann, P. (1965). Throwing and kicking ability of 8-year-old boys and girls. *Research Quarterly, 35,* 464–471.

Douglas, J. H., & Miller, J. A. (1977, September). Record breaking women. *Science News,* pp. 172–174.

Drage, J. S., Kennedy, C., Berendes, H., Schwartz, B. K., & Weiss, W. (1966). The Apgar score as an index of infant morbidity. *Developmental Medicine and Child Neurology, 8,* 141–148.

Duda, M. (1986). Prepubescent strength training gains support. *The Physician and Sports Medicine, 14,* 157–161.

du Plessis, M. P., Smit, P. J., du Plessis, L. A. S., Geyer, H. J., & Mathews, G. (1985). The composition of muscle fibers in a group of adolescents. In R. A. Binkhorst, H. C. G. Kemper, & W. H. M. Saris (Eds.), *Children*

and exercise XI. Champaign, IL: Human Kinetics.

Dworetzky, J. P., & Davis, N. J. (1989). *Human development: A lifespan approach.* St. Paul, MN: West.

Dzewaltowski, D. A. (1989). A social cognitive theory of older adult exercise motivation. In A. C. Ostrow (Ed.), *Aging and motor behavior.* Indianapolis: Benchmark Press.

Easton, T. A. (1972). On the normal use of reflexes. *American Scientist, 60,* 591–599.

Eaton, W. O., & Enns, L. R. (1986). Sex differences in human motor activity level. *Psychological Bulletin, 100,* 19–28.

Eckert, H. M. (1987). *Motor development* (3rd ed.). Indianapolis, IN: Benchmark.

Edleman, G. M. (1987). *Neural darwinism.* New York: Basic.

Edleman, G. M. (1989). *The remembered present.* New York: Basic.

Einkauf, D. K., Gohdes, M. L., Jensen, G. M., & Jewell, M. J. (1987). Changes in spinal mobility with increasing age in women. *Physical Therapy, 67,* 370–375.

Eisenberg, R. B. (1976). *Auditory competence in early life.* Baltimore: University Park Press.

Elkind, D. (1975). Perceptual development in children. *American Scientist, 63,* 533–541.

Elliot, R. (1972). Simple reaction time in children: Effects of incentive, incentive shift, and other training variables. *Journal of Experimental Child Psychology, 13,* 540–557.

Era, P., & Heikkinen, E. (1985). Postural sway during standing and unexpected disturbance of balance in random examples of men of different ages. *Journal of Gerontology, 40,* 287–295.

Erbaugh, S. J. (1984). The relationship of stability

performance and the physical growth characteristics of preschool children. *Research Quarterly for Exercise and Sport, 55,* 8–16.

Erhardt, R. P. (1982). *Developmental hand dysfunction.* Laurel, MD: RAMSCO.

Erikson, E. H. (1963). *Childhood and society* (2nd ed.). New York: W. W. Norton.

Espenschade, A. (1960). Motor development. In W. R. Johnson (Ed.), *Science and medicine of exercise and sport.* New York: Harper & Row.

Espenschade, A. (1963). Restudy of relationships between physical performances of school children and age, height, and weight. *Research Quarterly, 34,* 2540–2545.

Eveleth, P. B., & Tanner, J. M. (1990). *Worldwide variation in human growth* (2nd ed.). Cambridge, England: Cambridge University Press.

Everitt, A. V., & Huang, C. Y. (1980). The hypothalamus, neuroendocrine, and autonomic nervous systems in aging. In J. E. Birren & R. B. Sloane (Eds.), *Handbook of mental health and aging* (pp. 100–133). Englewood Cliffs, NJ: Prentice-Hall.

Fagan, J. W. (1984). Infants' long-term memory for stimulus color. *Developmental Psychology, 20,* 435–440.

Fagot, B. I., & Leinbach, M. D. (1987). Socialization of sex roles within the family. In D. B. Carter (Ed.), *Current conceptions of sex roles and sex typing: Theory and research.* New York: Praeger.

Fahey, T. D. (1990). The development of respiratory capacity during exercise in children and adolescents. In E. Meisami & P. S. Timiras (Eds.), *Handbook of human growth and developmental biology* (Vol. III,

Part B, pp. 209–218). Boca Raton: CRS Press.

Finegan, J. K., Niccols, G. A., & Sitarenios, G. (1992). Relations between prenatal testosterone levels and cognitive abilities at 4 years. *Developmental Psychology, 28*(6), 1075–1089.

Fiorentino, M. R. (1981). *A basis for sensorimotor development: The influence of the primitive, postural reflexes on the development and distribution of tone.* Springfield, IL: Charles C Thomas.

Fischer, K., & Pipp, S. C. (1984). Processes of cognitive development: Optimal level and skill acquisition. In R. J. Sternberg (Ed.), *Mechanisms of cognitive development.* New York: W. H. Freeman.

Fischman, M. G., Moore, J. B., & Steele, K. H. (1992). Children's one hand catching as a function of age, gender, and ball location. *Research Quarterly for Exercise and Sport, 63*(4), 349–355.

Fitts, P. M. (1954). The information capacity of the human motor system in controlling the amplitude of movement. *Journal of Experimental Psychology, 47,* 381–391.

Flavell, J. H. (1984). Discussion. In R. J. Sternberg (Ed.), *Mechanisms of cognitive development.* New York: W. H. Freeman.

Flavell, J. H., Beach, D. H., & Chinsky, J. M. (1966). Spontaneous verbal rehearsal in memory tasks as a function of age. *Child Development, 37,* 283–299.

Flavell, T. H. (1985). *Cognitive development* (2nd ed.). Englewood Cliffs, NJ: Prentice-Hall.

Fleg, J. L. (1986). Alterations in cardiovascular structure and function with advancing age. *American Journal of Cardiology, 57,* 33c-44c.

Fleishman, E. A., & Parker, J. F. (1962). Factors in the retention and relearning of perceptual-motor skill. *Journal of Experimental Psychology, 64,* 215–226.

Foos, P. W. (1989). Adult age differences in working memory. *Psychology and Aging, 4*(3), 269–275.

Forfar, J. O., & Arneil, G. C. (Eds.). (1992). *Textbook of pediatrics.* (4th ed.). London: Churchill Livingstone.

Fortney, V. L. (1983). The kinematics and kinetics of the running pattern of two-, four-, and six-year-old children. *Research Quarterly for Exercise and Sport, 54,* 126–135.

Fozard, J. L. (1980). The time for remembering. In L. W. Poon, J. L. Fozard, L. S. Cermak, D. Arenberg, & L. W. Thompson (Eds.), *New directions in memory and aging.* Hillsdale, NJ: Erlbaum.

Fried, P. A., & Watkinson, B. (1990). 36- and 48-month neuro-behavioral follow-up of children prenatally exposed to marijuana, cigarettes, and alcohol. *Developmental and Behavioral Pediatrics, 11,* 49–58.

Frisch, R. E. (1976). Fatness of girls from menarche to age 18 years, with a nomogram. *Human Biology, 48,* 353–359.

Frisch, R. E. (1991). Puberty and body fat. In R. M. Lerner, A. C. Petersen, & J. Brooks-Gunn (Eds.), *Encyclopedia of adolescence.* New York: Garland.

Frisch, R. E., Gotz-Welbergen, A. V., McArthur, J. W., Albright, T., Witschi, J., Bullen, B., Birnholz, J., Reed, R. B., & Hermann, H. (1981). Delayed menarche and amenorrhea of college athletes in relation to age of onset and training. *Journal of the American Medical Association, 246,* 1559–1563.

Frostig, M. (1963). *Developmental test of visual perception.* Palo Alto, CA: Consulting Psychologists Press.

Frostig, M. (1966). *Developmental test of visual perception—revised.* Palo Alto, CA: Consulting Psychologists Press.

Furman, L. N., & Walden, T. A. (1990). Effects of script knowledge on preschool children's communicative interactions. *Developmental Psychology, 26,* 227–233.

Gabbard, C. (1982). Comparison and relationships of selected upper limb and anthropometric measurements to strength and endurance among children aged 4–6 years. *Journal of Human Movement Studies, 8,* 55–91.

Gabbard, C. (1984). Teaching motor skills to children: Theory into practice. *The Physical Educator, 41,* 69–71.

Gabbard, C. (1985). Children's striking force characteristics. *Journal of Human Movement Studies, 11,* 53–63.

Gabbard, C. (1993). Foot laterality during childhood: A review. *International Journal of Neuroscience, 72*(3–4), 175–182.

Gabbard, C., & Bonfigli, D. (1987). Foot laterality in four-year-olds. *Perceptual and Motor Skills, 67,* 943–946.

Gabbard, C., Dean, M., & Haensly, P. (1991). Foot preference behavior during early childhood. *Journal of Applied Developmental Psychology, 12*(1), 131–137.

Gabbard, C., & Gentry, V. (1994). *Foot preference behavior: A developmental perspective.* Paper presented at the National AAHPERD convention, Denver.

Gabbard, C., Hart, S., & Kanipe, D. (1993). Hand preference

consistency and fine motor performance in young children. *Cortex, 29,* 1–8.

Gabbard, C., LeBlanc, B., & Lowy, S. (1994). *Physical education for children* (2nd ed.). Englewood Cliffs, NJ: Prentice-Hall.

Gabbard, C., & Patterson, P. E. (1980). Relationship and comparison of selected anthropometric measures to muscular endurance and strength in children aged 3–5 years. *Annuals of Human Biology, 7*(6), 583–586.

Gallahue, D. L. (1989). *Understanding motor development* (2nd ed.). Indianapolis, IN: Benchmark Press.

Gardner, R. A., & Broman, M. (1979). The Purdue pegboard: Normative data on 1334 school children. *Journal of Clinical Child Psychology, 1,* 156–162.

Garn, S. M. (1980). Human growth. *Annual Review of Anthropology, 9,* 275–292.

Garn, S. M. (1982). Nutrition. In A. M. Rudolph (Ed.), *Pediatrics* (17th ed., pp. 198–199). Norwalk, CT: Appleton-Century-Crofts.

Gasser, T., Kneip, A., Binding, A., Prader, A., & Molinasi, C. (1991). The dynamics of linear growth in distance, velocity, and acceleration. *Annuals of Human Biology, 18,* 187–205.

Gentry, V., & Gabbard, C. (1994). Foot preference behavior: A developmental perspective. *Abstracts of Research Consortium-AAHPERD,* Denver.

Geschwind, N., & Galaburda, A. S. (1985). Cerebral lateralization: Biological mechanisms, associations, and pathology: A hypothesis and program for research, I, II, III. *Archives of Neurology, 42,* 426–457, 521–552, 634–654.

Gesell, A. (1925). *The mental growth of the preschool child.* New York: Macmillan.

Ghatala, E. S., Carbonari, J. P., & Bobele, L. Z. (1980). Developmental changes in incidental memory as a function of processing level, congruity and repetition. *Journal of Experimental Child Psychology, 29,* 74–87.

Gibson, E. J. (1969). *The principles of perceptual learning and development.* New York: Appleton-Century-Crofts.

Gibson, E. J. (1982). The concept of affordances in perceptual development: The renascence of functionalism. In W. A. Collins (Ed.), *Minnesota symposia on child psychology* (Vol. 15, pp. 55–80). Hillsdale, NJ: Erlbaum.

Gibson, E. J. (1988). Exploratory behavior in the development of perceiving, acting, and the acquiring of knowledge. *Annual Review of Psychology, 39,* 1–41.

Gibson, E. J., & Spelke, E. J. (1983). The development of perception. In P. H. Mussen (Ed.), *Handbook of child psychology.* New York: Wiley.

Gibson, E. J., & Walk, R. D. (1960). The "visual cliff." *Scientific American, 202,* 64–71.

Gibson, J. J. (1979). *An ecological approach to perception.* Boston: Houghton Mifflin.

Gillberg, C., Rasmussen, P., & Wahlstrom, J. (1982). Minor neurodevelopment disorders in children born to older mothers. *Developmental Medicine and Child Neurology, 24,* 437–447.

Glick, P. C., & Lin, S. (1987). Remarriage after divorce: Recent changes and demographic variations. *Sociological Perspectives, 30,* 162–179.

Goldman-Rakic, P. S., Isseroff, A., Schwartz, M. L., & Bugbee, N. M. (1983). The neurobiology of cognitive development. In

P. H. Mussen (Ed.), *Handbook of child psychology* (Vol. 2, 4th ed.). New York: Wiley.

Gortmaker, S. L., Dietz, W. H., Jr., Sobol, A. M., & Wehler, C. A. (1987). Increasing pediatric obesity in the United States. *American Journal of Diseases of Children, 141,* 535–540.

Gottfried, A. W., & Bathurst, K. (1983). Hand preference across time is related to intelligence in young girls, not boys. *Science, 221,* 1074–1076.

Gould, D., Feltz, D. L., Hort, T., & Weiss, M. (1982). Reasons for discontinuing involvement in competitive youth swimming. *Journal of Sport Behavior, 5,* 155–165.

Gould, D., & Horn, T. (1984). Participation motivation in young athletes. In J. M. Silva & R. S. Weinberg (Eds.), *Psychological foundations of sport.* Champaign, IL: Human Kinetics.

Grattan, M. P., DeVos, E., Levy, J., & McClintock, M. K. (1992). Asymmetric action in the human newborn: Sex differences in patterns of organization. *Child Development, 63,* 273–289.

Greendorfer, S. L. (1976, September). *A social learning approach to female sport involvement.* Paper presented at the annual convention of the American Psychological Association, Washington, D. C.

Greendorfer, S. L. (1977). Role of socializing agents in female sport involvement. *Research Quarterly, 48,* 304–310.

Greendorfer, S. L. (1979). Children sport socialization influences of male and female track athlete. *Arena Review, 3,* 39–53.

Greendorfer, S. L., & Ewing, M. E. (1981). Race and gender differences in children's socialization into sport.

Research Quarterly for Exercise and Sport, 52, 301–310.

Greendorfer, S. L., & Lewko, J. H. (1978). Role of family members in sport socialization of children. *Research Quarterly, 49,* 146–182.

Greene, P. H. (1972). Problems of organization of motor systems. In R. Rosen & F. M. Snell (Eds.), *Progress in theoretical biology* (Vol. 2). New York: Academic Press.

Greenspan, E. (1983). *Little winners: Inside the world of the child sport star.* Boston: Little Brown.

Greulich, W. W., & Pyle, S. I. (1959). *Radiographic atlas of skeletal development of the hand and wrist* (2nd ed.). Stanford: Stanford University Press.

Guttentag, R. E. (1984). The mental effort requirement of cumulative rehearsal: A developmental study. *Journal of Experimental Child Psychology, 37,* 92–106.

Guyton, A. C. (1991). *Basic neuroscience* (2nd ed.). Philadelphia: Saunders.

Haaland, K. Y., Harrington, D. L., & Grice, J. W. (1993). Processing speed and aging. *Psychology and Aging, 8*(4), 617–632.

Hagen, J. W., Hargrave, S., & Ross, W. (1973). Prompting and rehearsal in short-term memory. *Child Development, 44,* 201–204.

Hager, A., Sjostrom, L., & Arvidsson, B. (1977). Body fat and adipose tissue cellularity in infants: A longitudinal study. *Metabolism, 26,* 607–614.

Hahn, W. K. (1987). Cerebral lateralization of function: From infancy through childhood. *Psychological Bulletin, 101,* 376–392.

Haith, M. H. (1991). *Setting a path for the 90's: Some goals and challenges in infant sensory and perceptual development.* Paper

presented at the Society for Research in Child Development meeting, Seattle.

Haith, M. M. (1966). The response of the human newborn to visual movement. *Journal of Experimental Child Psychology, 3,* 235–243.

Hale, S. (1990). A global developmental trend in cognitive processing speed. *Child Development, 61,* 653–663.

Hall, E. G., Durborow, B., & Progen, J. (1986). Self-esteem of female athletes and nonathletes relative to sex role type and sport type. *Sex Roles, 15,* 379–390.

Halverson, L. E., & Roberton, M. A. (1979). The effects of instruction on overhand throwing development in children. In K. Newell & G. Roberts (Eds.), *Psychology of motor behavior and sport—1978* (pp. 258–269). Champaign, IL: Human Kinetics.

Halverson, L. E., Roberton, M. A., & Harper, C. J. (1973). Current research in motor development. *Journal of Research Development Education, 6,* 56.

Halverson, L. E., Roberton, M. A., & Langendorfer, S. (1982). Development of the overarm throw: Movement and ball velocity changes by seventh grade. *Research Quarterly for Exercise and Sport, 53,* 198–205.

Halverson, L. E., Roberton, M. A., Safrit, M. J., & Roberts, T. W. (1977). Effect of guided practice on overhand throw ball velocities of kindergarten children. *Research Quarterly, 48,* 311–318.

Halverson, L. E., & Williams, K. (1985). Developmental sequences for hopping distance: A prelongitudinal screening. *Research Quarterly for Exercise and Sport, 56,* 37–44.

Hanson, M. (1965). *Motor performance testing of elementary school age children.* Unpublished doctoral dissertation, University of Washington, Seattle.

Harris, O., & Hunt, L. (1984). *Race and sports involvement: Some implications of sports for black and white youth.* Paper presented at the annual meeting of the American Alliance for Health, Physical Education, Recreation and Dance, Anaheim, CA.

Harter, S. (1981). A model of intrinsic mastery motivation in children: Individual differences and developmental change. In W. A. Collins (Ed.), *Minnesota symposium on child psychology* (Vol. 14, pp. 215–255.) Hillsdale, NJ: Erlbaum.

Harter, S. (1982). A cognitive-developmental approach to children's use of affect and trait levels. In F. Serafice (Ed.), *Sociocognitive development in context.* New York: Guilford Press.

Harter, S. (1987). The determinants and mediational rule of global self-worth in children. In N. Eiserberg (Ed.), *Contemporary topics in developmental psychology.* New York: Wiley.

Hasher, L., & Zacks, R. T. (1979). Automatic and effortful processes in memory. *Journal of Experimental Psychology: General, 108,* 356–388.

Hasselkus, B. R., & Shambes, G. M. (1975). Aging and postural sway in women. *Journal of Gerontology, 30,* 661–667.

Hata, E., & Aoki, K. (1990). Age at menarche and selected menstrual characteristics in young Japanese athletes. *Research Quarterly for Exercise and Sport, 61*(2), 178–183.

Haubenstricker, J., Branta, C., Seefeldt, V., Forsblom, L., & Kiger, J. (1990).

Prelongitudinal screening of a developmental sequence of skipping. Paper presented at the annual meeting of the North American Society for the Psychology of Sport and Physical Activity, Houston, TX.

Haubenstricker, J., & Sapp, M. (1985). A brief review of the Bruininks-Oseretsky test of motor proficiency. *Motor Development Academy Newsletter* (NASPE), *5*(1), 1–2.

Haubenstricker, J., & Seefeldt, V. (1986). Acquisition of motor skills during childhood. In V. Seefeldt (Ed.), *Physical activity and well-being* (pp. 41–102). Reston, VA: AAHPERD.

Haubenstricker, J., Seefeldt, V., Branta, C., & Wilson, D. (1992). The relationship of performance on selected motor tasks during childhood and adolescence to performance when adult stature is attained. *Abstracts of North American Society for Sport Psychology and Physical Activity,* Pittsburgh.

Havighurst, R. J. (1972). *Developmental tasks and education* (3rd ed.). New York: McKay.

Head, H. (1926). *Aphasia and kindred disorders of speech.* Cambridge: Cambridge University Press.

Hecean, H., & deAjuriaguerra, J. (1964). *Left-handedness: Manual superiority and cerebral dominance.* New York: Grune & Stratton.

Hensley, L. D., East, W. B., & Stillwell, J. L. (1982). Body fatness and motor performance during preadolescence. *Research Quarterly for Exercise and Sport, 53,* 133–140.

Heyward, V. H., Johannes-Ellis, S. M., & Romer, J. F. (1986). Gender differences in strength. *Research Quarterly for Exercise and Sport, 57,* 154–159.

Himes, J. H. (1988). Racial variation in physique and body composition. *Canadian Journal of Sport Science, 13*(2), 117–126.

Hirata, K. (1979). *Selection of olympic champions* (Vol. 1). Santa Barbara: Institute of Environmental Stress.

Hochberg, J. E. (1978). *Perception* (2nd ed.). Englewood Cliffs, NJ: Prentice-Hall.

Hodgkins, J. (1962). Influence of age on the speed of reaction and movement in females. *Journal of Gerontology, 17,* 385–389.

Hoffman, L. W. (1989). Effects of maternal employment in the two-parent family. *American Psychologist, 44,* 283–292.

Hoffman, S. J., Imwold, C. H., & Koller, J. A. (1983). Accuracy of prediction in throwing: A taxonomic analysis of children's performance. *Research Quarterly for Exercise and Sport, 54,* 33–40.

Hofsten, C. (1991). Structuring of early reaching movements: A longitudinal study. *Journal of Motor Behavior, 23,* 280–292.

Hofsten, C., & Rönnqvist, L. (1993). The structuring of neonatal arm movements. *Child Development, 64*(4), 1046–1057.

Hogue, C. C. (1982). Injury in late life: Prevention. *Journal of the American Geriatric Society, 30,* 276–280.

Hood, K. E. (1991). Menstrual cycle. In R. M. Lerner, A. C. Petersen, & J. Brooks-Gunn (Eds.), *Encyclopedia of adolescence.* New York: Garland.

Horn, T. S. (1985). Coach's feedback and changes in children's perceptions of their physical competence. *Journal of Educational Psychology, 77,* 174–186.

Horn, T. S., & Hasbrook, C. A. (1987). Psychological characteristics and the criteria

children use for self-evaluation. *Journal of Sport Psychology, 9,* 208–221.

Horn, T. S., & Weiss, M. R. (1991). A developmental analysis of children's self ability judgements in the physical domain. *Pediatric Exercise Science, 3,* 310–326.

Hughes, F. P., & Noppe, L. D. (1991). *Human development across the lifespan.* New York: Macmillan.

Hughes, J. E. (1979). *Manual for the Hughes basic motor assessment.* Golden, CO: University of Colorado.

Hughes, J. E., & Riley, A. (1981). Tools for use with children having minor motor dysfunction. *Physical Therapy, 61*(4), 503–510.

Humphrey, D. E., & Humphrey, G. K. (1987). Sex differences in infant reaching. *Neuropsychologia, 25,* 971–975.

Hurst, J. W. (Ed.). (1982). *The heart.* New York: McGraw-Hill.

Huston, A. C., Wright, J. C., Rice, M. C., Kerkman, D., & St. Peters, M. (1990). Development of television viewing patterns in early childhood: A longitudinal investigation. *Developmental Psychology, 26,* 409–420.

Isaacs, L. D. (1980). Effects of ball size, ball color, and preferred color on catching by young children. *Perceptual and Motor Skills, 51,* 583–586.

Jaffe, M., & Kosakov, C. (1982). The motor development of fat babies. *Clinical Pediatrics, 27,* 619–621.

James, W. ([1890]1950). *The principles of psychology.* New York: Dover.

Jersild, A. T., & Bienstock, S. F. (1935). Development of rhythm in young children. *Child Development Monographs* (No. 22).

Johnson, R. (1962). Measurements of achievement in fundamental skills of elementary school children. *Research Quarterly, 33,* 94–103.

Jones, M., Buis, J., & Harris, D. (1986). Relationships of race and sex to physical and motor measures. *Perceptual and Motor Skills, 63,* 169–170.

Kail, R. (1984). *The development of memory in children.* New York: W. H. Freeman.

Kail, R. (1991). Developmental change in speed of processing during childhood and adolescence. *Psychological Bulletin, 109*(3), 490–501.

Kaluger, G., & Kaluger, M. (1984). *Human development* (3rd ed.). St. Louis, MO: Mosby.

Kannel, W. B. (1967). Habitual level of physical activity and risk of coronary heart disease: The Framingham study. *Canadian Medical Association Journal, 96,* 811–812.

Kaplan, R. M., & Saccuzzo, D. P. (1993). *Psychological testing.* Pacific Grove, CA: Brooks/Cole.

Kasch, F. W., Wallace, J. P., Van Camp, S. P., & Verity, L. (1988). A longitudinal study of cardiovascular stability in active men aged 45 to 65 years. *The Physician and Sportsmedicine, 16*(1), 117–123.

Kavale, K., & Matson, P. D. (1983). One jumped off the balance beam: Meta analysis of perceptual motor training. *Journal of Learning Disabilities, 16,* 165–173.

Kellgren, J. H., & Lawrence, J. S. (1957). Radiological assessment of osteoarthrosis. *Annals of Rheumatic Diseases, 16,* 494–502.

Kellogg, R. (1969). *Analyzing children's art.* Palo Alto, CA: National Press Books.

Kelso, J. A. S. (1982). Epilogue: Two strategies for investigating action. In J. A. S. Kelso (Ed.), *Human motor behavior. An introduction* (pp. 283–287). Hillsdale, NJ: Erlbaum.

Kelso, J. A. S., Holt, K. G., Kugler, P. N., & Turvey, M. T. (1980). On the concept of coordinative structures as dissipative structures. II. Empirical lines of convergence. In G. E. Stelmach & J. Requin (Eds.), *Tutorials in motor behavior.* Amsterdam: North-Holland.

Kelso, J. A. S., & Norman, P. E. (1978). Motor schema formation in children. *Developmental Psychology, 14,* 153–156.

Kelso, J. A. S., Southard, D. L., & Goodman, D. (1979). On the nature of human interlimb coordination. *Science, 203,* 1029–1031.

Kelso, J. A. S., Tuller, B. H., Vatikiotis-Bateson, E., & Fowler, C. A. (1984). Functionally specific articulatory cooperation following jaw perturbations during speech: Evidence for coordinative structures. *Journal of Experimental Psychology: Human Perception and Performance, 10,* 812–832.

Kenyon, G. S., & McPherson, B. D. (1973). Becoming involved in physical activity and sport: A process of socialization. In G. L. Rarick (Ed.), *Physical activity: Human growth and development* (pp. 301–332). New York: Academic Press.

Keogh, J. (1965). *Motor performance of elementary school children* (USPHS grants MH08319-01 and HD01059). Los Angeles: University of California, Dept. of Physical Education.

Keogh, J. (1968). *Developmental evaluation of limb movement tasks* (Tech. Rep. No. 1–68). Los Angeles: University of California, Dept. of Physical Education.

Keogh, J., & Sugden, D. (1985). *Movement skill development.* New York: Macmillan.

Kephart, N. C. (1971). *The slow learner in the classroom* (2nd ed.). Columbus, OH: Charles E. Merrill.

Kerr, R. (1982). *Psychomotor learning.* Philadelphia: Saunders.

Kerr, R., & Booth, B. (1977). Skill acquisition in elementary school children and schema theory. In D. M. Landers & R. W. Christina (Eds.), *Psychology of motor behavior and sport* (Vol. 2, pp. 243–247). Champaign, IL: Human Kinetics.

Kerr, R., & Booth, B. (1978). Specific and varied practice of motor skill. *Perceptual and Motor Skills, 46,* 385–401.

Kingsley, P., & Hagen, J. (1969). Induced versus spontaneous rehearsal in short-term memory in nursery school children. *Developmental Psychology, 1,* 40–46.

Kinsbourne, M. (1975). The ontogeny of cerebral dominance. *Annals of the New York Academy of Sciences, 263,* 244–250.

Kirshnit, C. E., Ham, M., & Richards, M. H. (1989). The sporting life: Athletic activities during early adolescence. *Journal of Youth and Adolescence, 18,* 601–615.

Kirshnit, C. E., Richards, M. H., & Ham, M. (1988). *Athletic participation and body image during early adolescence.* Paper presented at the meeting of the American Psychological Association, Atlanta, GA.

Klatzky, R. L. (1984). *Memory and awareness.* New York: W. H. Freeman.

Kleinginna, P. A., & Kleinginna, A. M. (1988). Current trends toward convergence of the behavioristic, functional, and cognitive perspectives in experimental psychology. *The Psychological Record, 38,* 369–392.

Klinger, A., Masataka, T., Adrian, M., & Smith, E. (1980, April). *Temporal and spatial characteristics of movement patterns of women over 60.* (Cited in Adrian, 1982). American Alliance for Health, Physical Education, Recreation and Dance Research Consortium Symposium, Detroit.

Knobloch, H., & Pasamanick, B. (Eds.). (1974). *Gesell and Amatruda's developmental diagnosis* (3rd ed.). New York: Harper & Row.

Kobasigawa, A. (1974). Utilization of retrieval cues by children in recall. *Child Development, 45,* 127–134.

Krahenbuhl, G. S., & Martin, S. L. (1977). Adolescent body size and flexibility. *Research Quarterly, 48,* 797–799.

Krahenbuhl, G. S., Skinner, J. S., & Kohrt, W. M. (1985). Developmental aspects of maximal aerobic power in children. In R. L. Terjung (Ed.), *Exercise Sport Science Reviews* (Vol. 13, pp. 503–538). New York: Macmillan.

Kreutzer, M. A., Leonard, C., & Flavell, J. H. (1975). An interview study of children's knowledge about memory. *Monographs of the Society for Research in Child Development, 40* (1, Serial No. 159).

Kreutzfeldt, J., Haim, M., & Bach, E. (1984). Hip fracture among the elderly in mixed urban and rural population. *Age and Aging, 13,* 111–119.

Krmpotic-Nemanic, J. (1969). Presbycusis and retrocochlear structures. *International Audiology, 8,* 210–220.

Krotkiewski, M., Kral, J. G., & Karlsson, J. (1980). Effects of castration and testosterone substitution on body composition and muscle metabolism in rats. *Acta Physiologica Scandinavica, 109,* 233–237.

Kuczaj, S. A., & Maratsos, M. P. (1975). On the acquisition of front, back and side. *Child Development, 46,* 202–210.

Kugler, P. N., Kelso, J. A. S., & Turvey, M. T. (1980). On the concept of coordinative structures as dissipative structures. I. Theoretical lines of convergence. In G. E. Stelmach & J. Requin (Eds.), *Tutorials in motor behavior* (pp. 1–47). New York: North-Holland, Amsterdam.

Kugler, P. N., Kelso, J., & Turvey, M. (1982). On the control and coordination of naturally developing systems. In J. A. S. Kelso & J. E. Clark (Eds.), *The development of movement control and coordination* (pp. 5–78). New York: Wiley.

Kuhn, D., Nash, S. C., & Brucken, L. (1978). Sex role concepts of two- and three-year-olds. *Child Development, 49,* 445–451.

Kuypers, J. A., & Bengston, V. L. (1973). Social breakdown and competence: A model of normal aging. *Human Development, 16,* 181–201.

Labouvie-Vief, G. (1985). Intelligence and cognition. In J. E. Birren & K. W. Schaie (Eds.), *Handbook of the psychology of aging* (pp. 500–530). New York: Van Nostrand Reinhold.

Ladd, G. W., & Mize, J. A. (1983). A cognitive social learning model of social-skill training. *Psychological Review, 90,* 127–157.

Lamb, M. E. (1986). *The father's role: Applied perspectives.* New York: Wiley.

Landreth, C. (1958). *The psychology of early childhood.* New York: Alfred A. Knopf.

Lane, D. M., & Pearson, D. A. (1982). The development of selective attention. *Merrill-Palmer Quarterly, 28,* 317–345.

Lang, G. J., & Luff, A. R. (1990). Skeletal muscle growth, hypertrophy, repair, and regeneration. In E. Meisami & P. S. Timiras (Eds.), *Handbook of human growth and developmental biology* (Vol. 3). Boca Raton, FL: CRS Press.

Langendorfer, S. (1986). Books and media review. Test of gross motor development. *Adapted Physical Activity Quarterly, 3,* 186–190.

Larroche, J. C., & Amakawa, H. (1973). Glia of myelination and fat deposit during early myelogenesis. *Biology of the Neonate, 22,* 421–435.

Lasky, R. E. (1977). The effect of visual feedback of the hand on the reaching and retrieval behavior of young infants. *Child Development, 48,* 112–119.

Latchaw, M. (1954). Measuring selected motor skills in fourth, fifth, and sixth grades. *Research Quarterly, 25,* 439–449.

Lefford, A., Birch, H. G., & Green, G. (1974). The perceptual and cognitive bases for finger localization and selective finger movement in preschool children. *Child Development, 45,* 335–343.

Leme, S., & Shambles, G. (1978). Immature throwing patterns in normal adult women. *Journal of Human Movement Studies, 4,* 85–93.

Lennenberg, E. (1967). *Biological foundations of language.* New York: Wiley.

Leon, A. S., Connett, J., Jacobs, D. R., Jr., & Rauramaa, R. (1987). Leisure-time physical activity levels and risk of coronary heart disease and death: The multiple risk factor trial. *Journal of American*

Medical Association, 258, 2388–2395.

Leporé, F., Ptito, M., & Jasper, H. H. (Eds.). (1986). *Two hemispheres, one brain: Functions of the corpus callosum.* New York: Alan R. Liss.

Leudke, G. C. (1980). *Range of motion as the focus of teaching the overhand throwing pattern to children.* Unpublished doctoral dissertation, Indiana University, Bloomington.

Lewko, J. H., & Ewing, M. E. (1980). Sex differences and parental influences in sport involvement of children. *Journal of Sport Psychology, 2,* 62–68.

Liberty, C., & Ornstein, P. A. (1973). Age differences in organization and recall: The effects of training in categorization. *Journal of Experimental Child Psychology, 15,* 169–186.

Light, K. L., & Spirduso, W. W. (1990). Effects of adult aging on the movement complexity factor or response programming. *Journal of Gerontology, 45*(3), 107–109.

Light, L. L., & Anderson, P. A. (1983). Memory for scripts in young and older adults. *Memory and Cognition, 11,* 435–444.

Lindberg, M. A. (1980). Is knowledge base development a necessary and sufficient condition for memory development? *Journal of Experimental Child Psychology, 30,* 401–410.

Little, G. A. (1992). The fetus at risk. In R. A. Hookelman, S. B. Friedman, N. M. Nelron, & A. M. Seidel (Eds.), *Primary pediatric care* (2nd ed.). St. Louis, MO: Mosby Yearbook.

Lockman, J. J., & Thelen, E. (1993). Developmental biodynamics: Brain, body and behavior connections. *Child Development, 64*(4), 953–959.

Lohman, T. G. (1989). Assessment of body composition in children.

Pediatric Exercise Science, 1, 19–30.

Lohman, T. G., Boileau, R. A., & Slaughter, M. H. (1984). Body composition in children and youth. *Advances in Pediatric Sport Sciences:* Biological Issues (Vol. 1). Champaign, IL: Human Kinetics.

Londeree, B. R., & Moeschberger, M. L. (1982). Effects of age and other factors on maximal heart rate. *Research Quarterly for Exercise and Sport, 53,* 297–304.

Long, A. B., & Looft, W. R. (1972). Development of directionality in children: Ages six through twelve. *Developmental Psychology, 6,* 375–380.

Loovis, E. M., & Ersing, W. F. (1979). *Assessing and programming gross motor development for children* (2nd ed.). Loudonville, OH: Mohican Textbook Publishing.

Lorton, J. W., & Lorton, E. L. (1984). *Human development through the lifespan.* Monterey, CA: Brooks/Cole.

Lowrey, G. H. (1986). *Growth and development of children* (8th ed.). Chicago: Year Book Medical Publishers.

Loy, J. W., McPherson, B. D., & Kenyon, G. (1978). *Sport and social systems.* Reading, MA: Addison-Wesley.

Lund, J., & Lund, R. (1972). The effects of varying periods of visual deprivation on synaptogenesis in the superior colliculus of the rat. *Brain Research, 42,* 21–32.

Lund, R., & Lund, J. (1972). Development of synaptic patterns in the superior colliculus of the rat. *Brain Research, 42,* 1–20.

Lund, R. D. (1978). *Development and plasticity of the brain.* New York: Oxford University Press.

Lupinacci, N. S., Rikli, R. E., Jones, J., & Ross, D. (1993). Age and

physical activity effects on reaction time and digit symbol substitution performance in cognitively active adults. *Research Quarterly for Exercise and Sport, 64*(2), 144–150.

Maccoby, E. E. (1990). Gender and relationships: A developmental account. *American Psychologist, 45,* 513–520.

Macfarlane, A. (1977). *The psychology of childbirth.* Cambridge: Harvard University Press.

Macfarlane, A., Harris, P., & Barnes, I. (1976). Central and peripheral vision in early infancy. *Journal of Experimental Child Psychology, 21,* 532–538.

Magill, R. A. (1993). *Motor learning: Concepts and applications* (4th ed.). Dubuque, IA: Wm. C. Brown.

Malina, R. M. (1975). *Growth and development: The first twenty years.* Minneapolis: Burgess.

Malina, R. M. (1978). Adolescent growth and maturation: Selected aspects of current research. *Yearbook of Physical Anthropology, 21,* 63–94.

Malina, R. M. (1983). Menarche in athletes: A synthesis and hypothesis. *Annuals of Human Biology, 10,* 1–24.

Malina, R. M. (1984). Human growth, maturation, and regular physical activity. In R. A. Boileau (Ed.), *Advances in Pediatric Sport Sciences: Biological Issues* (Vol. 1). Champaign, IL: Human Kinetics.

Malina, R. M. (1990). Growth, exercise, fitness, and later outcomes. In C. Bouchard, R. J. Shephard, T. Stephens, J. R. Sutton, & B. D. McPherson (Eds.), *Exercise, fitness, and health* (pp. 637–653). Champaign, IL: Human Kinetics.

Malina, R. M., & Bouchard, C. (1991). *Growth, maturation,*

and physical activity. Champaign, IL: Human Kinetics.

Manchester, D., Woollacott, M., Zederbauer-Hylton, N., & Marin, O. (1989). Visual, vestibular and somatosensory contributions to balance control in the older adult. *Journal of Gerontology, 44*(4), 118–127.

Mandel, P., Bierth, R., Jacob, M., & Judes, C. (1974). Changes in nucleic acids during brain maturation. In S. R. Berenberg, Caniaris, & N. P. Masse (Eds.), *Pre- and postnatal development of the human brain.* New York: S. Karger.

Mandler, J. M. (1983). Representation. In P. H. Mussen (Ed.), Handbook of child psychology (Vol. 3, 4th ed.). New York: Wiley.

Manis, F. R., Keating, D. P., & Morrison, F. J. (1980). Developmental differences in the allocation of processing. *Journal of Experimental Child Psychology, 29,* 156–169.

Marker, K. (1981). Influence of athletic training on the maturity process. In J. Borms, M. Hebbelinck, & A. Venerando (Eds.), *The female athlete* (pp. 117–126). Basel, Switzerland: Karger.

Marshall, W. A., & Tanner, J. M. (1969). Variation in the pattern of pubertal changes in girls. *Archives of Diseases in Children, 44,* 291–303.

Marteniuk, R. G. (1976). *Information processing in motor skills.* New York: Holt, Rinehart and Winston.

Martens, R. (1978). *Joy and sadness in children's sports.* Champaign, IL: Human Kinetics.

Marx, J. J. M. (1979). Normal iron absorption and decreased red cell iron uptake in the aged. *Blood, 53,* 204–211.

McArdle, W. D., Katch, F. I., & Katch, V. L. (1991). *Exercise physiology: Energy, nutrition, and human performance* (3rd ed.). Philadelphia: Lea & Febiger.

McClenaghan, B. A., & Gallahue, D. L. (1978). *Fundamental movement: A developmental and remedial approach.* Philadelphia: W. B. Saunders.

McDonnell, P. M. (1979). Patterns of eye-hand coordination in the first year of life. *Canadian Journal of Psychology, 33,* 253–267.

McDonnell, P. M., & Abraham, W. C. (1981). Adaptation to displacing prisms in infants. *Perception, 8,* 175–185.

McFarlan, D. (Ed.). (1989). *Guinness book of world records.* New York: Bantam Books.

McGraw, M. B. (1966). *The neuromuscular maturation of the human infant.* New York: Hafner.

McManus, I. C., Sik, G., Cole, D. R., Mellon, A. F., Wong, J., & Kloss, J. (1988). The development of handedness in children. *British Journal of Developmental Psychology, 6,* 257–273.

McPherson, B. D. (1978). The child in competitive sport: Influences of the social milieu. In R. A. Magill, M. J. Ash, & F. L. Smoll (Eds.), *Children in sport: A contemporary anthology* (pp. 219–249). Champaign, IL: Human Kinetics.

Meeuwsen, H. J., Tesi, J. M., & Goggin, N. C. (1992). Psychophysics of arm movements and human aging. *Research Quarterly for Exercise and Sport, 63*(1), 19–24.

Meltzoff, A. N. (1988). Imitation, objects, tools, and the rudiments of language in human ontogeny. *Human Evolution, 3,* 45–64.

Meltzoff, A. N., & Moore, M. K. (1977). Imitation of facial and manual gestures by human neonates. *Science, 198,* 75–78.

Meltzoff, A. N., & Moore, M. K. (1983). The origins of imitation in infancy: Paradigm, phenomena, and theories. In L. P. Lisitt & C. K. Rovee-Collier (Eds.), *Advances in infancy research* (Vol. 2). Norwood, NJ: Ablex.

Metheny, E. (1941). The present status of strength testing for children of elementary and preschool age. *Research Quarterly, 12,* 115–130.

Meyers, J. (1967). Retention of balance coordination learning as influenced by extended lay-offs. *Research Quarterly, 38,* 72–78.

Mihalik, B. J., O'Leary, J. J., McGuire, F. A., & Dottavio, F. D. (1989). Sports involvement across the life of man: Expansion and contraction of sports activities. *Research Quarterly for Exercise and Sport, 60*(4), 396–398.

Milani-Comparetti, A. (1981). The neurophysiologic and clinical implications of studies on fetal motor behavior. *Seminars in Perinatology, 5,* 183–189.

Milani-Comparetti, A., & Gidoni, E. A. (1967). Pattern analysis of motor development and its disorders. *Developmental Medicine and Child Neurology, 9,* 625.

Miller, G. A. (1956). The magical number seven, plus or minus two: Some limits on our capacity for processing information. *Psychological Review, 63,* 81–97.

Mischel, W. (1984). Convergences and challenges in the search for consistency. *American Psychologist, 39,* 251–364.

Molliver, M., Kostovic, I., & Van Der Loos, H. (1973). The development of synapses in the cerebral cortex of the human fetus. *Brain Research, 50,* 403–407.

Molnar, G. (1978). Analysis of motor disorder in retarded infants and young children. *American Journal of Mental Deficiency, 83,* 213–222.

Moore, K. C. (1988). *The developing brain: Chemically oriented embryology* (4th ed.). Philadelphia: W. B. Saunders.

Morgan, D. G. (1992). Neurochemical changes with aging: The dispositions toward age-related mental disorders. In T. E. Birren, R. B. Sloane, & D. G. Cohen (Eds.), *Handbook of mental health and aging* (2nd ed., pp. 175–200). San Diego, CA: Academic Press.

Morris, A. M., Williams, J. M., Atwater, A. E., & Wilmore, J. H. (1982). Age and sex differences in motor performance of 3 through 6 year old children. *Research Quarterly for Exercise and Sport, 53,* 214–221.

Morris, E. K. (1988). Contextualism: The world view of behavior analysis. *Journal of Experimental Child Psychology, 46,* 289–323.

Morris, J. N., Pollard, R., & Everitt, M. G. (1980). Vigorous exercise in leisure time: Protection against coronary heart disease. *Lancet, 2,* 1207–1210.

Mounoud, P., & Bower, T. G. R. (1974). Conservation of weight in infants. *Cognition, 3,* 229–240.

Moxley, S. E. (1979). Schema: The variability of practice hypothesis. *Journal of Motor Behavior, 11,* 65–70.

Muir, D., & Field, J. (1979). Newborn infants' orientation to sound. *Child Development, 50,* 431–436.

Murray, M. P. (1985). Shoulder motion and muscle strength of normal men and women in two age groups. *Clinical Orthopedics and Related Research, 192,* 268–273.

Murray, M. P., Kory, R. C., & Sepic, S. B. (1970). Walking patterns of normal women. *Archives of Physical Medicine and Rehabilitation, 51,* 637–650.

Nachshon, I., Denno, D., & Aurand, S. (1983). Lateral preferences of hand, eye and foot: Relation to cerebral dominance. *International Journal of Neuroscience, 18,* 1–10.

National Children and Youth Fitness Study I. (1985). *Journal of Physical Education, Recreation and Dance, 56*(1), 43–90.

National Children and Youth Fitness Study II. (1987). *Journal of Physical Education, Recreation and Dance, 58*(9), 49–96.

Neisser, U. (1967). *Cognitive psychology.* New York: Appleton-Century-Crofts.

Nelson, C. J. (1981). *Locomotor patterns of women over 57.* Unpublished master's thesis, Washington State University, Pullman.

Nelson, K. (1977). Cognitive development and the acquisition of concepts. In R. C. Anderson, R. J. Spiro, & W. E. Montague (Eds.), *Schooling and the acquisition of knowledge.* Hillsdale, NJ: Erlbaum.

Neugarten, B. L., Havighurst, R. J., & Tobin, S. S. (1968). Personality and patterns of aging. In B. L. Neugarten (Ed.), *Middle age and aging.* Chicago: University of Chicago Press.

Neumann, E., & Ammons, R. B. (1957). Acquisition and long term retention of a simple serial perceptual motor skill. *Journal of Experimental Psychology, 53,* 159–161.

Newell, K. M. (1986). Constraints on the development of coordination. In M. G. Wade & H. T. A. Whiting (Eds.), *Motor development in children: Aspects of coordination and control* (pp. 341–361). Amsterdam: Martinus Nijhoff Publishers.

Norval, M. A. (1947). Relationships of weight and length of infants at birth to the time at which they begin to walk alone. *The Journal of Pediatrics, 30,* 676–678.

Nowakowski, R. S. (1987). Basic concepts of CNS development. *Child Development, 58,* 568–595.

Oliver, M. (1980). Race, class, and the family's orientation to mobility through sport. *Sociological Symposium, 30,* 62–86.

Olson, G. M., & Strauss, M. S. (1984). The development of infant memory. In M. Moscovitch (Ed.), *Infant memory: Its relation to normal and pathological memory in humans and other animals.* New York: Plenum.

Ornstein, P. A., Stone, B. P., Medlin, R. G., & Naus, M. J. (1985). Retrieving for rehearsal: An analysis of active rehearsal in children's memory. *Developmental Psychology, 21,* 633–641.

Oscai, L. B., Babirak, S. P., McGarr, J. A., & Spirakis, C. N. (1974). Effect of exercise on adipose tissue cellularity. *Federal Proceedings, 33,* 1956–1958.

Oscai, L. B., Spirakis, C. N., Wolff, C. H., & Beck, R. J. (1972). Effects of exercise and of food restriction on adipose tissue cellularity. *Journal of Lipid Research, 13,* 588–592.

Ostrow, A. C., Jones, D. C., & Spiker, D. D. (1981). Age role expectations and sex role expectations for selected sport activities. *Research Quarterly for Exercise and Sport, 52,* 216–227.

Overstall, P. W., Exton-Smith, A. N., Imms, F. J., & Johnson, A. L. (1977). Falls in the elderly related to postural imbalance. *British Medical Journal, 1,* 261–264.

Oxedine, J. B. (1984). *Psychology of motor learning.* Englewood Cliffs, NJ: Prentice-Hall.

Paffenbarger, R. S., Jr., Hyde, R. T., Wing, A. L., & Hsieh, C. C. (1986). Physical activity, all-cause mortality, and longevity of college alumni. *New England Journal of Medicine, 314,* 605–613.

Paikoff, R. L., Buchanan, C. M., & Brooks-Gunn, J. (1991). Hormone-behavior links at puberty, methodological links in the study of. In R. M. Lerner, A. C. Petersen, & J. Brooks-Gunn (Eds.), *Encyclopedia of adolescence.* New York: Garland.

Painter, M. A. (1994). Developmental sequences for hopping as assessment instruments: A generalized analysis. *Research Quarterly for Exercise and Sport, 65*(1), 1–10.

Palmer, C. F. (1989). The discriminating nature of infant's exploratory actions. *Developmental Psychology, 25,* 885–893.

Panton, L. B., Graves, J. E., & Pollock, M. L. (1990). Effect of aerobic and resistance training on fractionated reaction time and speed of movement. *Journal of Gerontology, 45*(1), 26–31.

Parcel, G. S., Simons-Morton, B. G., O'Hara, N. M., Baranowski, T., Kolbe, L. J., & Bee, D. E. (1987). School promotion of healthful diet and exercise behavior: An integration of organizational change and social learning theory intervention. *Journal of School Health, 57*(4), 150–156.

Parizkova, J. (1968). Longitudinal study of the development of body composition and body build in boys of various physical activity. *Human Biology, 40,* 212–225.

Parizkova, J., & Carter, J. E. L. (1976). Influence of physical activity on stability of somatotypes in boys. *American Journal of Physiology and Anthropology, 44,* 327–339.

Parker, S. J., & Barrett, D. (1992). Maternal type A behavior during pregnancy, neonatal crying, and infant temperament: Do type A women have type A babies? *Pediatrics, 89,* 474–479.

Parnell, R. W. (1958). *Behavior and physique: An introduction to practical and applied somatometry.* London: Arnold.

Parten, M. (1932). Social play among preschool children. *Journal of Abnormal and Social Psychology, 27,* 243–269.

Payne, V. G. (1985). Effects of object size and experimental design on object reception by children in the first grade. *Journal of Human Movement Studies, 11,* 1–9.

Payne, V. G., & Isaacs, L. D. (1987). *Human motor development: A lifespan approach.* Mountain View, CA: Mayfield Publishing Company.

Peabody Developmental Motor Scales. (1983). Allen, TX: DLM Teaching Resources.

Perelle, I. B. (1975). Difference in attention to stimulus presentation mode with regard to age. *Developmental Psychology, 11,* 403–404.

Perlmutter, M. (1980). An apparent paradox about memory and aging. In L. W. Poon, J. L. Fozard, L. S. Cermak, D. Arenberg, & D. W. Thompson (Eds.), *New directions in memory and aging: Proceedings of the George Talland memorial conference.* Hilldale, NJ: Erlbaum.

Perlmutter, M. (1983). Learning and memory through adulthood. In M. W. Riley, B. B. Hess, & K. Bond (Eds.), *Aging in society: Selected reviews of recent research.* Hillsdale, NJ: Erlbaum.

Perlmutter, M., & Hall, E. (1985). *Adult development and aging.* New York: John Wiley & Sons.

Peters, M. (1988). The size of the corpus callosum in males and females: Implications of a lack of allometry. *Canadian Journal of Psychology, 42* (3), 313–324.

Peters, M. (1990). Footedness: Neuropsychological identification of motor problems: Can we learn something from the feet and legs that hands and arms will not tell us? *Neuropsychology Review, 1,* 165–183.

Petrofsky, J. S., & Lind, A. R. (1975). Aging, isometric strength and endurance, and cardiovascular responses to static effort. *Journal of Applied Physiology, 38,* 91–95.

Pettersen, L., Yonas, A., & Fisch, R. O. (1980). The development of blinking in response to impending collision in preterm, full term, and postterm infants. *Infant Behavior and Development, 3,* 155–165.

Piaget, J. (1952). *The origins of intelligence in children.* New York: International Universities Press.

Piaget, J. (1963). *The origins of intelligence in children.* (M. Cook, Trans.). New York: W. W. Norton.

Piaget, J. (1971). *The construction of reality by the child.* New York: Ballentine.

Piaget, J. (1985). *The equilibration of cognitive structures: The central problem of intellectual development.* Chicago: University of Chicago Press.

Pick, A. D. (Ed.). (1979). *Perception and its development: A tribute to Eleanor J. Gibson.* Hillsdale, NJ: Erlbaum.

Pick, H. L., & Pick, A. D. (1970). Sensory and perceptual development. In P. H. Mussen (Ed.), *Carmichael's manual of child psychology* (Vol. 1, 3rd ed., pp. 773–847). New York: Wiley.

Pieper, A. (1963). *Cerebral function in infancy and childhood.* New York: Consultants Bureau.

Pierson, I. M., & Montoye, H. J. (1958). Movement time, reaction time, and age. *Journal of Gerontology, 13,* 418–421.

Plagenhoef, S. (1971). *Patterns of human motion: A cinematographic analysis.* Englewood Cliffs, NJ: Prentice-Hall.

Poe, A. (1976). Description of the movement characteristics of two-year-old children performing the jump and reach. *Research Quarterly, 47,* 260–268.

Pontius, A. A. (1973). Neuro-Ethics of "walking" in the newborn. *Perceptual and Motor Skills, 37,* 235–245.

Poon, L. W. (1985). Differences in human memory with aging: Nature, causes, and clinical implications. In J. E. Birren & K. Warner Schaie (Eds.), *Handbook of the psychology of aging* (2nd ed.). New York: Van Nostrand Reinhold.

Porac, C., & Coren, S. (1981). *Lateral preferences and human behavior.* New York: Springer-Verlag.

Porac, C., Coren, S., & Duncan, P. (1980). Life-span age trends in laterality. *Journal of Gerontology, 35,* 715–721.

Powell, K. E. (1987). Physical activity and the incidence of coronary heart disease. *Annual Review of Public Health, 8,* 253.

Prechtl, H. F. R., & Hopkins, B. (1986). Developmental transformations of spontaneous movements in early infancy. *Early Human Development, 14,* 233–238.

Pressley, M., & Levin, J. R. (1977). Task parameters affecting the efficacy of a visual imagery learning strategy in younger and older children. *Journal of Experimental Child Psychology, 24,* 53–59.

Pressley, M., & MacFadyen, J. (1983). Mnemonic mediator retrieval at testing by preschool and kindergarten children. *Child Development, 54,* 474–479.

Previc, F. H. (1991). A general theory concerning the prenatal origins of cerebral lateralization in humans. *Psychological Review, 98*(3), 299–334.

Provins, K. A. (1992). Early infant motor asymmetries and handedness: A critical evaluation of the evidence. *Developmental Neuropsychology, 8*(4), 325–365.

Quirion, A., deCareful, D., Laurencell, L., Method, D., Vogelaere, P., & Dulac, S. (1987). The physiological response to exercise with special reference to age. *Journal of Sports Medicine and Physical Fitness, 27,* 143–149.

Rabinowiz, T. (1974). Some aspects of the maturation of the human cerebral cortex. In S. R. Berenberg & N. P. Masse (Eds.), *Pre- and postnatal development of the human brain.* New York: S. Karger.

Rarick, G. L. (1973). *Physical activity: Human growth and development.* New York: Academic Press.

Revised Denver Prescreening Developmental Questionnaire. (1986). Denver: DDM.

Rice, T., & Plomin, R. (1983). Hand preferences in the Colorado adoption project. *Behavior Genetics, 13,* 550.

Richter, C. P. (1934). The grasp reflex of the newborn infant. *American Journal of Developmental Children, 48,* 327–332.

Rikli, R., & Busch, S. (1986). Motor performance of women as a function of age and physical activity level. *Journal of Gerontology, 41,* 645–649.

Rikli, R. E., & Edwards, D. J. (1991). Effects of a three year

exercise program on motor function and cognitive processing speed in older women. *Research Quarterly for Exercise and Sport, 62*(1), 61–67.

Roach, E. G., & Kephart, N. C. (1966). *Purdue perceptual motor survey.* Columbus, OH: Merrill.

Roberton, M. A. (1977). Stability of stage categorizations across trials: Implications for the "stage theory" of overarm throw development. *Journal of Human Movement Studies, 3,* 49–59.

Roberton, M. A. (1978). Longitudinal evidence for developmental stages in the forceful overarm throw. *Journal of Human Movement Studies, 4,* 167–175.

Roberton, M. A. (1984). Changing motor patterns during childhood. In J. R. Thomas (Ed.), *Motor development during childhood and adolescence* (pp. 48–85). Minneapolis: Burgess.

Roberton, M. A. (1990). Interlimb timing changes in the development of hopping. NASPSPA abstracts.

Roberton, M. A., & Halverson, L. E. (1984). *Developing children—their changing movement.* Philadelphia: Lea & Febiger.

Roberton, M. A., & Halverson, L. E. (1988). The development of locomotor coordination: Longitudinal change and invariance. *Journal of Motor Behavior, 20*(3), 197–241.

Roberton, M. A., Halverson, L. E., Langendorfer, S., & Williams, K. (1979). Longitudinal changes in children's overarm throw ball velocities. *Research Quarterly for Exercise and Sport, 50,* 256–264.

Roberton, M. A., Williams, K., & Langendorfer, S. (1980). Prelongitudinal screening of motor development sequences.

Research Quarterly for Exercise and Sport, 51, 724–731.

Roosa, M. W. (1984). Maternal age, social class, and the obstetric performance of teenagers. *Journal of Young and Adolescence, 13,* 365–374.

Rosenhall, U. (1973). Degenerative patterns in the aging human vestibular neuroepithelia. *Acta Otolarngolica, 76,* 208–220.

Ross, A. O. (1976). *Psychological aspects of learning disabilities and reading disorders.* New York: McGraw-Hill.

Ross, J. G., & Gilbert, G. G. (1985). The national children and youth fitness study: A summary of findings. *Journal of Physical Education, Recreation and Dance, 56*(1), 45–50.

Ross, J. G., & Pate, R. R. (1987). The national children and youth fitness study II: A summary of findings. *Journal of Physical Education, Recreation and Dance, 58*(9), 51–56.

Rudman, D., Feller, A. G., Nagraj, H. S., Gergans, G. A., Lalitha, P. Y., Goldberg, A. F., Schlenker, R. A., Cohn, L., Rudman, I. W., & Mattson, D. E. (1990). Effects of human growth hormone in men over 60 years old. *The New England Journal of Medicine, 323*(1), 1–6.

Rudolph, R. S. (1982). Aspects of child health. In A. M. Rudolph (Ed.), *Pediatrics* (17th ed. pp. 1–7). Norwalk, CT: Appleton-Century-Crofts.

Ruff, H. A. (1980). The development of perception and recognition objects. *Child Development, 51,* 981–992.

Ryan, E. D. (1962). Retention of stabilometer and pursuit motor skills. *Research Quarterly, 33,* 593–598.

Ryan, E. D. (1965). Retention of stabilometer performance over extended periods of time. *Research Quarterly, 36,* 46–51.

Sabatino, D. A. (1985). Review of Bruininks-Oseretsky test of motor proficiency. In J. V. Mitchell, Jr. (Ed.), *Ninth mental measurement yearbook.* Lincoln, NE: Buros Institute of Mental Measurements.

Sage, G. H. (1984). *Motor learning and control.* Dubuque, IA: Wm. C. Brown.

Salapatek, P., Benchtold, A. G., & Bushnell, B. W. (1976). Infant acuity as a function of viewing distance. *Child Development, 47,* 860–863.

Salthouse, T. A. (1985). Motor performance and speed of behavior. In J. E. Birren & K. W. Schaie (Eds.), *Handbook of the psychology of aging* (2nd ed.). New York: Van Nostrand Reinhold.

Salvia, J. Z., & Ysseldyke, J. E. (1991). *Assessment* (5th ed.). Boston: Houghton Mifflin.

Sameroff, A. J., & Chandler, M. J. (1975). Reproductive risk and the continuum of caretaking casualty. In F. D. Horowitz (Ed.), *Review of child development research* (Vol. 4). Chicago: University of Chicago Press.

Santrock, J. W. (1989). *Life-span development* (3rd ed.). Dubuque, IA: Wm. C. Brown.

Santrock, J. W. (1995). *Life-span development* (5th ed.). Dubuque, IA: Wm. C. Brown.

Scanlan, T. (1984). Competitive stress and the child athlete. In J. M. Silva & R. S. Weinberg (Eds.), *Psychological foundations of sport.* Champaign, IL: Human Kinetics.

Schachter, D. L., & Moscovitch, M. (1984). Infants, amnesiacs, and dissociable memory systems. In M. Moscovitch (Ed.), *Infant memory: Its relation to normal and pathological memory in humans and other animals.* New York: Plenum.

Schaie, K. W., & Willis, S. L. (1991). *Adult development and aging* (3rd ed.). New York: HarperCollins.

Schank, R., & Abelson, R. (1976). *Scripts, plans, goals and understanding.* Hillsdale, NJ: Erlbaum.

Scheibel, A. B. (1992). Structural changes in the aging brain. In J. E. Birren, R. B. Sloane, & G. D. Cohen (Eds.), *Handbook of mental health and aging* (2nd ed., pp. 147–174). San Diego, CA: Academic Press.

Scheinfeld, A. (1939). *You and heredity.* New York: Frederick A. Stokes.

Schmidt, R. A. (1975). A schema theory of discrete motor skill learning. *Psychological Review, 82,* 225–260.

Schmidt, R. A. (1977). Schema theory: Implications for movement education. *Motor Skills: Theory into Practice, 2,* 36–48.

Schmidt, R. A. (1988). *Motor control and learning* (2nd ed.). Champaign, IL: Human Kinetics.

Schulman, J. L., Buist, C., Kasper, J. C., Child, D., & Fackler, E. (1969). An objective test of speed and fine motor function. *Perceptual and Motor Skills, 29,* 243–255.

Schultz, R., & Curnow, C. (1988). Peak performance and age among superathletes. *Journal of Gerontology, 43,* 113–120.

Schultz, R., Musa, D., Staszewski, J., & Siegler, R. S. (1994). The relationship between age and major league baseball performance: Implications for development. *Psychology and Aging, 9*(2), 274–286.

Schutte, J. E., Townsend, E. J., Hugg, J., Shoup, J., Molina, R. M., & Blomquist, C. G. (1984). Density of lean body mass is greater in blacks than in whites. *Journal of Applied Physiology, 56,* 1647–1649.

Schwartz, R. S., Shuman, W. P., Bradbury, V. L., Cain, K. C., Fellingham, G. W., Beard, J. C., Kahn, S. E., Stratton, J. R., Cerqueina, M. D., & Abrass, I. B. (1990). Body fat distribution in healthy young and older men. *Journal of Gerontology, 45*(6), 181–185.

Seefeldt, V. (1972). *Developmental sequences of catching skills.* Paper presented at the American Alliance for Health, Physical Education, Recreation and Dance, Houston.

Seefeldt, V. (1974). Perceptual-motor programs. In J. H. Wilmore (Ed.), *Exercise and sports science reviews* (Vol. 2). New York: Academic Press.

Seefeldt, V., & Branta, C. F. (1984). Patterns of participation in children's sport. In J. Thomas (Ed.), *Motor development during childhood and adolescence.* Minneapolis: Burgess.

Seefeldt, V., & Haubenstricker, J. (1976). *Developmental sequences of fundamental motor skills.* Unpublished research, Michigan State University, East Lansing.

Seefeldt, V., & Haubenstricker, J. (1982). Patterns, phases, or stages: An analytical model for the study of developmental movement. In J. A. Kelso & J. E. Clark (Eds.), *The development of movement control and coordination.* New York: Wiley.

Seeley, R. R., Stephens, T. D., & Tate, P. (1992). *Anatomy and physiology.* St. Louis: Mosby.

Seils, L. (1951). The relationship between measure of physical growth and gross motor performance of primary-grade children. *Research Quarterly, 22,* 244–260.

Sekuler, R., & Blake, R. (1994). *Perception* (3rd ed.). New York: Alfred A. Knopf.

Shannon, C. E., & Weaver, W. (1949). *The mathematical theory of communication.* Urbana: University of Illinois Press.

Shapiro, D. C., & Schmidt, R. A. (1982). The schema theory: Recent evidence and developmental implications. In J. A. Kelso & J. Clark (Eds.), *The development of movement control and coordination.* New York: Wiley.

Shea, C., Shebilske, W., & Worchel, S. (1993). *Motor learning and control.* Englewood Cliffs, NJ: Prentice-Hall.

Sheehan, F. (1954). *Baseball achievement scales for elementary and junior high school boys.* Unpublished master's thesis, University of Wisconsin, Madison.

Sheldon, W. H., Dupertuis, C. W., & McDermott, E. (1954). *Atlas of men: A guide for somatotyping the adult male of all ages.* New York: Harper.

Shephard, R. J. (1978). *Physical activity and aging.* Chicago: Yearbook Medical.

Shephard, R. J. (1982). *Physical activity and growth.* Chicago, IL: Year Book Medical Publishers.

Shock, N. W. (1962). The physiology of aging. *Scientific American, 206*(1), 100–110.

Skinner, B. F. (1974). *About behaviorism.* New York: Alfred A. Knopf.

Skre, H. (1972). Neurological signs in a normal population. *Acta Neurological Scandanavia, 48,* 575–606.

Smith, A. D. (1980). Age differences in encoding, storage, and retrieval. In L. W. Poon, J. L. Fozard, L. S. Cermak, D. Arenberg, & L. W. Thompson (Eds.), *New directions in memory and aging.* Hillsdale, NJ: Erlbaum.

Smith, E. L., & Serfass, R. C. (Eds.). (1981). *Exercise and aging: The scientific basis.* Hillside, NJ: Enslow.

Smith, L. B., Kemler, D. G., & Aronfried, J. (1975). Developmental trends in voluntary selective attention: Differential aspects of source distinctiveness. *Journal of Experimental Child Psychology, 20,* 353–362.

Smith, R. E., Smoll, F. L., & Curtis, B. (1979). Coach effectiveness training: A cognitive-behavioral approach to enhancing relationship skills in youth sport coaches. *Journal of Sport Psychology, 1,* 59–75.

Smith, S. (1983). Performance differences between hands in children on the motor accuracy test—revised. *The American Journal of Occupational Therapy, 37*(2), 96–102.

Smoll, F. L. (1974). Development of spatial and temporal elements of rhythmic ability. *Journal of Motor Behavior, 6,* 53–58.

Smoll, F. L., & Schutz, R. W. (1990). Quantifying gender differences in physical performance: A developmental perspective. *Developmental Psychology, 26*(3), 360–369.

Snyder, E. E., & Spreitzer, E. A. (1973). Family influence and involvement in sport. *Research Quarterly, 44,* 249–255.

Snyder, E. E., & Spreitzer, E. A. (1983). *Social aspects of sport* (2nd ed.). Englewood Cliffs, NJ: Prentice-Hall.

Spelke, E. S. (1979). Exploring audible and visible events in infancy. In A. D. Pick (Ed.), *Perception and its development: A tribute to Eleanor J. Gibson.* Hillsdale, NJ: Erlbaum.

Spelke, E. S., & Owsley, C. J. (1979). Intermodal exploration and knowledge in infancy. *Infant Behavior and Development, 2,* 13–27.

Spirduso, W. W. (1975). Reaction and movement time as a

function of age and physical activity level. *Journal of Gerontology, 30,* 435–440.

Sporns, O., & Edelman, G. M. (1993). Solving Bernstain's problem: A proposal for the development of coordinated movement by selection. *Child Development, 64*(4), 960–981.

Sports Illustrated for Kids. (1992). Annual survey of kid's favorite sports.

Stennet, R. G., Smythe, P. C., & Hardy, H. (1972). Developmental trends in letter printing skills. *Perceptual and Motor Skills, 34,* 182–186.

Sterns, H., Barrett, G. V., & Alexander, R. A. (1985). Accidents and the aging individual. In J. E. Birren & K. W. Schaie (Eds.), *Handbook of the psychology of aging* (pp. 703–724). New York: Van Nostrand Reinhold.

Stratton, R. K. (1978a). Information processing deficits in children's motor performance: Implications for instruction. *Motor Skills: Theory and Practice, 3,* 49–55.

Stratton, R. K. (1978b). *Selective attention deficits in children's motor performance: Can we help?* Paper presented at the annual conference of the North American Society for Psychology of Sport and Physical Activity, Tallahassee, FL.

Strohmeyer, H. S., Williams, K., & Schaub-George, D. (1991). Developmental sequences for catching a small ball: A prelongitudinal screening. *Research Quarterly for Exercise and Sport, 62*(3), 257–266.

Sugden, D. A. (1980). Developmental strategies in motor and visual motor short-term memory. *Perceptual and Motor Skills, 51,* 146.

Sullivan, M. W. (1982). Reactivation: Priming forgotten memories in human infants. *Child Development, 53,* 516–523.

Surwillo, W. W. (1977). Developmental changes in the speed of information processing. *Journal of Psychology, 97,* 102.

Sutherland, D. H. (1984). *Gait disorders in childhood and adolescence.* Baltimore: Williams and Wilkins.

Swearingen, J. J., Braden, G. E., Badgley, J. M., & Wallace, T. F. (1969). *Determination of centers of gravity in children.* Washington, D.C.: Federal Aviation Administration.

Tan, L. E. (1985). Laterality and motor skills in four-year-olds. *Child Development, 56,* 119–124.

Tanner, J. M. (1978). *Fetus into man: Physical growth from conception to maturity.* Cambridge, MA: Harvard University Press.

Tanner, J. M., Whitehouse, R. H., Marshall, W. A., Healy, M. J., & Goldstein, H. (1975). *Assessment of skeletal maturity and prediction of adult height (TWZ method).* New York, NY: Academic Press.

Temple, I. G., Williams, H. G., & Bateman, N. J. (1979). A test battery to assess intrasensory and intersensory development of young children. *Perceptual and Motor Skills, 48,* 643–659.

Thelen, E. (1979). Rhythmical stereotypes in normal human infants. *Animal Behavior, 27,* 699–715.

Thelen, E. (1985). Developmental origins of motor coordination: Leg movements in human infants. *Developmental Psychology, 18,* 11.

Thelen, E., & Bradley, N. S. (1988). Motor development: Posture and locomotion. In E. Meisami & P. S. Timiras (Eds.), *Handbook of human growth and developmental biology.* Boca Raton, FL: CRC Press.

Thelen, E., & Cooke, D. W. (1987). The relation between newborn stepping and later locomotion: A new interpretation. *Developmental Medicine and Child Neurology, 29,* 380–393.

Thelen, E., Corbetta, D., Kamm, K., Spencer, J., Schneider, K., & Zernicke, R. F. (1993). The transition to reaching. Mapping intention and intrinsic dynamics. *Child Development, 64,* 1058–1098.

Thelen, E., & Fisher, D. M. (1982). Newborn stepping: An explanation for a "disappearing reflex." *Developmental Psychology, 18,* 760.

Thelen, E., & Fisher, D. M. (1983). From spontaneous to instrumental behavior: Kinematic analysis of movement changes during very early learning. *Child Development, 54,* 129–140.

Thelen, E., & Ulrich, B. D. (1991). Hidden skills: A dynamical systems analysis of treadmill stepping during the first year. *Monographs of Society for Research in Child Development, 56*(1), serial #223.

Thelen, E., Ulrich, B. D., & Jensen, J. C. (1989). The developmental origins of locomotion. In M. Woollacott & A. Shumway-Cook (Eds.), *The development of posture and gait across the lifespan* (pp. 25–47). Columbia: University of South Carolina Press.

Thomas, J. R., Thomas, K. T., Lee, A. M., Testerman, E., & Ashy, M. (1983). Age differences in use of strategy for recall of movement in a large scale environment. *Research Quarterly for Exercise and Sport, 54,* 264–272.

Thomas, R. J., & French, K. (1985). Gender differences across age in motor performance: A meta-analysis. *Psychological Bulletin, 98*(2), 260–282.

Thomas, R. J., Nelson, J. K., & Church, G. (1991). A developmental analysis of gender differences in health-related physical fitness. *Pediatric Exercise Science, 3,* 28–42.

Thompson, J. E. (1990). Maternal stress, anxiety, and social support during pregnancy: Possible directions for prenatal intervention. In I. R. Merkatz & J. E. Thompson (Eds.), *New perspectives on prenatal care.* New York: Elsevier.

Till, R. E. (1985). Verbatim and inferential memory in young and elderly adults. *Journal of Gerontology, 40,* 316–323.

Timiras, P. S., Vernadakis, A., & Sherwood, N. M. (1968). Development and plasticity of the nervous system. In N. S. Assali (Ed.), *Biology of gestation.* New York: Academic Press.

Toglia, J. U. (1975). Dizziness in the elderly. In W. Fields (Ed.), *Neurological and sensory disorders in the elderly.* New York: Grune & Stratton.

Touwen, B. C. L. (1976). Neurological development in infancy. In *Clinics in Developmental Medicine* (No. 58). London: William Heinemann.

Touwen, B. C. L. (1988). Motor development: Developmental dynamics and neurological examination in infancy. In E. Meisami & P. S. Timitas (Eds.), *Handbook of human growth and developmental biology* (Vol. 1). Boca Raton, FL: CRS Press.

Trevarthen, C. (1983). Development of the cerebral mechanisms for language. In V. Kirk (Ed.), *Neuropsychology of language, reading and spelling.* New York: Academic Press.

Trevarthen, C. (1984). How control of movement develops. In

H. D. A. Whiting (Ed.), *Human motor actions-Bernstein reassessed* (pp. 223–261). Amsterdam: North Holland.

Tulving, E. (1985). How many memory systems are there? *American Psychologist, 40,* 385–398.

Turvey, M. T. (1977). Preliminaries to a theory of action with reference to vision. In R. Shaw & J. Bransford (Eds.), *Perceiving, acting, and knowing* (pp. 211–265). Hillsdale, NJ: Erlbaum.

Ulrich, B. D. (1989). Development of stepping patterns in human infants: A dynamical systems perspective. *Journal of Motor Behavior, 21,* 392–406.

Ulrich, B. D., & Ulrich, D. A. (1985). The role of balancing in performance of fundamental motor skills in 3-, 4-, and 5-year-old children. In J. E. Clark & J. H. Humphrey (Eds.), *Motor development: Current selected research* (Vol. 1). Princeton, NJ: Princeton Book Co.

Ulrich, D. A. (1985). *Test of gross motor development.* Austin, TX: Pro-Ed.

Van Camp, S. P., & Boyer, J. L. (1989). Cardiovascular aspects of aging. *The Physician and Sportsmedicine, 17*(4), April, 120–129.

Van Duyne, H. J. (1973). Foundations of tactical perception in three to seven year olds. *Journal of the Association for the Study of Perception, 8,* 1–9.

Vogel, P. G. (1986). Effects of physical education programs on children. In V. Seefeldt (Ed.), *Physical activity and well-being.* Reston, VA: American Alliance for Health, Physical Education, Recreation and Dance.

Von Hofsten, C. (1979). Development of visually directed reaching: The approach

phase. *Journal of Human Movement Studies, 5,* 160–178.

Von Hofsten, C. (1980). Predictive reaching for moving objects by human infants. *Journal of Experimental Child Psychology, 30,* 369–382.

Wagner, S., Winner, E., Cicchetti, D., & Gardner, H. (1981). "Metaphorical" mapping in human infants. *Child Development, 52,* 728–731.

Walsh, R. P. (1983). Age differences in learning and memory. In D. S. Woodruff & J. E. Birren (Eds.), *Aging: Scientific perspectives and social issues* (2nd ed.). Monterey, CA: Brooks/Cole.

Wankel, L. (1985). Personal and situational factors affecting exercise involvement: The importance of enjoyment. *Research Quarterly for Exercise and Sport, 56,* 275–282.

Warren, M. P. (1980). The effects of exercise on pubertal progression and reproductive function in girls. *Journal of Clinical Endocrinology and Metabolism, 51,* 1150–1156.

Watson, J. B. (1928). *Psychological care of infant and child.* New York: Norton.

Weale, R. A. (1975). Senile changes in visual acuity. *Transactions of the Opthalmological Societies of the United Kingdom, 95,* 36–38.

Weil, W. B., Jr. (1982). Specific dietary needs. In A. M. Rudolph (Ed.), *Pediatrics* (17th ed., pp. 181–183). Norwalk, CT: Appleton-Century-Crofts.

Weiss, M. R. (1983). Modeling and motor performance: A developmental perspective. *Research Quarterly for Exercise and Sport, 54,* 190–197.

Weiss, M. R. (1987). Self-esteem and achievement in children's sport and physical activity. In D. Gould & M. R. Weiss (Eds.), *Advances in pediatric sport sciences: Vol. 2. Behavioral issues* (pp. 87–119). Champaign, IL: Human Kinetics.

Weiss, M. R. (1993). Psychological effects of intensive sport participation on children and youth: Self-esteem and motivation. In B. R. Cahill & A. J. Pearl (Eds.), *Intensive participation in children's sports* (pp. 39–69). Champaign, IL: Human Kinetics.

Weiss, M. R., & Knoppers, A. (1982). The influence of socializing agents on female collegiate volley players. *Journal of Sport Psychology, 4,* 267–279.

Weiss, M. R., & Sisley, B. (1984). Where have all the coaches gone? *Sociology of Sport Journal, 1*(4), 322–347.

Welford, A. T. (1977). Motor performance. In J. E. Birren & K. W. Schaie (Eds.), *Handbook of the psychology of aging.* New York: Van Nostrand Reinhold.

Welford, A. T. (1980). Motor skills and aging. In C. H. Nadeau, W. R. Halliwell, K. M. Newell, & G. C. Roberts (Eds.), *Psychology of motor behavior and sport—1979* (pp. 253–268). Champaign, IL: Human Kinetics.

Welford, A. T. (1984). Between bodily changes and performance: Some possible reasons for slowing with age. *Experimental Aging Research, 10,* 73–88.

Welford, A. T., Norris, A. H., & Shock, N. W. (1969). Speed and accuracy of movement and their changes with age. *Acta Psychologica, 30,* 3–15.

Wellman, B. L. (1937). Motor achievements in preschool children. *Childhood Education, 13,* 311–316.

Weltman, A. (1989). Weight training in prepubertal children: Physiologic benefit and potential damage. In O. Bar-Or (Ed.),

Advances in pediatric sport science (Vol. 3). Champaign, IL: Human Kinetics.

Weltman, A., Janney, C., Rains, C. B., Strand, K., Berg, B., Tippitt, S., Wise, J., Cahill, B. R., & Katch, F. I. (1986). The effects of hydraulic resistance strength training in prepubertal males. *Medicine and Science in Sports Exercise, 18*(6), 629–638.

Werner, E., & Smith, R. (1982). *Vulnerable but invincible: A study of resilient children.* New York: McGraw-Hill.

West, J. R. (1986). *Alcohol and brain development.* London: Oxford University Press.

Whitbourne, S. K. (1985). *The aging body: Physiological changes and psychological consequences.* New York: Springer-Verlag.

White, B. L. (1971). *Human infants: Experience and psychological development.* Englewood Cliffs, NJ: Prentice-Hall.

Wickens, C. D., & Benel, D. C. R. (1982). The development of time-sharing skills. In J. A. S. Kelso & J. E. Clark (Eds.), *The development of movement control and co-ordination.* New York: Wiley.

Wickstrom, R. L. (1980). *Acquisition of ball-handling skill.* Paper presented at the Research Section of the American Alliance for Health, Physical Education, Recreation and Dance, Detroit.

Wickstrom, R. L. (1983). *Fundamental motor patterns* (3rd ed.). Philadelphia: Lea & Febiger.

Wild, M. (1938). The behavior pattern of throwing and some observations concerning its course of development in children. *Research Quarterly, 9,* 20–24.

Wilkinson, R. T., & Allison, S. (1989). Age and simple reaction

time: Decade differences for 5,325 subjects. *Journal of Gerontology, 44*(2), 29–35.

Williams, H., Clement, A., Logsdon, B., Scott, S., & Temple, I. (1970). *A study of perceptual-motor characteristics of children in kindergarten through sixth grade.* Unpublished paper, University of Toledo.

Williams, H., Temple, I., & Bateman, J. (1978). Perceptual-motor and cognitive learning in young children. *Psychology of motor behavior and sport II.* Champaign, IL: Human Kinetics.

Williams, H. G. (1983). *Perceptual and motor development.* Englewood Cliffs: Prentice-Hall.

Williams, H. G. (1990). Aging and eye-hand coordination. In H. G. Williams (Ed.), *Development of eye-hand coordination.* Columbia: University of South Carolina Press.

Williams, H. G., & Breihan, S. K. (1979). *Motor control tasks for young children.* Unpublished manuscript, University of Toledo.

Williams, K., Haywood, K., & VanSant, A. (1990). Movement characteristics of older adult throwers. In J. E. Clark & J. H. Humphery (Eds.), *Advances in motor development research* (Vol. 3, pp. 29–44). New York: AMS Press.

Williams, K., Haywood, K., & VanSant, A. (1991). Throwing patterns of older adults: A follow-up investigation. *International Journal of Aging and Human Development, 33,* 279–294.

Willmott, M. (1986). The effect of vinyl floor surface and carpeted floor surface upon walking in elderly hospital inpatients. *Age and Ageing, 15,* 119–120.

Winterhalter, C. (1974). *Age and sex trends in the development of* selected balancing skills. Unpublished master's thesis, University of Toledo, Toledo, OH.

Wodrich, D. L. (1990). *Children's psychological testing: A guide for non-psychologists* (2nd ed.). Baltimore: Paul H. Brooks.

Woollacott, M. (1993). Age related change in posture and movement. *The Journals of Gerontology, 48,* 56–60.

Woollacott, M. H., Shumway-Cook, A., & Nasher, L. (1982). Postural reflexes and aging. In F. J. Pirozzlo & G. J. Maletta (Eds.), *The aging motor system.* New York: Praeger.

Woollacott, M. H., Shumway-Cook, A., & Williams, H. (1989). The development of posture and balance control in children. In M. H. Woollacott, M. & A. Shumway-Cook (Eds.), *Development of posture and gait across the life span.* Columbia: University of South Carolina Press.

Wright, R. E. (1981). Aging, divided attention, and processing capacity. *Journal of Gerontology, 36,* 605–614.

Wulf, G., & Schmidt, R. A. (1988). Variability in practice: Facilitation in retention and transfer through schema formation or context effects? *Journal of Motor Behavior, 20*(2), 133–149.

Wyke, B. (1975). The neurological basis of movement. In K. S. Hold (Ed.), *Clinics in Developmental Medicine* (No. 55). London: William Heinemann.

Zauner, C. W., Maksud, M. G., & Melichna, J. (1989). Physiological considerations in training young athletes. *Sports Medicine, 8*(1), 15–31.

Zelazo, P. R. (1983). The development of walking, new findings, and old assumptions. *Journal of Motor Behavior, 15,* 99–137.

Zelazo, P. R., Zelazo, N. A., & Kolb, S. (1972). Walking in the newborn. *Science, 176,* 314–315.

Zittel, L. L. (1994). Gross motor assessment of preschool children with special needs: Instrument selection considerations. *Adapted Physical Activity Quarterly, 11,* 245–260.

Ziviana, J. (1983). Qualitative changes in dynamic tripod grip between seven and fourteen years of age. *Developmental Medicine and Child Neurology, 25,* 778–782.

Zuckerman, B. (1989). Effects of maternal marijuana and cocaine use on fetal growth. *New England Journal of Medicine, 320,* 762–768.

Zwiren, L. D. (1989). Anaerobic and aerobic capacities in children. In T. Rowland (Ed.), *Pediatric Exercise Science 1*(4), 31–44.

Credits

Chapter 1

Figure 1.4: From Claire B. Kopp and Joanne B. Krakow, *The Child,* (pg. 648), © 1982 by Addison-Wesley Publishing Company, Inc. Reprinted by permission of the publisher.

Chapter 2

Figure 2.1: From Ross M. Durham, *Human Physiology,* Copyright 1989 Wm. C. Brown Communications, Inc., Dubuque, Iowa. All Rights Reserved. Reprinted by permission. **Figure 2.2:** From L. L. Langley, I. R. Telford and J. B. Christensen, *Dynamic Anatomy and Physiology,* 3rd edition. Copyright © 1969 McGraw-Hill, Inc., New York, NY. Reprinted by permission. **Figure 2.3:** From John W. Hole, *Human Anatomy and Physiology,* third edition. Copyright © 1984 Wm. C. Brown Communications, Inc., Dubuque, Iowa. All Rights Reserved. Reprinted by permission. **Figure 2.4:** From George H. Sage, *Motor Learning and Control,* Copyright © 1984 Wm. C. Brown Communications, Inc., Dubuque, Iowa. All Rights Reserved. Reprinted by permission. **Figure 2.5:** From L. L. Langley, I. R. Telford and J. B. Christensen, *Dynamic Anatomy and Physiology,* 3rd edition. Copyright © 1969 McGraw-Hill, Inc., New York, NY. Reprinted by permission.

Figure 2.7: From J. L. Conel (1939–1963), *Postnatal Development of the Human Cerebral Cortex,* Vol. I–VI. Copyright © Harvard University Press, Cambridge, MA. Reprinted by permission. **Figure 2.9:** Adapted from Rorke & Riggs, 1969; Dekaban, 1970; Yakolev & Lecours, 1976. Reprinted by permission of Blackwell Scientific Publications, Oxford, England.

Chapter 3

Figure 3.1: From J. M. Tanner, *Fetus Into Man,* Copyright © 1978 by J. M. Tanner. Reprinted by permission of Harvard University Press, Cambridge, MA. **Figure 3.2:** From R. M. Malina, *Growth and Development: The First Twenty Years,* Copyright © 1975 Burgess Publishing Company, Minneapolis, MN. Reprinted by permission. **Figure 3.3:** From John W. Santrock, *Life-Span Development,* 2nd edition. Copyright © 1986 Wm. C. Brown Communications, Inc., Dubuque, Iowa. All Rights Reserved. Reprinted by permission. **Figure 3.5:** From John W. Santrock, *Children,* 4th edition. Copyright © 1989 Wm. C. Brown Communications, Inc., Dubuque, Iowa. All Rights Reserved. Reprinted by permission. **Figure 3.7:** Reprinted with permission of Ross Laboratories, Columbus, OH 43216. **Figure 3.8:**

Reprinted with permission of Ross Laboratories, Columbus, OH 43216. **Figure 3.16:** From J. M. Tanner, *Fetus Into Man,* Copyright © 1978 by J. M. Tanner. Reprinted by permission of Harvard University Press, Cambridge, MA. **Figure 3.22:** From "Exercise and Osteoporosis" by H. J. Montoye. In *Exercise and Osteoporosis,* by H. M. Eckert and H. J. Montoye (Eds.), 1984. Champaign, IL: Human Kinetics. Copyright 1984 by the American Academy of Physical Education. Reprinted by permission. **Figure 3.27:** From "Body Composition in Children and Youth" by Timothy G. Lohman, Richard A. Boileau, and Mary H. Slaughter. In *Advances in Pediatric Sport Sciences I,* page 43 by Richard A. Boileau (Ed.), 1984. Champaign, IL: Human Kinetics. Copyright 1984 by Human Kinetics Publishers. Reprinted by permission. **Figure 3.34:** From A. F. Roche, "Secular Trends in Stature, Weight, and Maturation" in *Monographs of the Society for Research in Child Development,* 44, Serial No. 179, 1977. Copyright © 1977 The Society for Research in Child Development, Inc., Chicago, Il. Reprinted by permission.

Chapter 4

Figure 4.1: From Ross M. Durham, *Human Physiology,* Copyright

Index